BANANA

NEW INNOVATIONS

H.P. Singh
M.M. Mustaffa

WESTVILLE
WESTVILLE PUBLISHING HOUSE
NEW DELHI

H.P. Singh and M.M. Mustaffa (2009) Banana – New Innovations, pp 335
Westville Publishing House, New Delhi

First Edition 2009

ISBN 978-81-85873-49-7

Published by Mrinal Goel, Westville Publishing House
47, B-5, Paschim Vihar, New Delhi – 110 063
Tel: 011-25284742, Fax: 011-25267469
Email: westville_2002@yahoo.co.in
Printed at Salasar Imaging Systems, New Delhi

Foreword

Banana and plantain are food for millions of people in the world and recognized as the third global food commodity in terms of gross value of production. Banana is known for its antiquity in India, from its mention in the *Balmiki Ramayana* (2029 BC), *Kautillya's Arthashastra* (300-400 BC) and its presence in paintings and sculpture in the Ajanta and Allora. Research has intensified around the globe to improve production and productivity, considering the threat of many pests and diseases in addition to management constraints. Bioversity International through its network on banana and plantain improvement has played the role of catalyst in research and development across the world.

India has emerged as the largest producer of banana, with the production of about 22 million tonnes of fruits per year. This achievement can be attributed to research and developmental strategies for banana. The National Research Centre for Banana established at Trichy in 1993, with a mandate for improving production and productivity of banana through basic and strategic research, and also the Association for the Improvement in Production and Utilization of Banana (AIPUB) have done appreciable work in diffusion of technologies, which is evident in terms of improved production and productivity. Dissemination of technologies is a key to success. Therefore, publication of a book ***'Banana — New Innovations'*** edited by Dr H.P. Singh and Dr M.M. Mustaffa is a valuable effort. The book has broad coverage of genetic resources and their utilization for further breeding programmes, biotechnology, production system management, irrigation technologies, biotic and abiotic stresses, pests, post-harvest management and value-addition. I am sure that the book will be extremely useful for students, researchers, scientists, farmers, policy-planners and research and development managers.

Emile Frison
Director-General
Biodiversity International

Preface

Banana and plantain play a pivotal role in nutritional and livelihood security of the people across the world. There has been international initiatives, more intensely during the last 5 decades, which resulted in science- based technologies and focussed attention in improvement for sustainable production. In the scenario of declining land and water resources, threat of biotic and abiotic stresses for food and livelihood security, are the challenges for increasing population. In this context, banana could be one of the options for livelihood security.

Banana has its antiquity, and is interwoven in our cultural heritage, playing a significant role in socio-economic development of the country. It is filling, easy-to-digest, nearly fat-free, rich source of carbohydrate with high calorific value, and free from sodium, making it salt-free diet. Small holders and marginal farmers are highly benefited by growing banana as it provides regular income. Being a preferred fruit, during last decades, it has been witnessed a significant growth in the area and production of banana. However, there are many constraints which required to be addressed systematically with utilization of latest information and improved technologies. Therefore, to draw the vast resource of scientific talent available in the country and abroad, and to develop strategies for the improvement in production and utilization of banana, the book ***'Banana — New Innovations'*** has been brought out.

This book is a compilation of scientific information which covers national and international scenario of banana and plantain, genetic resource management, breeding and biotechnology, production system, nutrition and water management, insect pest and disease management, post-harvest management, product diversification and trade. Considering the past achievements and challenges ahead, we are sure, this compilation of scientific information will have a way for fine tuning of research, development and policies, to guide Indian banana industry on the road to success.

We are grateful to all fellow colleagues who have contributed their valuable chapters in this book. We sincerely appreciate the support of all the contributors in bringing out this book. Last but not the least, I am thankful to all those, who have helped in compilation, editing and publication of this book. Financial support provided by the NHB, ICAR and others are thankfully acknowledged.

H.P. Singh 18/5

(H.P. Singh)

About the Editors

Born on 2 July 1950, Dr H.P. Singh obtained his Doctorate in Horticulture from the University of Agricultural Sciences, Bangalore. Starting his career as a Scientist at the Central Horticulture Experiment Station, Chethalli (Coorg) in Karnataka in 1972, after post graduation in Horticulture, with Gold Medal, Dr Singh has contributed significantly to horticultural research and development. His professional career has earned 3 international and 19 national awards and 5 fellowships including the Fellow of National Academy of Agricultural Sciences (NAAS). In the career spanning of 37 years, Dr Singh has contributed significantly to horticultural research, management and development and also human capital formation. Dr Singh held various positions which include, Project Coordinator (Fruits), Director, NRC for Banana, Horticulture Commissioner, Govt. of India, Chairman, Coconut Development Board, Dean (Agriculture), GBPUA&T, Pantnagar, Vice–Chancellor, Rajendra Agricultural University, Pusa, and now Deputy Director–General (Horticulture), ICAR. He was also Chairman of APCC, Jakarta, National Director for internationally-aided projects and currently he was in a Chairperson of Steering Committee of BAPNET of Bioversity International.

Dr H.P. Singh has developed and released several varieties, technologies and products. He has contributed significantly to microirrigation development. He provided new dimension of mission mode approach for horticultural development. His contribution as an academician also created an impact on agricultural education. The efforts of Dr Singh has provided a strong foundation for horticultural development, leading to **Golden Revolution** in the country. He is also responsible for the establishment of National Research Centre for Banana at Trichy and creation of national field repository of banana and is a founder General Secretary and Editior-in-Chief of AIPUB and now the President. He has been closely associated with activities of Asia and Pacific Network ASPNET(INIBAP) and is member of several professional societies. Dr Singh has been instrumental in organizing several national and international conferences for the cause of horticulture and has headed committees as Chairman both in India and aborad. Mission mode approach for horticultural development, starting from planning to execution has been due to his insight. He has provided leadership to horticulture and is credited to have authored and edited 39 books and 30 bulletins, besides 175 research papers. He is widely traveled in India and abroad.

Dr M.M. Mustaffa, Director, National Research Centre for Banana, ICAR, New Delhi, has obtained his Doctorate in Horticulture from the University of Agricultural Sciences, Bangalore. Born on 01.08.1953, he did his B. Sc in Agriculture and M. Sc degree in Horticulture from Tamil Nadu Agricultural University, Coimbatore.

Starting his career as a Scientist at Indian Institute of Horticultural Research, Hessaraghatta, Bangalore, he has contributed significantly to horticultural research, management and development. He has specialized in Tropical Fruits, especially in banana, citrus and pineapple in the areas, *viz.* genetic conservation, canopy management, integrated nutrient management, irrigation management and use of growth regulators.

As a scientist, he has developed and released a banana variety, many cost-effective, improved production technologies and value-added products. Has assembled the largest collection of banana from India and also from other centers of diversity. He has developed many production based technologies like canopy management, modified high-density planting, integrated nutrient management and post-harvest technologies including package for export in banana. He is instrumental in the promotion of tissue culture plants, efficient water and fertigation management and precision farming in banana. He has developed many eco-friendly technologies for the control of many pests and diseases in banana.

He has been closely associated with the activities of Association for the Improvement in Production and Utilization of Banana as Vice-President and Banana Asia Pacific Network (BAPNET) as India's Representative. Associated with Bioversity, France in banana related international collaborative activities. He has publication in National and International Journals, authored book chapters, research papers in various National and International seminars, symposium and workshop and member of several professional societies.

During his career, he was awarded an International **"Pisang Raja Award"** by Banana Asia Pacific Network (BAPNET), Philippines and one national award "**Kadali Puraskar"** by Association for the Improvement in Production and Utilization of Banana (AIPUB). He is a "Fellow of Association" for the Improvement in Production and Utilization of Banana. Under his leadership, the National Research Centre for Banana, Trichy, was awarded the **'Plaque of Honour**" Recognition by Bioversity International- Commodities for Livelihood, France and Banana Asia Pacific Network (BAPNET).

Contents

Acronyms

AIPUB	:	Association for Improvement, Production and Utilization of Banana
BDA	:	Biological Diversity Act
CBD	:	Convention on Biological Diversity
HGRM	:	Horticultural Genetic Resources Management
IARCs	:	Intenational Agricultural Research Centre
ICAR	:	Indian Council of Agricultural Research
INIBAP	:	International Network for the Improvement of Banana and Plantation (Bioversity)
IPR	:	Intellectual Property Rights
ITPGRFA	:	International Trusty on Plant Genetic Resources for Food and Agriculture
NAGS	:	National Active Germplasm Sites
NBA	:	National Biodiversity Authority
NBPGR	:	National Bureau of Plant Genetic Resources
NHB	:	National Horticulture Board
NHM	:	National Horticulture Mission
PGR	:	Plant Genetic Resources
WTO	:	World Trade Organisation

Contributors

Ambika D.S.
Division of Fruit Crops
Indian Institute of Horticultural Research
Hessaraghatta – 560 089
Karnataka

Auxcillia J.
Department of Fruit Crops
Horticultural College and Research Institute
Tamil Nadu Agricultural University
Coimbatore – 641 003
Tamil Nadu

Balamohan T.N
Department of Fruit Crops
Horticultural College and Research Institute
Tamil Nadu Agricultural University
Coimbatore – 641 003
Tamil Nadu

Balasubramanian V.
National Research Centre for Banana
Tiruchirapalli – 620 102
Tamil Nadu

Balasubramanyan S.
Horticultural College and Research Institute
Tamil Nadu Agricultural University
Periyakulam East – 625 604
Tamil Nadu

Bandyopadhyay B.
AICRP Tropical Fruits
Banana Research Centre
Bidhan Chandra Krishi Viswavidyalaya
Kalyani, Nadia
West Bengal – 741 234

Byadgi A.S.
Department of Plant Pathology
University of Agricultural Sciences
Dharwad – 580 005
Karnataka

Chih Ping Chao
Taiwan Banana Research Institute
Chiuji, Taiwan
Republic of China

Chowdhury R. Ray
Department of Fruits and Orchard Management
Faculty of Horticulture
Bidhan Chandra Krishi Viswavidyalaya
Mohanpur, Nadia
West Bengal – 741 252

Daksha Bhatt
Sun Agrigenetics Pvt. Ltd
Vadodara – 390021
Gujarat

Das A.
Department of Plant Physiology
Faculty of Agriculture
Bidhan Chandra Krishi Viswavidyalaya
Mohanpur, Nadia
West Bengal – 741 252

Dattatreya M. Kadam
Central Institute of Post Harvest Engineering & Technology
Ludhiana – 141 004
Punjab

Deb P.
Department of Pomology and Post Harvest Technology
Faculty of Horticulture
Uttar Banga Krishi Viswavidyalaya
Pundibari, Cooch Behar
West Bengal – 736 165

Devinder Dhingra
Central Institute of Post Harvest Engineering & Technology
Ludhiana – 141 004
Punjab

Dinesh Kumar
Central Horticultural Experiment Station (IIHR)
Bhubaneswar – 751 019
Orissa

Dinesh M.R.
Division of Fruit Crops
Indian Institute of Horticultural Research
Hessaraghatta – 560 089
Karnataka

Firke P.V.
Director of Horticulture
Pune
Maharashtra

George V Thomas
Central Plantation Crops Research Institute
Kasaragod – 671 124
Kerala

Ghosh S. K.
Department of Pomology and Post Harvest Technology
Faculty of Horticulture
Uttar Banga Krishi Viswavidyalaya
Pundibari, Cooch Behar
West Bengal – 736 165

Gohil S.N.
Department of Horticulture
Mahatama Phule Krishi Vidyapeeth
Rahuri – 413 722
Maharashtra

Gupta S.N.
Department of Plant Physiology
Faculty of Agriculture
Bidhan Chandra Krishi Viswavidyalaya
Mohanpur, Nadia
West Bengal – 741 252

Hasan M.A.
Department of Fruits and Orchard Management
Faculty of Horticulture
Bidhan Chandra Krishi Viswavidyalaya
Mohanpur, Nadia
West Bengal – 741 252

Hiremath S.C.
Department of Botany
Karnataka University
Dharwad – 580 003
Karnataka

Indumathi S.
National Research Centre for Banana
Tiruchirapalli – 620 102
Tamil Nadu

Jai Oberoi
Vice President - Advisor
Fresh Bharti WalMart Pvt Ltd
Sector 53, Gurgaon
Haryana

Jegadeeswari V.
Department of Fruit Crops
Horticultural College and Research Institute
Tamil Nadu Agricultural University
Coimbatore – 641 003
Tamil Nadu

Jeyakumar P.
Department of Fruit Crops
Horticultural College and Research Institute
Tamil Nadu Agricultural University
Coimbatore – 640 003
Tamil Nadu

Jim Thomas
All India Network Project on Agricultural Ornithology
Department of Entomology, College of Horticulture
Kerala Agricultural University
Thrissur – 680 656
Kerala

Jonathan E. I.
Department of Nematology
Tamil Nadu Agricultural University
Coimbatore – 641 003
Tamil Nadu

Kadam U. S.
College of Agricultural Engineering
Mahatama Phule Krishi Vidyapeeth
Rahuri – 413 722
Maharashtra

Kavitha M.
Department of Fruit Crops
HC & RI, Tamil Nadu Agricultural University
Coimbatore – 641 003
Tamil Nadu

Kavitha P. S.
Department of Fruit Crops
HC & RI, Tamil Nadu Agricultural University
Coimbatore – 641 003
Tamil Nadu

Kotur S. C.
Indian Institute of Horticultural Research
Hessaraghatta
Bangalore – 560 089
Karnataka

Krishnan P.N.
Division of Biotechnology and Bioinformatics
Tropical Botanic Garden and Research Institute
Palode, Thiruvananthapuram – 695 562
Kerala

Kumar N.
Department of Fruit Crops
Horticultural College and Research Institute
Tamil Nadu Agricultural University
Coimbatore – 641 003
Tamil Nadu

Kumar V.
National Research Centre for Banana
Tiruchirapalli – 620 102
Tamil Nadu

Kumudha R.
National Research Centre for Banana
Tiruchirapalli – 620 102
Tamil Nadu

Maicykutty
College of Horticulture
Kerala Agricultural University
Vellanikkara
Kerala

Majumder D.
Department of Agricultural Statistics
Faculty of Horticulture
Bidhan Chandra Krishi Viswavidyalaya
Mohanpur, Nadia
West Bengal – 741 252

Mali M.D.
Department of Horticulture
MPKV, Rahuri – 413 722
Maharashtra

Mani Chellappan
All India Network Project on Agricultural Ornithology
Department of Entomology, College of Horticulture
Kerala Agricultural University
Thrissur – 680 656
Kerala

Mathew P.
College of Horticulture
Kerala Agricultural University
Vellanikkara
Kerala

Mathhew Prasad
Central Institute of Post Harvest Engineering & Technology
Ludhiana – 141 004
Punjab

Mohana S.
National Research Centre for Banana
Tiruchirapalli – 620 102
Tamil Nadu

Mukunthakumar S.
Division of Biotechnology and Bioinformatics
Tropical Botanic Garden and Research Institute
Palode, Thiruvananthapuram – 695 562
Kerala

Mustaffa M.M.
National Research Centre for Banana
Tiruchirapalli – 620 102
Tamil Nadu

Nache Gowda V.
Division of Horticulture
University of Agricultural Sciences
GKVK, Bangalore
Karnataka

Nalina L.
Department of Fruit Crops
Horticultural College and Research Institute
Tamil Nadu Agricultural University
Coimbatore – 641 003
Tamil Nadu

Nandhini P.
NRC for Banana
Thogamalai Road, Thayanur post
Trichy – 620102
Tamil Nadu

Narayana C.K.
National Research Centre for Banana
Tiruchirapalli – 620 102
Tamil Nadu

Narayana Swamy P.
Division of Horticulture
University of Agricultural Sciences
GKVK, Bangalore
Karnataka

Pandey V.
Central Horticultural Experiment Station (IIHR)
Bhubaneswar – 751 019
Orissa

Patil R. T.
Central Institute of Post Harvest Engineering & Technology
Ludhiana – 141 004
Punjab

Pitam Chandra
Asst. Director General (PE)
Indian Council of Agricultural Research
New Delhi – 110 012

Pitswati Buara
Horticulture Research Institute
Bangkok 10900
Thailand

Poornima K.
Department of Fruit Crops
HC & RI, Tamil Nadu Agricultural University
Coimbatore – 641 003
Tamil Nadu

Prashant Bhatt
Sun Agrigenetics Pvt. Ltd
Vadodara – 390 021
Gujarat

Pratibha Brahmi
National Bureau of Plant Genetic Resources
Pusa Campus
New Delhi – 110 012

Praveen G.
Division of Biotechnology & Bioinformatics
Tropical Botanic Garden & Research Institute
Palode, Thiruvananthapuram – 695 562
Kerala

Pushpalatha P.B.
College of Horticulture
Kerala Agricultural University
Vellanikkara, Thrissur
Kerala

Raghavendra Prasad T.
Division of Horticulture, University of Agricultural Sciences
GKVK, Bangalore
Karnataka

Rajamanickam C.
Department of Fruit Crops and PHT
HC & RI, Tamil Nadu Agricultural University
Periyakulam East – 625 604
Tamil Nadu

Rajendran L.
Department of Fruit Crops
HC & RI, Tamil Nadu Agricultural University
Coimbatore – 641 003
Tamil Nadu

Rajesh T.
National Research Centre for Banana
Tiruchirapalli – 620 102
Tamil Nadu

Rajmohan K.
Department of Pomology and Floriculture
College of Agriculture
Vellayani, Thiruvananthapuram – 695 522
Kerala

Rajmohan R.
National Research Centre for Banana
Tiruchirapalli – 620 102
Tamil Nadu

Raju V.K.
College of Horticulture
Kerala Agricultural University
Vellanikkara, Thrissur
Kerala

Ramakrishnappa K.
Director of Horticulture
Lalbagh
Bangalore – 560 004
Karnataka

Ravi I.
NRC for Banana
Thogamalai Road, Thayanur post
Trichy – 620 102
Tamil Nadu

Reddy B.M.C.
Central Institute for Subtropical Horticulture
Rehmankhera, Lucknow – 227 107
Uttar Pradesh

Rekha A.
Division of Fruit Crops
Indian Institute of Horticultural Research
Hessaraghatta – 560 089
Karnataka

Rema Menon
Banana Research Station, Kerala Agricultural University
Kannara, Thrisssur
Kerala

Reni M.
College of Horticulture
Kerala Agricultural University
Vellanikkara, Thrissur
Kerala

Roy D.R.
Central Institute of Post Harvest Engineering & Technology
Ludhiana – 141 004
Punjab

Santhanalakshmi K.
National Research Centre for Banana
Trichy – 620 102
Tamil Nadu

Sarkar S.
Department of Fruits and Orchard Management
Bidhan Chandra Krishi Viswavidyalaya
Mohanpur, Nadia
West Bengal – 741 252

Seeni S.
Division of Biotechnology and Bioinformatics
Tropical Botanic Garden and Research Institute
Palode, Thiruvananthapuram – 695 562
Kerala

Selvarajan R.
National Research Centre for Banana
Tiruchirapalli – 620 102
Tamil Nadu

Senthamizh Selvi
Department of Fruit Crops
HC & RI, Tamil Nadu Agricultural University
Coimbatore – 641 003
Tamil Nadu

Sharma S.K.
National Bureau of Plant Genetic Resources
Pusa Campus
New Delhi – 110 012

Sheela K.B.
College of Horticulture
Kerala Agricultural University
Vellanikkara, Thrissur
Kerala

Shiddlingeswara G.C.
Division of Horticulture
University of Agricultural Sciences
GKVK
Bangalore – 560 065
Karnataka

Singh H. P.
Deputy Director General (Horticulture)
Indian Council of Agricultural Research
New Delhi – 110 012

Sivakumar S. D.
Tamil Nadu Agricultural University
Coimbatore
Tamil Nadu

Smitha M.S.
College of Horticulture
Kerala Agricultural University
Vellanikkara
Kerala

Soorianathasundaram K.
Department of Fruit Crops
HC & RI, Tamil Nadu Agricultural University
Coimbatore – 641 003
Tamil Nadu

Suganya Devi P.
National Research Centre for Banana
Tiruchirapalli – 620 102
Tamil Nadu

Sukhada Mohandas
Indian Institute of Horticultural Research
Hessaraghatta
Bangalore – 560 089
Karnataka

Suma
Banana Research Station
Kerala Agril. University
Kannara, Thrisssur
Kerala

Sundararaju P.
National Research Centre for Banana
Thogamalai Road, Thayanur post
Tiruchirapalli – 620 102
Tamil Nadu

Suresh C.P.
Department of Pomology & Post Harvest Technology
Faculty of Horticulture
Uttar Banga Krishi Viswavidyalaya
Pundibari, Cooch Behar
West Bengal – 736 165

Thangavelu R.
National Research Centre for Banana
Tiruchirapalli – 620 102
Tamil Nadu

Uma S.
NRC for Banana
Thogamalai Road, Thayanur post
Tiruchirapalli – 620 102
Tamil Nadu

Vasudeva Rao V.
Project Coordinating Cell
ANGRAU
Hyderabad – 500 030
Andhra Pradesh

Veeraragavathatham D.
HC & RI, Tamil Nadu Agricultural University
Coimbatore – 641 003
Tamil Nadu

Vishal Nath
Central Horticultural Experiment Station (IIHR)
Bhubaneswar – 751 019
Orissa

1

Research and Development in Banana and Plantain — National and International Scenario

H.P. Singh

Horticulture, consisting of fruits, vegetables, spices, floriculture, medicinal and aromatic plants and plantation crops, a pleasantry before independence, has emerged as a core sector in agriculture, passing through various phases. Investments made in horticulture during Ninth and Tenth Five-Year Plans have been highly productive in transforming an agrarian economy in many states, which has provided insight for reversing the trend of ever-declining farmers' income, and above all, addressing the nutritional security and environmental concern. Thus, horticulture has emerged as the best option for diversification in agriculture for effective land-use planning. Growing economic importance of horticultural crops, especially of fruits and vegetables could be attributed to an increasing demand arising from domestic as well as overseas market. Initiatives taken by the government and other stakeholders have been rewarding in terms of increased production and productivity. Most significant developments in last decade happened is that horticulture has moved from rural confine to commercial production, and this changing scenario has encouraged private sector investment in production system management.

MEDICINAL AND THERAPEUTIC VALUES

Banana is known for its antiquity and it is interwoven with Indian heritage and culture. It is considered as the symbol of 'prosperity and fertility'. Owing to its greater socio-economic significance and multifaceted uses, banana is referred as '*Kalpatharu*' (plant of virtues) and *Kalpavriksh*. Botanically, banana fruit is a wonder berry, which forms the staple food of millions of people across the globe, providing more balanced food than any other fruit or vegetable. As a dessert, banana is more filling, easy-to-digest, fat-free, rich source of carbohydrates with a calorific value of 90 Kcal per 100 g fruit and free from sodium, making it a salt-free food fruit. It contains various vitamins and therapeutic values for the treatment of many diseases. In North-Eastern India, banana powder is used as a baby food. Being a good source of potassium, banana is good for heart patients. Stem core juice is a well-recognized medicine for dissolving kidney stones. Plantains are rich in vitamin A (beta carotenes) and they aid in digestion. Ripe fruits are being used in the treatment of asthma and bronchitis. Nectar from it flower buds is rich in vitamin, hence given for strengthening of babies. Juice from male buds provides effective remedy for all stomach problems. The powdered peel of ripe banana has antiseptic properties, hence used in healing of cuts and wounds. Banana skin is also a rich source of vitamin and is curative to retina related disorders.

Uses of Banana

There has been ample diversification in products. Frozen banana purees are being used in manufacturing of yoghurts, milk shakes, ice-creams, breads, cakes, banana flavored drinks, baby food and sauces. Banana flour obtained by drying and grinding the green fruits is easily digestible than the cereal flour. Banana juice is being used in production of beer with low alcohol content and rich in vitamin $B_{6,}$ can help to relieve stress and anxiety. Unripe banana and plantain can be dried and made into a meal, which can be used to substitute up to 70-80% of the grains in pig and dairy diets. Starch can be extracted from banana pseudostems and corms, which is used for making glue, and in manufacturing of cartons for exporting fresh bananas. Banana seeds are used for making necklaces and other ornaments. Use of sap as a dye, fruit as a meat tenderizer and ash in manufacturing of soap and anti-dandruff shampoos. Sap can be used as a natural film forming resin, which along with isopropyl alcohol produces better quality ink for inkjet printers. Dried banana peel contains 30-40% tannin, making it suitable for leather processing. The production of floor wax and shoe polish from banana peels are also being explored. Under non-edible sector of banana products, fibre plays a major role. Banana yields a fibre, which is used extensively in manufacturing of certain papers, especially where more strength is required. The papers are also being used for making tea bags and currency notes like those of Japanese Yen. It has use in textile industry, for making cordages and handicrafts.

NATIONAL AND INTERNATIONAL SCENARIO

Globally, banana (*Musa* sp.) is fourth most important commodity after rice, wheat and corn. It is produced in tropical and subtropical regions of developing economies. It is grown in more than 130 countries across the world in a 8.84 milliom ha, producing 97.15 million tonnes of banana and plantain. India is largest producer of banana in the world, contributing 21.0 per cent to the global production of banana, the total production being 23.2 million tonnes from 0.646 million ha. However, total production is largely consumed domestically and has negligible share in the global trade. Brazil is next largest producer of banana followed by the Philippines, Indonesia, China and Ecuador. There has been a growth in production of banana in China. The Philippines continue to lead in area and export among Asian countries. Ecuador produces 6.13 million tonnes from an area of 227 thousand ha, which has declined as compared to 2001-02 but has substantial share in export. In tropical zones of Colombia, plantain is not only an important part of the human diet but fruits and plants are indispensable feed for domestic animals as well. The total area under banana and plantain in Colombia has declined considerably to 62 thousand ha with a total production of 1.58 million tonnes. Mexico grows in one-sixth of total area, of which 35% is under irrigation. The total production of banana in the country is around 2.36 million tonnes from an area of 78 thousand ha. In most of these countries, banana and plantain are important from socio-economic point of view. Banana is largely traded from Latin American countries, South America and the Philippines in Asia, while plantain is largely consumed locally. Productivity is highest in Cost Costa Rica (Table 1.1).

Table 1.1. Area, production and productivity of banana and plantain in major banana producing countries

Country	*Area ('000ha)*		*Production (tonnes)*		*Productivity (tonnes/ha)*	
	2001-02	*2005-06*	*2001-02*	*2005-06*	*2001-02*	*2005-06*
India	466	565	14.20	19.19	30.50	34.00
Brazil	513	491	5.74	6.58	11.18	13.41
Philippines	400	415	5.06	5.63	12.65	13.56
Indonesia	285	315	3.60	4.87	8.14	15.49
China	259	269	5.40	6.25	20.84	23.20
Ecuador	298	227	8.03	6.13	26.94	27.07
Cameroon	312	082	2.25	0.79	7.21	9.72
Mexico	074	078	1.97	2.36	26.62	29.99
Columbia	444	062	4.20	1.58	9.45	25.15
Costa Rica	054	042	2.32	2.22	42.96	52.54
Others	5,705	5,705	41.77	41.77	7.32	7.32
Total	8,810	8,251	94.56	97.378	203.81	251.45

India has remained the largest banana producer in the world for the past one decade. Present production of 23.2 million tonnes is three times the production of next leading banana-producing country, Brazil. Almost all leading countries have showed growth in banana industry both in terms of area and production. This is mainly attributed to conversion of virgin soil into banana field and cultivating high-yielding Cavendish clones like Williams, Valery, Grand Naine *etc.* A decreasing trend in area and production is observed in Cameroon and Columbia, whereas in Cameroon, the reduction is attributed to shifting to perennial crops like oil palm and pest and disease problem in perennial banana orchards. Costa Rica remained the leading country in productivity (52.54 tonnes/ha) due to contract farming by main export giants like Dole and Chiquita. Irrespective of polyclonal situation with varying productivity, India remained the next best with 34 tonnes/ha followed by Mexico. Brazil has a productivity of only 13.5 tonnes/ ha owing to polyclonal situation and large area being under Pome groups. Though Philippines has the third largest area (0.42 million ha), it ranks only eighth in production next to China and Indonesia. Conversely, Indonesia and China which rank fourth and fifth in area under banana, are the third and sixth largest producers, indicating their higher productivity (Fig. 1.1).

In India, banana is well adopted in the regions varying from humid tropics to humid subtropics and semi-arid subtropics, and up to 2,000 m above mean sea-level. Among horticultural crops, contribution of banana to agricultural Gross Domestic Product (GDP) is highest (1.99 % (Fig 1.2). Availability of wide genetic diversity, varying production system, its suitability on wide range of agroclimatic conditions are attributed for a wide range adoption of banana. It is grown as homestead garden as well as commercial plantation. Commercial production system is largely irrigated, while subsistence cultivation is under rainfed farming. Interesting feature observed in the country is that commercial production has largely concentrated

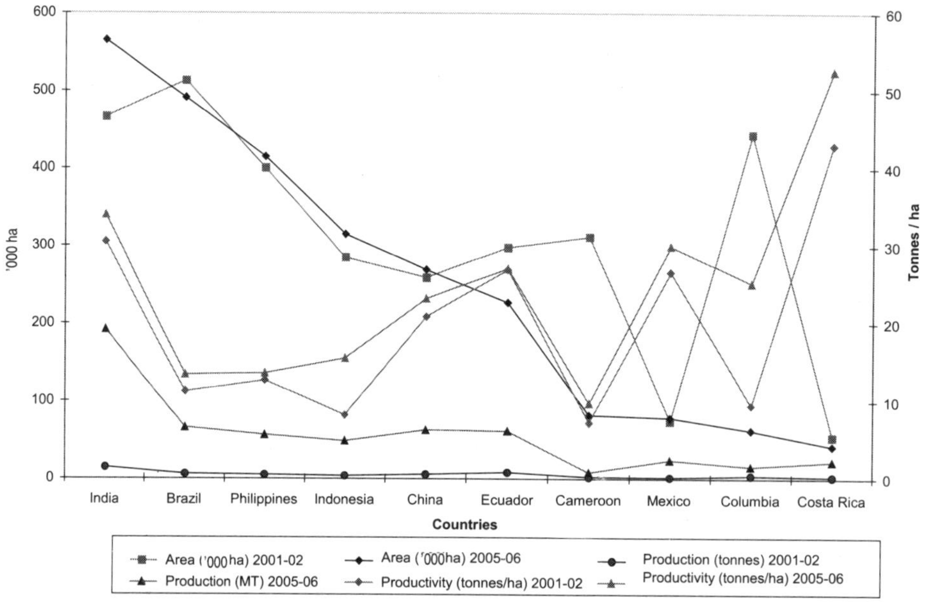

Fig.1.1. Area, production and productivity of major banana-growing countries

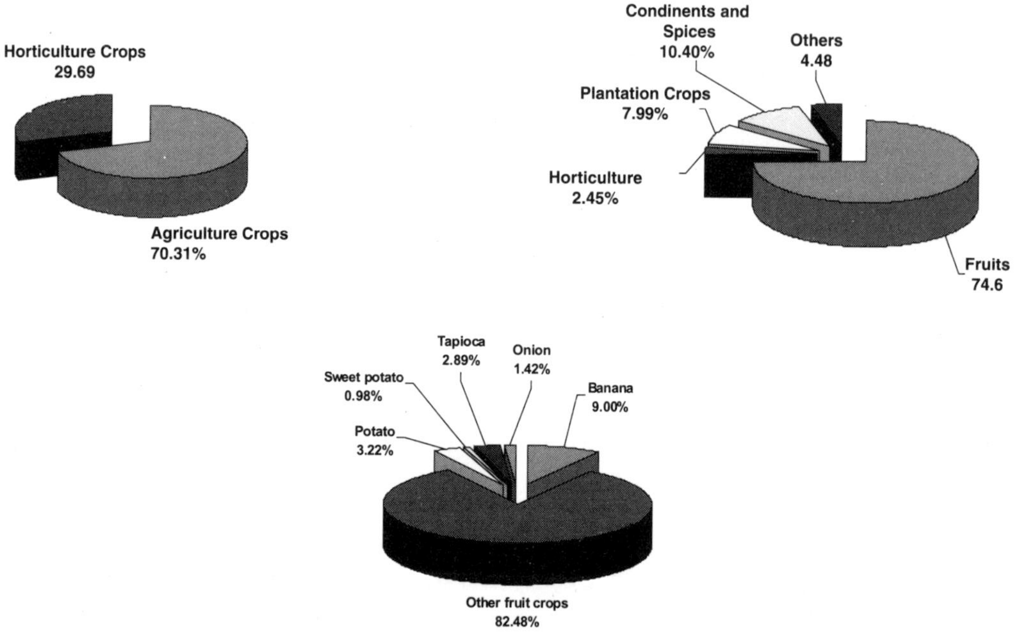

Fig. 1.2. Share of horticultural crops and banana in agricultural GDP

along the river. In Bihar, it is grown in Gangetic plains, while in Tamil Nadu, major concentration is in Cauvery basin. In Andhra Pradesh, concentration of production is around Godavari. In Maharashtra, it is more concentrated around the Tapti River. This phenomenon is attributed to soil and availability of water. However, its cultivation has also been extended to areas where water sources have been created. Banana has assumed more significance nowadays, as it is the source of regular income for small and marginal farmers, and responds very well to crop management system.

An interesting trend is observed in growth in production and productivity. The production has increased from 7.79 million tonnes in 1991-92 to 23.20 million tonnes in 2006-07. The productivity has increased from 20.30 to 38.75 tonnes/ha during the same period, while the area has increase from 383.9 to 646 thousand ha. A decline in production has been noticed during 2001-04 due to sever drought (Fig. 1.3).

DISTRIBUTION PATTERN IN INDIA

The major banana-growing states are Andhra Pradesh, Assam, Bihar, Gujarat, Karnataka, Kerala, Madhya Pradesh, Maharashtra, Orissa and West Bengal. Tamil Nadu has the largest area (1.14 lakh ha), followed by Maharashtra (0.80 lakh ha) and Andhra Pradesh (0.75 lakh ha). Tamil Nadu also ranks first in production (6.12 million tonnes), followed by Maharashtra (4.96 million tonnes). Gujarat and Andhra Pradesh are the next, with a total production of 3.15 and 2.25 million tonnes respectively. However, highest productivity is recorded in Maharashtra, followed by Gujarat and Tamil Nadu (Table 1.2). Cultivar and production system vary in these

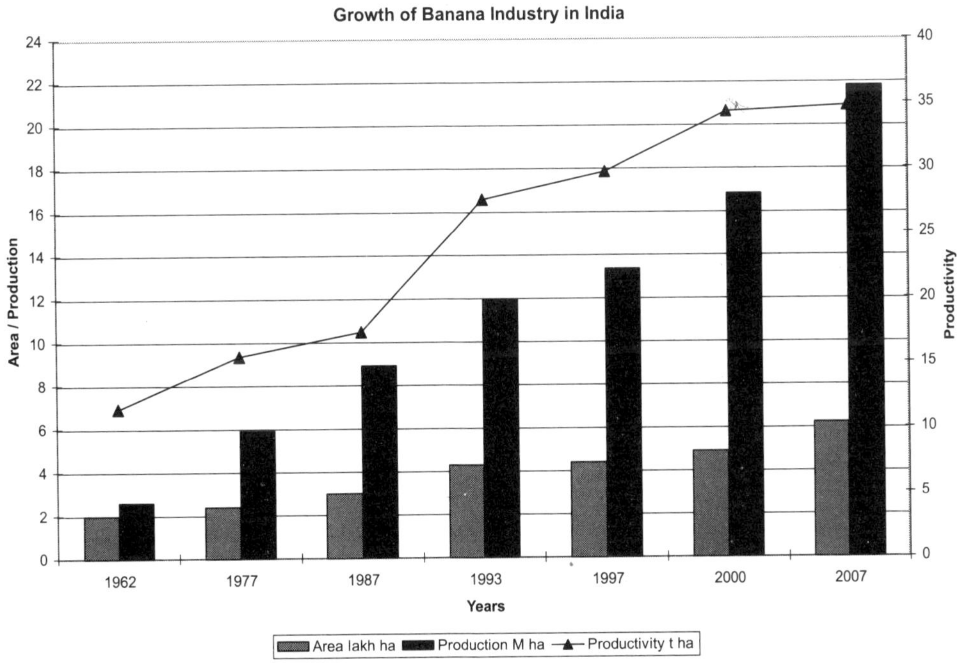

Fig. 1.3. Trend in area, productionand productivity in India

regions, which influence the productivity. Interestingly, variation in productivity ranges from 13.5 to 63.6 tonnes/ha, which is attributed to cultivars, production system and management strategies. Maharashtra excels in productivity owing to monoculture of high-yielding Cavendish and Grand Naine clones coupled with adoption of improved technologies, i.e. microirrigation, fertigation and use of tissue culture plants besides targeted management. Strategically, adoption of improved technologies in other states has potential to improve productivity. Production has increased largely due to increase in productivity during the decade (Table 1.2).

Table 1.2. State-wise area, production and productivity of banana in India

State	*Area ('000ha)*	*Production ('000 million tonnes)*	*Productivity (tonnes/ha)*
Andhra Pradesh	75.1	2254.3	30.01
Assam	44.1	610.9	13.85
Bihar	30.5	1329.4	43.58
Gujarat	57.7	3157.7	54.73
Karnataka	59.9	1513.3	25.26
Kerala	61.5	493.9	8.03
Madhya Pradesh	15.2	788.2	51.85
Maharashtra	80.0	4962.9	62.03
Tamil Nadu	114.1	6116.5	53.60
West Bengal	37.4	892.2	23.85
Others	71.4	1085.5	15.2
Total	646.9	23204.8	35.87

Source: FAO statistics 2006

VARIETIES GROWN IN INDIA

India is endowed with diverse agroclimatic conditions which have encouraged the development and sustenance of a large number of varieties catering to the local needs. Though, more than 20 varieties have assumed the status of commercial cultivation, Dwarf Cavendish forms the mainstay of the Indian banana industry, owing to its high yield, wide market acceptability, short crop duration and high economic returns per unit area. But this cultivar is not common in coastal region due to its high susceptibility to Sigatoka leaf spot disease. Poovan is another cultivar grown commercially in different regions for its wider adaptability and tolerance to drought and diseases. Njali Poovan (Elakki Bale or Ney Poovan or Safed Velchi) is also grown commercially in different regions of the country. Rasthali is significant in commercial production and its success largely depends on higher price it fetches. Virupakshi, Monthan, Karpuravalli and Chakia are also important in some regions of the country. Nendran or Rajali are grown in Kerala and Tamil Nadu. The clonal situation prevailing in different regions indicates that varietal need cannot be unified and would need regional consideration. Interestingly, Cavendish group of banana has largest area and production. Its contribution is more than 50% to the production (Figs. 1.4 and 1.5). The cultivars grown in different states are presented in Table 1.3.

Table 1.3. Important cultivars of banana grown in different states

State	*Cultivars*
Andhra Pradesh	Grand Naine, Dwarf Cavendish, Robusta, Thellachakkarakeli (Cavendish), Amrithpani, Monthan, Yenagu Bantha, Karpura-chakkarakeli (Mysore)
Assam	Jahaji, Borjahaji, Manjahaji, Robusta, Honda, Chenichampa (Mysore), Jatikol (Mysore), Digj (Silk), Kulpait (Silk), Manohar, Bhimkol, Kanchkol, Attikol, Bharatmoni
Bihar	Dwarf Cavendish, Chinia, Chenichampa, Alpan (Mysore), Malbhog (Silk), Muthia (Bluggoe), Kothia (Bluggoe), Gauria (Bluggoe), Kanthali (Pisang Awak)
Gujarat	Grand Naine, Dwarf Cavendish, Lacatan, Harichal, Gandevi Selection
Karnataka	Grand Naine, Elakki Bale, Dwarf Cavendish, Robusta, Poovan (Mysore), Karibale (Monthan), Rasabale (Silk), Jwaribale (Pome)
Kerala	Njali Poovan, Red banana, Nendran (Plantain), Palayankodan (Mysore), Poovan (Silk), Monthan
Maharashtra	Grand Naine, Safed velchi (AB), Dwarf Cavendish, Sindhuri, Ardhapuri, Lalvelchi (Mysore), Rajeli (Plantain)
Tamil Nadu	Grand Naine, Matti, Sevvazhai (Red Banana), Ney Poovan, Robusta, Poovan (Mysore), Rasthali (Silk), Virupakshi (Pome), Pachanadan (Pome), Nendran, Karpuravalli (Pisang Awak), Sakkai (Bluggoe), Monthan, Peyan
West Bengal	Amrit Sagar, Giant Governor, Lacatan, Mortman (Silk), Champa (Mysore), Kanthali (Pisang Awak)

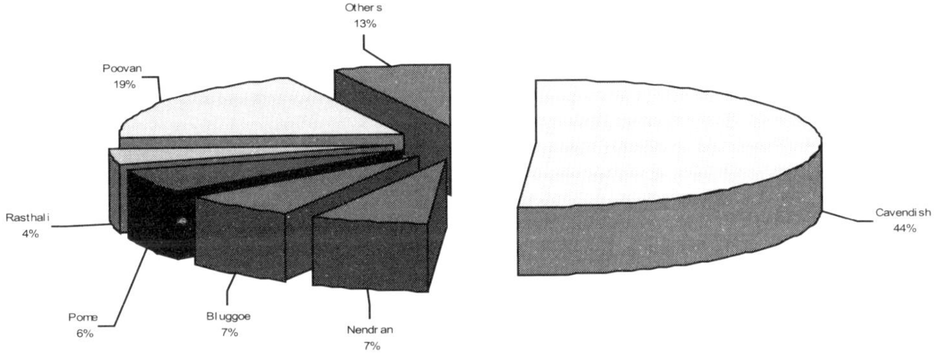

Fig. 1.4. Area (%) under different banana cultivars

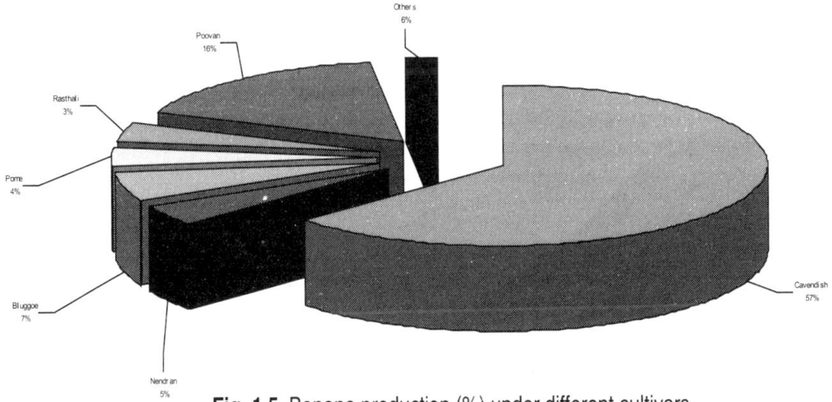

Fig. 1.5. Banana production (%) under different cultivars

Dwarf Cavendish (AAA)

Dwarf Canvendish group of cultivars constitutes 57% of the total production owing to its high yield, ability to withstand strong winds, short crop cycle, and good response to microirrigation, and high economic returns/unit area. This group of banana is known by different names having different cultivars, i.e. Basrai, Sinduri, Bhusawal, Robusta, Giant Governor *etc.* Gandevi selection known as Hanuman or Padarse has gained popularity. A selection Madhukar is becoming popular in Karnataka, owing to very high bunch weight (50-60 kg) in light red soil under optimum nutrient and water management. Grand Naine, introduced during 1995, has also been well-adopted for commercial production and has covered considerable area.

Red Banana (AAA)

Red Banana, a choice variety, is popular in Tamil Nadu, Kerala, Karnataka, Andhra Pradesh and to some extent in western and central India. Its plant is very tall and robust in stature (2.5-3.5 m tall and 70-80 cm girth). The pseudostem, petiole, midrib and fruit skin are purplish-red in colour. Crop duration varies from 15 to 18 months. The bunch weight is 15-20 kg with 7-8 hands having 60-80 fruits. The length of fruits is 15-18 cm and girth 15-16 cm. Fruit is less sweet and pulp is orange-yellow with a characteristic flavour. The fruit skin is thick and purplish-red. It is highly susceptible to leaf spot diseases, Fusarium wilt (race-1), weevils, nematodes apart from viruses like Banana Bunchy Top, Mosaic Virus and Banana Bract Mosaic Virus. Red Banana often reverts to a green type known an Venkadali, Pacha Kappa, Sandhana Vazhai and Vella vazhai with green colour of pseudostem, petiole and fruit skin instead of red. Fruit skin turns to attractive yellow on ripening. Rest of the characters are similar to that of Red Banana. Red Banana is an unstable mutant of its green counterpart. It is believed to have many medicinal properties. This cultivar is cultivated in some parts of South India. It is low-yielding and takes longer duration with bunches weighing 20-22 kg each and 70-90 fruits.

Rasthali (AAB)

It is the choicest table banana for its tasty, crisp, good sweet-blended and pleasant flavoured fruits. Plant is medium-statured, takes 13-15 months for harvest with bunches weighing 15-18 kg each. When grown in acidic soils, its fruits have hard lumps. *Fusarium* wilt is a major threat to this cultivar resulting in restricted cultivation. There has been a significant achievement in selecting a cultivar under the group, which can tolerate infestation.

Poovan (Mysore, AAB)

A popular cultivar grown all over the country in a perennial cropping system, it is a leading commercial cultivar of southern and north-eastern states. It bears heavy bunches weighing 20-24 kg each, with closely packed short and stout fruits having a conspicuous beak. Its fruits are slightly acidic and crop duration being 13-14 months. Its ease of cultivation and hardiness make Poovan a popular cultivar. It is known by different names, *viz.* Mysore, Poovan, Palayankodan, Chiru Champa, Chakkarkeli *etc.*

Nendran (French Plantain, AAB)

It is most prized cooking variety used in Kerala, fetching a premium price during festive occasions. It is also cultivated in Tamil Nadu for marketing in Kerala or for processing. Nendran is generally short-duration having bunch weight of 8-15 kg, with 30-50 fingers/bunch. The fruits remain starchy even on ripening. A number of ecotypes of Nendran, *viz.* Zanzibar, Otta Moongil, Moongil, Kali Ethan, Valiettan, Manjeri Nendran, Chengalikodan, Nedu Nendran, Chenganacheri Nendran, Attu Nendran, Myndoli and Padali Murian are known. Moongil is considered to be a mutant of Nendran with only one hand with large fruits and is grouped as plantain.

Hill Banana (Pome, AAB)

An elite banana in south India, it is preferred for its fruits having unique aroma and taste. It fetches high price over other banana varieties. Bunches weigh 11-13 kg having about 60 stout fruits, which have a very good keeping quality. Pulp is not so juicy but rather dry and sweeter. It is grown mainly under perennial system of cultivation on hills of Tamil Nadu and Karnataka. The cultivar is known as Virupakshi, Sirumalai, Ladan *etc.* There is variability within the cultivar. Interestingly, this group develops better flavour when grown in hills. There has been threat to this cultivar due to banana bunchy top. However, rehabilitation using disease-free plants has helped in its recovery.

Monthan (ABB)

This vegetable banana is becoming popular due to its therapeutic value. It is grown commercially. There are a number of varieties in this group but there is a regional preference. Yenugo Bontha has been selected and promoted for cultivation, which has wider acceptability. Popularly known cultivars are Karibonth, Mannan, Monthan, Batheera Kothia, Muthia, Gauria etc. There is an emerging potential for widening its commercial cultivation. Karpuravalli (ABB) belonging to Pisang Awak group is highly tolerant to drought, salt and wind. It is cultivated for its earliness in production. The cultivar is popular in Tamil Nadu and other states. Its fruit is starchy but very sweet when fully ripe.

Ney Poovan (AB)

Commonly known as Elakkibale, it is grown in Karnataka. It is becoming choicest cultivar due to its excellent aroma and better post-green life. Although, yield is poor as compared to Cavendish, but farmers get compensated by higher price realisation. There is a potential for widening market base and extend its cultivation.

Udhayam (ABB)

A new variety released by NRC for Banana, Trichy, average bunch weight is 37 kg having a potential up to 45-50 kg, which is 40% higher than local Karpuravalli. Crop duration is 13 months and produces high yield in ratoons also. Exhibits field tolerance to Sigatoka leaf spot diseases and nematode infestations. Its cylindrical bunch with well-spaced hands are suitable for long distance transportation and less prone to fruit damage during transit. Its fruits are high in sugar with 31° Brix with good blend of acidity and sweetness, hence suitable for processing into value-added products like figs, banana juice, wine, etc. It performs well in Tamil Nadu, Kerala, Andhra Pradesh, Karnataka, Bihar and Tripura.

HYBRIDS

Co 1

With typical acid/apple flavour it is a substitute for Hill banana, which develops aroma when grown at high altitude. However, its commercial cultivation has remained restricted due to small-sized bunch and low yield.

BRS 1 (Agniswar × Pisang Lilin)

A hybrid from BRS, Kannara, it has a very short cropping cycle. Plants are resistant to leaf spot, *Fusarium* wilt and burrowing nematode. Its medium statured plants produce 14-16 kg bunches and do not need support. Elongated fruits turn attractive golden yellow on ripening. Slightly acidic fruits, upon full ripening develop high sugar content. H-1 has faster ratoonability and completes 4 crop cycles in 3 years.

BRS 2 (Vannan × Pisang Lilin)

Developed at BRS, Kannara, its plants grow up to 7-8 feet. Crop cycle is short (11-12 months). On an average, a bunch weigh from 15 to 20 kg with short, stout, dark green Poovan like fruits arranged very compactly. Fruits are slightly acidic with pleasant sweet and sour aroma. It is tolerant to leaf spot diseases and nematodes, and suitable for subsistence cultivation.

FHIA 01

A hybrid derivative of AAB – Pome cv. Prata Ana × Prata Ana (Dwarf Prata), SH 3142 from FHIA, Honduras, introduced in India has good apple flavour and is resistance to Sigatoka, burrowing nematodes, and races 1 and 4 of Fusarium wilt. The cultivar appeared to be a substitute to Ladan cultivar. This cultivar produces a bunch weighing 32-35 kg. With excellent pulp quality, it is an excellent cultivar for processing. This cultivar needs to be popularised for large-scale production.

SABA (ABB)

An exotic introduction from the Philippines, it has a yield potential of 40kg bunch with an average weight of 35-37 kg. Highly suitable to saline sodic soils even up to pH 8.8-9.0, it is resistant to Septoria leaf spot disease. It is best substitute to local Adukku Monthan, Chakkia clones.

FHIA 18 (AAB)

It is a Pome type banana. It is a dessert type with sweet acid taste and apple flavour. It is resistant to black Sigatoka and provides bunches of 25-30 kg each.

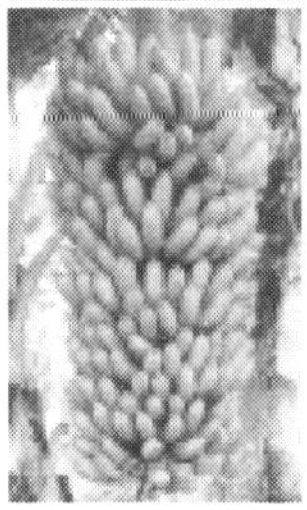

FHIA 23 (AAAA)

This is another important hybrid developed from the cross AAA – Gros Michel × Highgate × SH 3362. The promising features are tolerance to black Sigatoka and resistance to race 1 of *Fusarium* wilt. It is cultivated extensively in Cuba where it is twice as productive as Cavendish because of its tolerance to black Sigatoka.

FHIA 25 (AAAB)

It is a hybrid SH 3648 (4x) × SH 3142 (2x). It produces very large bunches of 47 kg. Its plants are dwarf, robust with high resistance to black Sigatoka, long green life, easy to peel, fast to cook with excellent texture and flavour when both boiled and fried.

RESEARCH AND DEVELOPMENT

Institutional Efforts

Recognizing the importance of banana and plantain at global level in terms of food security to millions of small farmers throughout the developing countries of tropics and subtropics, the International Network for the Improvement of Banana and Plantain (INIBAP) was established in 1985 and now this organisation is known as Biodiversity International. Of late, improvement in cultivars having resistance to biotic and abiotic stresses is receiving focus globally. In this context, management of *Musa* genetic resources is a key activity of INIBAP. A germplasm is basic resource to develop improved varieties. It underpins the efforts by national programme to identify the varieties that best suits to the needs of local farmers and consumers. The BAPNET, a networking activity between South Asia and South East Asian countries with headquarters at the Philippines, is also involved in coordination of *Musa* activities. In addition, PROMUSA, a global programme for banana and plantain improvement, has successfully demonstrated how the network and partnership approaches could enhance the deliverables, through catalyzing the activities of different national programmes.

A global *Musa* genomics consortium launched by INIBAP, reflect the strength of the networking approach. In recent years, *Musa* genome offers new perspectives in the monocotyledons due to its small size and availability of different ploidy levels and species within *Musa*. The combination of parthenocarpy with sterility and vegetative propagation, place the banana among one of the few opportunities to study the specific gene expression in different chromosomal environment. The Fundacion Hondurena de Investigation Agricola (FHIA), Honduras, has developed many banana hybrids, which are resistant to multiple diseases and pests having high and consistent yield. This has demonstrated that there exist a potential to develop new cultivars through tireless efforts and commitments.

India is one of the major centres of *Musa* diversity in South-East Asia with more than 8 species and 120 distinct clones. The *M. acuminata* s.sp. *burmannicoides* (AA) is one of the most important accessions which has become a major gene donor of black Sigatoka resistance. A total of 310 core accessions collected at NRC for Banana, Tiruchirapalli, have been described and documented; yet there is huge diversity, which need to be collected and conserved. Some success has been achieved through selection from the existing variability; identification and popularisation of superior clones have to be intensified. India being a source of high diversity in banana and plantain, conventional breeding in conjunction with molecular breeding approaches under network mode would be rewarding for banana in future. Mutation breeding attempted in past has not yielded the expected results but has widened the knowledge. With success in biotechnology, mutation breeding can be successfully used. With the advancement in banana improvement programme at global level having focus to sequence *Musa* genome to identify each gene, describe its function, map the gene expression, develop genetic markers for major traits and with better understanding of transgenic expression. There is all likelihood for a breakthrough in banana breeding by which the challenges of biotic and abiotic stresses could be faced and to produce more for the growing population. Thus, all efforts in future would be needed in breeding to achieve the target.

India is credited to have emerged as the largest producer of banana with unprecedented yield increase. This has been possible due to adoption of improved production technologies, which includes, high-density planting, use of *in-vitro* plants, fertigation and management of insect pests and diseases. These developments have lead to the innovation of precision farming technology in banana, to reap the maximum output. However, there is a regional disparity in adoption of technologies. Insect pests and diseases continue to be a major problem. There is a lot of scope for improving the post-harvest management and diversity. Network programme in participatory mode having horizontal and vertical integration could be a key to success.

Production System and Crop Management

Banana, a group of crops consisting of dessert banana, cooking banana, plantain, pome, silk and others, has different production systems and technologies. Depending upon resource availability, cultivars, traditions and marketing, different production systems have been adopted, which have their own merits. In Maharashtra and Gujarat, banana is grown in rotation with one plant crop, while tall cultivars are maintained for 4-5 years in other parts of the country. Under wetland banana system, deep trenches after every two rows are common in Tamil Nadu and parts of Karnataka. Garden land is widely adopted system of production in many banana-growing states, while planting on hill slope is common in Tamil Nadu, Karnataka and North-Eastern regions. Normally, cultivars adopted in this system need low inputs. Banana is also grown as subsistence crop in homestead or as shade or nurse crop in plantations. Mixed cropping with arecanut and coconut is grown in Kerala, coastal Karnataka, Tamil Nadu, West Bengal and Andhra Pradesh. In commercial planting, banana is invariably managed with high inputs.

Traditionally, banana has been propagated by suckers or rhizomes but realising the advantages of *in-vitro* propagation, mass multiplication through shoot tip culture has gained popularity for the production of disease-free plants. High multiplication rate per unit time, better establishment with hardening, uniformity at flowering and harvesting, higher returns with increased yields and freeness from nematode, fungal and bacterial pathogens are some of the advantages which have attracted many commercial firms in India for mass multiplication using micropropagation techniques.

Success in *in-vitro* propagation of banana is reported in many tissue culture units, which have established its credentials through effective quality control system. Currently, about 8-9 million tissue-cultured plants are planted annually. However, there is a need to focus on quality assurance more critically. To achieve this, quality standards have been developed for an ideal tissue-cultured plant and certification standards for banana tissue culture laboratories have also been formulated. In addition, to have an uniform certification of tissue-cultured banana plants produced by laboratories, the Department of Biotechnology (DBT), New Delhi, has accredited molecular biology and virology laboratories of NRC, Banana, for fidelity testing and virus indexing of tissue cultured plants of banana for its genetic purity and freeness from virus diseases. Thereby, quality of micropropagated plants is certified so as farmers are benefited. Biological hardening using consortium of mycorrhiza has further enhanced the utility of this technology.

Significant increase in yield due to high-density planting has been achieved. A plant population of 4,500 plant/ha is practised in Maharashtra on vertisol, while in other states 3,000-3,500 plants/ha are planted in three-crop cycle. Double-row system of planting advocated has been practised in Maharashtra and Gujarat. In this system, row-to-row distance is maintained at 1.2 m and distance between two rows is maintained at 1.8-2.0 m. Plant-to-plant distance is kept at 1.2 m. A plant population of 5,200 plants/ha can be accommodated in this system. In a modified high-density planting system, three suckers/pit at a spacing of 1.8m × 3.6m or 2.0 × 3.2m, accommodating 4,500-4,600 plants/ha is becoming popular in Tamil Nadu and Karnataka. This method increases the productivity by 50%, reducing the cost of production by saving 25% fertilizer and 30-40% water. The system provides space for growing intercrops in the initial stages. The system has also helped in reducing the cost of laterals where drip irrigation is followed.

Generally, banana needs 150-200g N, 40-60g P_2O_5 and 200-300 g K_2O/plant/year depending upon soil and cultivar. Full dose of P_2O_5 at planting, 3 splits of N and K are recommended. Application of one-fourth N and one-third K_2O at reproductive phase is also found beneficial. Application of N in reproductive phase delays the senescence of leaves, improving the bunch weight. Finger filling is better when one-third of K_2O is applied at shooting stage. Response to Zn, B and Mn sprays has been observed. The NRC for Banana, Trichy, and IIHR, Bangalore, have developed effective micronutrient mixtures to correct the deficiencies and to increasee yield and quality of fruits. To integrate the nutrient needs, plant analysis could be practised for efficient utilisation of nutrients. Azospirillum and VAM association increase yields in Poovan and Ney Poovan cultivars. With changing scenario in banana production, efficient nutritional management system needs emphasis for reduced cost of production and increased productivity.

Organic source of nutrients improves quality and soil health by increasing the rhizosphere microflora and improving quality and shelf-life of bananas. Banana responds very well to applied irrigation. Drip irrigation is becoming increasingly important. This system has resulted in increasing the productivity. Drip irrigation is widely adopted especially in low-rainfall area. Mulching using sugarcane trash (5-6 tonnes/ha) has been recommended for better production. Polythene mulch is also found beneficial for conserving moisture. Application of fertilizers through drip irrigation is effective for enhancing nutrient efficiency and increasing productivity. Fertigation increases the yield in Robusta and Grand Naine significantly, with a need-based application and economy in fertilizers. This helps in efficient utilization of nutrients. The popularisation of tissues cultured plants has necessitated higher requirement of fertilizers 300g N, 50-75g P_2O_5 and 400-450 g K_2O/ plant /year in 7-8 split applications at 30 days interval from 30 days after planting. This is recommended for getting higher yield. Wrapping of bunches is effective in enhancing the quality. Covering bunches with polythene sleeves improve the quality of bunches to export grade.

Pest and Disease Management

Banana weevil (*Cosmopolites sordidus*) and pseudostem-borer (*Odoiporus longicollis*) are widely distributed causing considerable damage. Though, mites (*Tetranychus*) and thrips cause huge damage to flowers and fruits, the situation is not alarming. Banana aphid (*Pentalonia nigronervosa*) is a cause of concern as a vector in transmitting Banana Bunchy Top Virus (BBTV), in areas under perennial cultivation. But in India, most of the commercial cultivars like Cavendish clones and plantain are under annual cultivation, while Pome, Silk, Mysore are under 1+2 production cycles with effective controls to check the build-up of aphids. Eco-friendly approaches using pseudostem traps with bioagents like *Beaveria* sp. and EPN *Heterorhabditis indica* are effective in controlling weevils in upland and different systems of cultivation. Transgenic plants with weevil resistance developed in Isreal and FABI, South Africa are under field-testing. Development of pheromones and semio-chemical based kairomones specific to these pests are also in progress. Scarring beetle (*Besilepta subcostatum*) is a serious pest only in northeast and eastern states (Table 1.4). The fruits affected by this pest have poor market acceptability.

Presence of *Radopholus similis* (burrowing nematode), *Pratylenchus coffeae* (root-lesion nematode), *Meloidogyne incognita* (root-knot nematode), *Helicotylenchus multicinctus, H. dihystera* (spiral nematode) and *Heterodera oryziocola* (cyst nematode) are in different parts of the country (Table 1.5). The *Radopholus similis* and *Pratytenchus coffeae* are widespread all over India. The *Helicotylenchus* is more serious in north-eastern region, while cyst nematode is a serious problem in Kerala only. Growing sunnhemp as an intercrop and application of neem cake invariably reduce the population of nematodes. Breeding for nematode resistance has been a thrust area and many diploids have been identified with tolerance traits. Effective local biocontrol agents like *Pseudomonas fluorescence*, *Paecilomyces lilacinus* and *Trichoderma viride* have been identified for the management of nematodes. Use of VAM (*Glomus* sp.) has also encouraging results in integrated management of nematodes.

Among fungal diseases, *Fusarium* wilt and leaf spot diseases and Banana Bract Mosaic Virus (BBrMV), Banana Streak Virus (BSV), Banana Bunchy Top Virus (BBTV) among viral diseases are important (Table 1.6). Fusarium wilt *(Fusarium oxysporum* f.sp. cubense*)* race 1 is serious in Rasthali (Silk AAB), Karpuravalli, Pachanadan and Virupakshi banana in the order of severity. Of late, tropical race 4 is becoming a serious threat to Cavendish group of bananas in Malaysia, Indonesia, and the Philippines which needs effective quarantine to contain the entry of this disease to India. Sigatoka *(Cercospora musae),* the imperfect stage of yellow sigatoka, is of importance only in humid tropics and coastal regions. Choice of cultivar, cropping season and production systems have effectively contributed towards reduced incidence of this disease. Effective fungicide and chemical management strategy has been developed to control post-harvest diseases. Heart rot (*Erwinia* sp.) is becoming a serious limiting factor in tissue culture banana cultivation for which effective chemical and biocontrol agents have been identified.

Among viral diseases, BBTV is a serious problem. It has devastated large plantations of Hill banana, an elite cultivar in Tamil Nadu. Collection of plants from disease-free areas, timely roguing and field sanitation are effective in checking its incidence. Aphid vector, *viz. Pentalonia nigronervosa* is present in all regions, especially in plantations with perennial cultivation and even virus-free planting material, fails to remain free for a long time due to secondary infection. Banana Streak Virus (BSV) in Poovan banana is found to be present in all banana-growing regions with varied degree of symptoms. This is becoming serious in other cultivars also. The BBrMV is also widespread, causing considerable damage. BBrMV is serious in Ilantain, Robusta and Ney Poovan banana spreading fast through infected suckers. The CMV or variegated chlorosis is also found in many banana-growing regions, which spread through infected planting material.

Since, there is no control measure for viral diseases, selection of disease-free planting materials and mother plants for micropropagation are very important. The planting material and micropropagated planting material should be diagnosed for presence of viruses. Accordingly effective molecular diagnostic techniques like ELISA, PCR and RTPCR have been developed at NRC for Banana, Trichy, and are being used for simultaneous detection of 2-3 viruses.

Table 1.4. Banana insect pests, distribution, symptoms and control measures

Pest	*Areas of occurrence*	*Symptoms*	*Susceptible cultivars*
Rhizome weevil *(Compolites sordidus)*	Kerala, Assam, Maharashtra, Bihar, Andhra Pradesh	Galleries of network in rhizome	Nendran, Poovan Malbhog
Pseudostem-borer *(Odoiporus longicollis)*	Andhra Pradesh, Bihar, Assam, Karnataka	Exudation of plan sap and drooping of faecal matter from the holes	Nendran, Kunnan, Poovan, Rasthali
Scarring-beetle *(Besilepta subcostatum)*	Bihar, Assam, West Bengal	Fingers have full of scars	All cultivars
Aphids *(Pentalonia nigronervosa)*	Kerala, Tamil Nadu, Karnataka	Bunchy appearance of crown and plant is unproductive	All cultivars
Mealy bugs *(Icerya aegyptiaca)*	Southern and North-Eastern states	Fruit loses consumer appeal	Pisang Awak and its ecotypes
Thrips *(Thrips florum)*	All banana-growing regions patches with scraped	Fruits have brown look	Cavendish group, Plantain group, Silk and Monthan

Table 1.5. Major nematodes and their distribution in India

Nematode	*Distribution*	*Susceptible cultivars*
Root-knot nematode *(Meloidogyne incognita)*	Tamil Nadu, Bihar, Andhra Pradesh, Karnataka	Nendran
Burrowing nematode *(Radopholus similis)*	Kerala, Bihar, Maharashtra, Andhra Pradesh, Nendran, Poovan, Rasthali, Karpuravalli	Robusta, Karnataka, Gujarat, Tamil Nadu
Root-lesion nematode *(Pratylenchus coffeae)*	Kerala, Tamil Nadu, Andhra Pradesh, Karnataka, Gujarat	Almost all commercial cultivars
Spiral nematode *(Helicotylenchus multicinctus)*	Kerala, Assam, Bihar, Maharashtra, Tamil Nadu	Almost all commercial cultivars
Cyst nematode *(Heterodera oryzicola)*	Kerala, Karnataka, Assam	

Table 1.6. Important diseases and their distribution

Disease	*Symptoms*	*Susceptible cultivars*
Fusarium wilt (Panama wilt) *Fusarium oxysporum* f.sp. cubense	Symptoms appear only 5-6 months after planting, yellowing of older leaves. Splitting of pseudostem at base	Silk, Ney PoovanPisang Awak, Pome, Bluggoe and Monthan
Sigatoka leaf spot *Mycosphaerella* sp.	At earlier stage, small lesions appear on leaves which become pale-yellow or greenish-yellow streaks	Cavendish and Mysore group
Heart rot *(Erwinia carotovara)*	Planted suckers start rotting and emit off-odour. In older plants rotting appears at collar regions and leaf base. Base of pseudostem gets swollen and split at later stages	Tissue cultured Cavendish, Robusta, Grand Naine
Cigar end rot	Black necrotic rot appears at tip of fingers and extends towards whole arrears of fruits	Bluggoe group, Nendran, Ney Poovan
Anthracnose *Colletotrichum musae*	Black lesions on fruit skin	Cavendish and all varieties

Post-harvest Management

Banana is highly perishable. The post-harvest losses are estimated around 25-30 %. At ambient temperature, banana has an average shelf-life of 1-5 days. The short storage life of banana is worsened by marketing system. The marketing system in India usually involves several retailers. Buying and selling takes time, leading to increased damage. Transportation is often delayed, and can fail altogether, because of poor conditions of vehicles and roads. A combination of high perishability, high ambient temperature, slow marketing systems, and poor market conditions lead to losses in fruit quality, and ultimately to post-harvest losses. Shelf-life could be extended by appropriate storage at 13-14 ºC. Effective pre and post-harvest management technology is available to extend the storage life and also export of banana. To maintain product quality throughout the market chain, whole system, from farm to consumer, should be considered. Banana and plantain can be processed into different byproducts as value-addition to tide over the glut and to get maximum profitability. The NRC for Banana has developed 12 value-added products and commercialized four of them which are available in market nowadays.

Improvements in one area may be ineffective if other areas of the retail system are not considered. Cultural practices also affect the post-harvest quality of banana. Timely cultivation, moderate use of nitrogen fertilizer, avoidance of drought, and control of fungal infection, all increase the storage potential. Fruits harvested at appropriate maturity ensure the highest quality. In commercial production of dessert banana, date of harvesting is based on size rather than maturity, because fruits can be ripened as required by exposure to ethylene. Market standards dictate the size at harvesting. It is essential to ensure the exact time to harvesting. For this, maturity indices could be followed. In banana, maturity indices for harvesting are degree of fullness of fingers, i.e. disappearance of angularity in a cross-section. Banana is harvested mature-green and ripen upon arrival at destination markets since fruits ripened on plant often split and have poor quality. Post-harvest management is essentially required to improve better return and to reduce losses.

Marketing and Supply of Banana

Banana supply and marketing are mainly governed by private sector except in some cases where cooperatives handle its marketing. Unlike in other commodities, cash marketing is most common, where money payment is done immediately after every transaction. Banana markets are specialized markets located in all big cities, and most of the marketing functions are followed. Primary wholesale market is major marketing yard for all commercial clones in the country. Irrespective of location of farm, bunches are transported by growers to nearest towns. In some cases, bunches are sold by farmers to pre-harvest contractors even before harvesting, at 50% of the cost in advance. After harvesting, bunches are transported to primary wholesale markets by growers or by pre-harvest contractors. Secondary wholesale markets are located at distant places from major area of production. Grading and sorting are mainly done in marketing yards for the convenience of price fixing. Though there is no definite yardstick for grading and sorting, size of bunches and external fruit appearance determine the quality of produce. In last 10 years, average price of banana has shown increasing trend while seasonal fluctuation is very wide. Pre-winter and summer prices are generally high while prices come down during October-December (Figs 1.6 and 1.7).

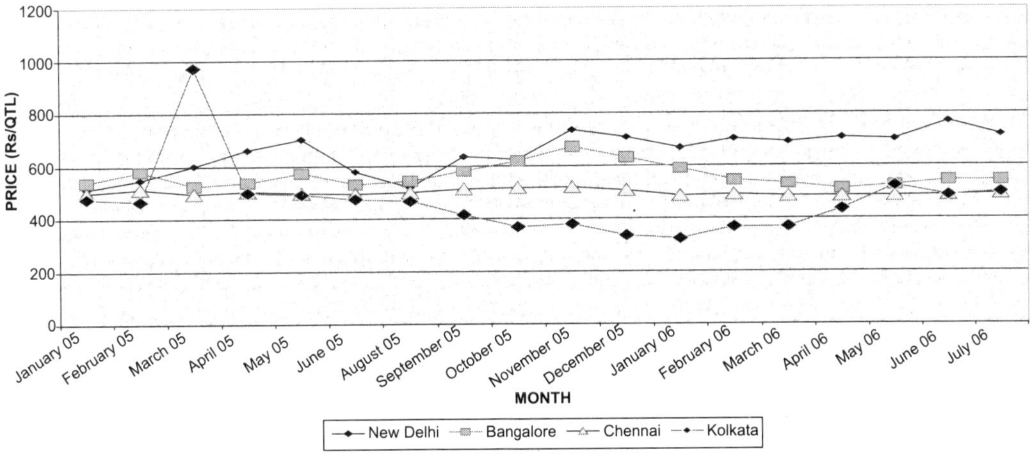

Fig. 1.6. Monthly average wholesale price in major markets (2005-2006)

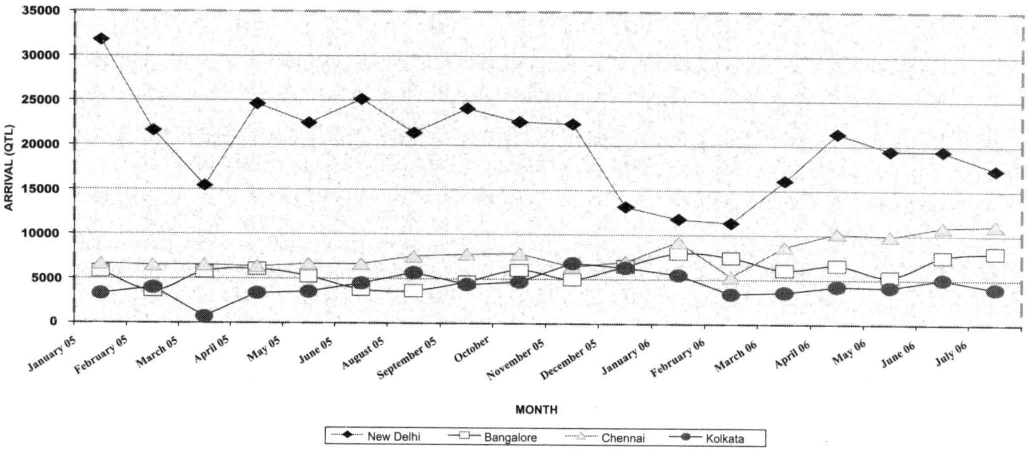

Fig. 1.7. Average arrivals of banana in major markets

Banana Export

Bananas are fifth most important agricultural commodity in world trade after cereals, sugar, coffee and cocoa of the 97 million tonnes of banana and plantain produced annually, only 14% are traded globally. The world's buyers of banana purchase a total volume of 14.65 million tonnes, valued at US$ 4.41billion (2000). The USA remains the major importer of banana with a share of 30% of global import worth US $ 1.50 billion. Other major importing countries of banana and plantain are Germany, Japan, Belgium and United Kingdom. The Latin American countries dominate the US export market for banana. Costa Rica and Ecuador are top exporters, capturing half of the US market for banana at a combined volume of 1.8 million tonnes. World exports of bananas also show a high level of concentration, with developing countries, accounting for the bulk of exports.

Only Latin America and the Caribbean supply about 71% of the total exports (2006). The four leading banana exporter countries (Ecuador, Costa Rica, Philippines and Colombia) in 2006 accounted for 65% of the world exports. Ecuador alone provided more than 30% of global banana exports. Nevertheless, participation of Latin American and Caribbean supply has been slightly reduced during the nineties, while participation of Asian exports has increased. The analysis of export income dependence on bananas provides also interesting results. For major exporting countries like Ecuador or Costa Rica, exports of bananas represented 20.2 and 25.4% of the total exports (2006). The highest levels of dependence on banana exports can be found in the Windward Islands countries: Saint Lucia, St. Vincent and Grenadines, Dominica and Grenada (in case of Saint Lucia, for example, the proportion of banana exports over total exports was 49.6%) (Fig. 1.8).

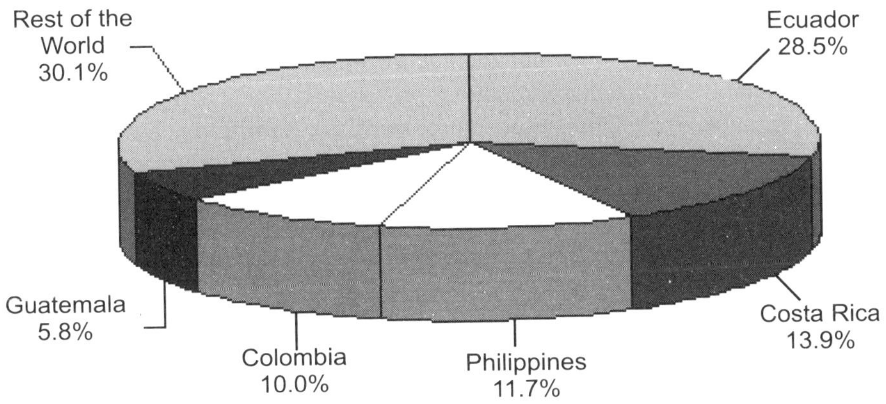

Fig. 1.8. Distribution of total world banana exports (2000-2004)

Banana Imports

Bananas are imported mainly by the European Union, the USA and Japan, which together accounted for about 67% of the world imports (2004), while the first 10 banana-importing countries represented more than 80% of total imports (considering the EU as a whole). Although geographical concentration of imports remains high, the analysis of import data shows a tendency towards diversification of destination markets for bananas, particularly during the nineties. This is shown by evolution of imports by the rest of the world group of countries (Fig.1.9). It would reflect the increasing importance of emerging markets as the Russian Federation, China or Eastern European countries as destinations for banana exports. While during the last decade, all importing areas have increased their banana imports, it is possible to appreciate certain stagnation in the volume of imports of European Union during the nineties.

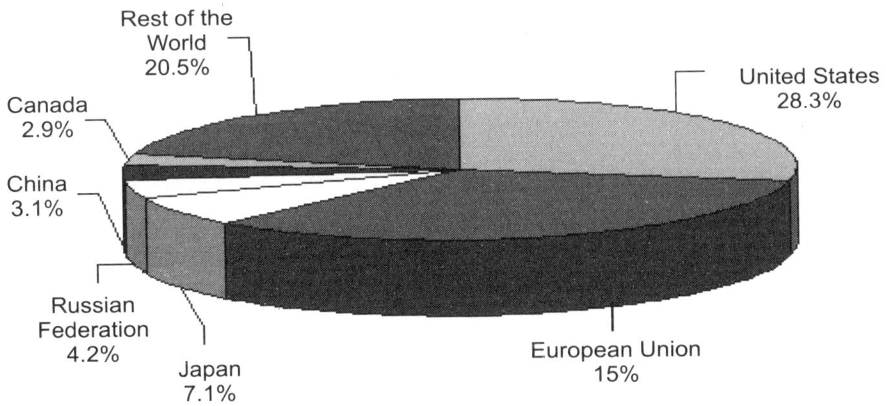

Fig. 1.9. Distribution of total world banana imports (2000-2004)

Trends of Banana Prices in Major Markets

In general, banana market prices at the international level are determined by demand and supply availability. Supply factors would include weather conditions, diversification of supply sources, area available for banana production, yields, *etc*. Demand from banana consumers would depend on income growth and economic evolution in different countries, population growth, consumer preferences, banana import regimes, *etc*.

Although banana market presents an oligopolistic structure, this does not actually mean that transnational marketing companies have a great power to set selling prices for bananas, particularly during the nineties with increasing competition existing among them. In addition, dominant position of these large banana marketing companies has been challenged by some other importing groups that have appeared as a result of efforts of producers to directly market their bananas outside the multinational channel. Therefore, banana price formation is most determinant at the retail stage, particularly due to increasing dominant position of distributors (supermarket chains) (Fig. 1.10).

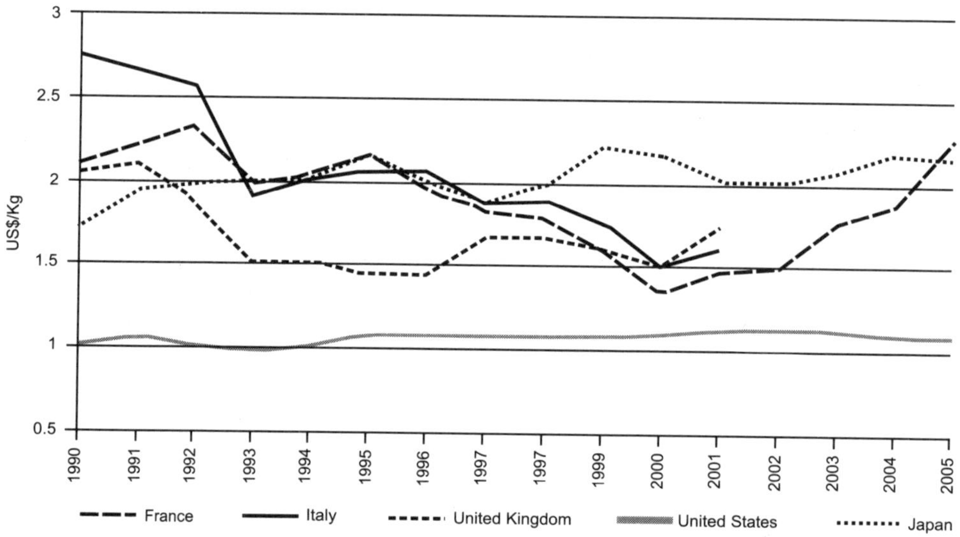

Fig. 1.10. Retail prices of banana in selected markets (US$/kg)

CONCLUSION

The role of banana and plantain has to play in livelihood security of about a billion people globally, there have been international initiatives, more intensely during the last 7-9 years, which resulted in use of science-based technologies and focussed attention on improvement for sustainable production. Development scenario in India is also highly appreciable, with 149% increase in production and 71% in productivity due to adoption of improved production technologies and efficient management of pests and diseases. The task ahead is to reduce regional imbalances. There is also an urgent need to exploit biotechnological tools for improving productivity and reducing post-harvest losses. Opportunities emerging due to growing market for organic banana requires to be harnessed. Emerging market in

Middle-East, East Europe and other developing economies could be exploited. To keep its leadership of the Indian banana in production and diversity, horizontal and vertical integration coupled with institutional support system would be essential, which shall improve the quality of life of people dependent on banana.

2

Gene Safeguard and Management of Plant Genetic Resources in IPR Regime

S.K. SHARMA AND PRATIBHA BRAHMI

PLANT genetic resources (PGR) are a subset of biodiversity of cultivated and related wild species containing genetic material of actual or potential value and PGR are one of the essential components that hold the key to food and nutritional security for the world. During the last three decades, rapid progress in agriculture has been witnessed, along with collection, conservation and sustainable utilization of PGR. Conservation of PGR is one of the most widely discussed issues in agriculture and environment today. In the recent past, ownership of genetic resources and equity issues have changed our outlook towards these resources from freely available for all to "valuable assets" with a price.

AGRO-BIODIVERSITY IN INDIA

Indian sub-continent has a rich and varied heritage of biodiversity encompassing a wide spectrum of habitats from tropical rain forests to alpine vegetation and from temperate forests to coastal wetlands. It is one of the eight centres of origin (Vavilov, 1951) and the 12 mega gene centres of the world. It possesses 11.9% of world flora and 33% of the country's recorded flora are endemic to the region and concentrated mainly in the North-East, Western Ghats, North-West Himalayas and the Andaman and Nicobar Islands. Of the 49,219 higher plant species, 5,725 are endemic belonging to 141 genera under 47 families (Nayer, 1980). Of these, 3,500 are found in the Himalayas and adjoining regions and 1,600 in the Western Ghats alone (Arora, 1991). Myers *et al.* (2000) have brought out an updated list of 25 global hot spots of biodiversity as areas featuring exceptional concentration of endemic species and experiencing exceptional loss of habitat. This classification also has two hot spots extending into India. These are Western Ghats/ Sri Lanka and Indo-Burma (covering the Eastern Himalayas) regions. These are included among most important eight hot spots. In addition, India has 26 recognized endemic centres that are home to nearly one-third of all the total flowering plants identified and described todate.

India is also homeland to 167 cultivated species and 329 wild relatives of crop plants (Arora, 1991). There are about 583 cultivated crops in India. Diversity in crops available in

India is indicated (Fig.2.1). It has 30,000-50,000 landraces of rice, pigeonpea, mango, turmeric, ginger, sugarcane, gooseberries etc. and ranks seventh in terms of contribution to world agriculture. Further, around 1,500 wild edible plant species are exploited by native tribes. These include 145 species of roots and tubers, 521 of leafy vegetables/ greens, 101 of buds and flowers, 647 of fruits and 118 of seeds and nuts (Arora and Pandey, 1996). In addition, nearly 9,500 plant species of ethno-botanical usage are reported from the country. Of which, around 7,500 are of ethno-medicinal importance and 3,900 are multipurpose edible species.

Primary Centre of Diversity
Rice, kodomillet, little millet, pigeonpea, *urdbean*, moth bean, cotton, Capsularis jute, smooth gourd, ridged gourd, pointed gourd, jack fruit, banana, mango, jamun, Bengal cardamom, black pepper, *Rauvolfia serpentina* and *Saussurea lappa*.

Regional Centre of Diversity (Expansion of Primary Centre)
Wheat, barley, proso millet, foxtail millet, sugarcane, bamboos, buckwheat, mungbean, chickpea, rice bean, cluster bean, *Brassica* spp., cucumber, bitter gourd, bottle gourd, snake gourd, citrus, taro, yams, ginger, turmeric and small cardamom

Secondary Centre of Diversity
African Crops: sorghum, pearl millet, finger millet, cowpea, niger, safflower, brinjal, okra, sesame and castor. Tropical American Crops: maize, amaranth, tomato, pumpkin, chilli and chayote

Fig. 2.1. Diversity in crops in India

The Indian Gene Centre has been further enriched by the Mughals, Spaniards and Portuguese who introduced pear, grape, date palm, garlic, spinach, maize, tomato, chilli, peas, cashew nut etc. while Arabs possibly introduced clove, coriander, cumin, fennel, coffee etc. and Britishers introduced cocoa, strawberry, litchi, potato, tea *etc*.

Plant Genetic Recourses

Genetic resources of actual or potential value are being lost at an alarming rate due to habitat destruction, land degradation, over-exploitation of water resources, forestry practices, urban expansion, changing social and cultural norms, and adoption of improved varieties/ other technologies of intensive agriculture by farmers. This erosion has caused losses of several desirable and potentially useful genes threatening the basic food security. The contradictions between modern agricultural practices and traditional ones can be overcome with sustainable farming practices, supported by adequate policies, institutional mechanism, and scientific management of genetic diversity. Extensive genetic diversity provides stability to farming systems and sustains changing environmental conditions that may occur in future (Gautam, 2004).

Horticultural Genetic Resources

India has a vast heritage of these resources and a large number of horticultural crops cultivated is due to different agroclimatic factors, socio-economic, cultural and ethnic needs. Horticulture Genetic Resources Management is more complex as compared to the field crops, requiring different management strategies (Fig.2.2). The conservation approach combining both the *in-situ* and *ex-situ* methods is the best option for horticultural crops.

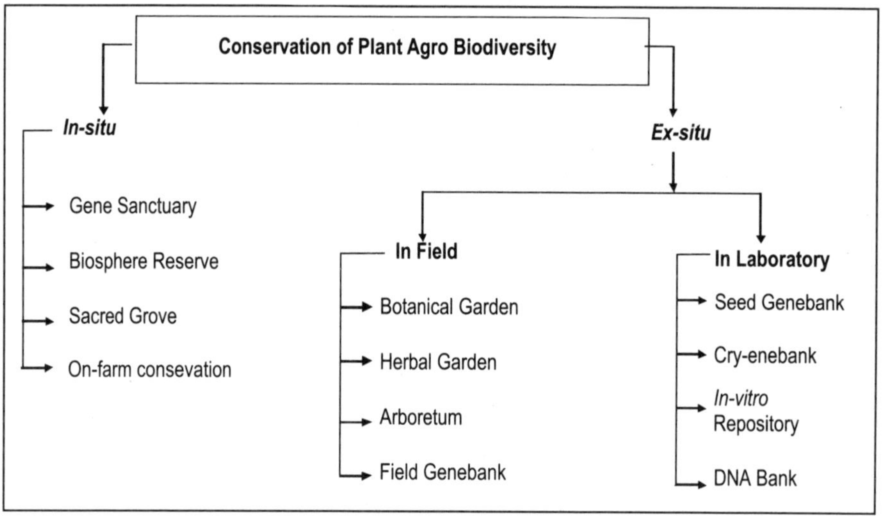

Fig. 2.2. Management practices for conservation of genetic resources

Over the years, India has developed sound and scientific management regime for *ex-situ* conservation and access to its genetic resources (Dhillon and Saxena, 2003). Groups of institutions, scientific societies and non-governmental organizations are addressing the task with NBPGR, New Delhi, as the nodal agency for its coordination. It aims at an efficient management of PGR by providing access to various crop improvement programmes. It also encompasses the National Gene Bank Network which is primarily responsible for conservation of germplasm on long-term basis. Ten regional stations of NBPGR are located in different agroclimatic zones of the country.

About 57 National Active Germplasm Sites (NAGSs) are the integral components of the network. The NAGS, are based at the premier institutes for specific crops or crop groups and are entrusted with the responsibility of multiplication, evaluation, conservation of active collections and their distribution to users both at national and international levels. Various other national institutes, All India Coordinated Crop Research Projects, State Agricultural Universities (SAUs) and other stakeholders are also linked to the network. International Agriculture Research Centres (IARCs) are involved in conservations and use of PGR is also effectively linked to the network.

Major categories of plant genetic resources which are required to be conserved are land races, farmers' varieties, obsolete cultivars, modern cultivars, breeding lines, genetic stocks, wild relatives, weedy races and potential domesticates or other wild species of potential value. The NBPGR, New Delhi, since its inception in 1976, has collected over two lakh accessions belonging to various categories from all over India. Over the years, the emphasis on collection of wild and weedy relatives of crop plants has increased. The proportion of these collections has increased to about 39% in 2005-06 from 6% during 1976-1998 (NBPGR, New Delhi 2007). In future, collection strategies would be based on the gaps in existing collections and on sources of tolerance to biotic and abiotic stresses including areas of high genetic erosion and depletion due to habitat loss.

Future Thrusts in Management of PGR

Ex-situ conservation methods including seed storage, cryobanking and *in-vitro* are costly and energy intensive technologies. Research on cost-effective conservation strategies is going on at NBPGR, New Delhi. Conventional storage of seeds for long conservation (more than 50 years) is done with a moisture content of seeds at 5%. The status of germplasm conserved by various methods is indicated (Fig.2.3).

Research on ultra desiccation of seeds to below 5% moisture has shown that seeds could be stored for more than 10 years without the need for refrigeration at an ambient condition in sealed aluminum foil pouches. Other strategies for low conservation cost are seeds stored under naturally occurring permafrost conditions such as higher reaches of the Himalayas or the arctic region at the global level. Svalbard Seed Vaults near Greenland in Norway is being developed as a safety duplicate site for all major collections of the world. The capacity of this facility is about 3 million seed samples, permafrost and thick rock cover in the area ensures that temperature never rises above -3.5 °C even after loss of refrigeration for one year. Similar site with permafrost conditions in the Indian Territory of the Himalayan region is being short listed as a possible safety duplicate site by NBPGR, New Delhi.

Genomic Resource Banks

Another approach for conservation of genetic resources is through 'gene' conservation. Genomic Resource Bank aims to conserve genomic DNA, gene constructs, promoters, cDNA libraries and DNA of rare and endangered species. Such banks cut across boundaries of plants, animals, fish or microbial species. This futuristic approach of 'gene' bank can help in the enhanced utilization of genetic resources through distribution of genetic and genomic resources for research on structural and functional genomics. These can also provide DNA markers for

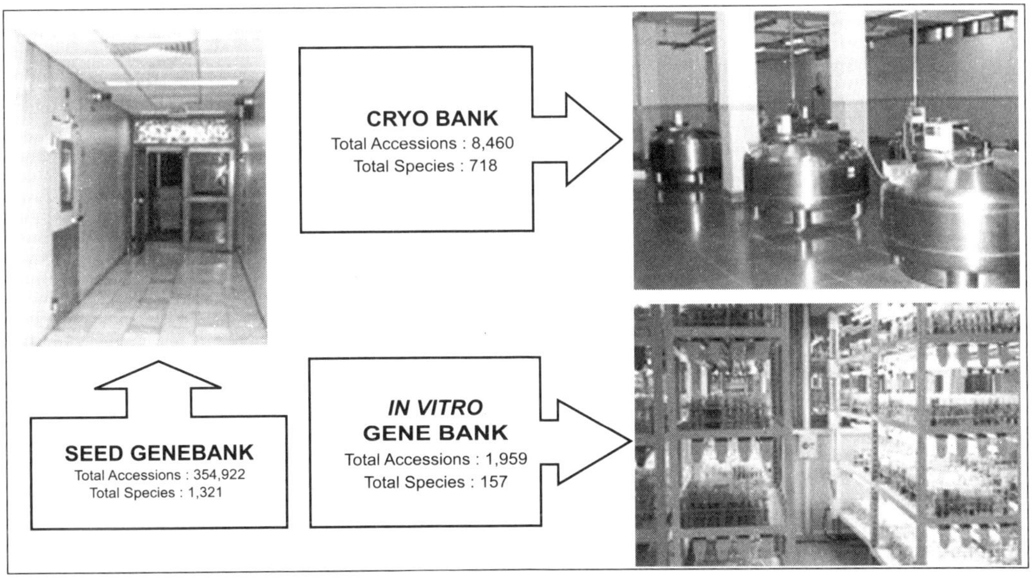

Fig. 2.3. Germplasm conservation at NBPGR, New Delhi

marker assisted selection in crop improvement programmes and help in discovery of novel genes/alleles, promoters, markers development and assessment of genetic diversity and possibilities are endless. A proposal for Genetic Resource Bank is under consideration by the ICAR during Eleventh Five Year Plan.

Optimum Utilization of Conserved Genetic Resources

Maintenance of a large number of germplasm accessions in field is an expensive affair and handling of a large number of accessions for selections of desirable material hinders its effective utilization. The concept of core collections has therefore been developed where about 10% of the size of all collections of a species is maintained which represents the possible variations within a species. At NBPGR, New Delhi, core collections have been developed in sesame, okra, green/gram and fingermillet. Similar efforts are also underway to develop core collections in brinjal, tomato and frenchbean.

GENETIC RESOURCES MANAGEMENT IN EMERGING IPR REGIME

Many international and national developments during last 15 years have directly or indirectly affected bio-resource management programmes. Efforts are being made to streamline the access and management of bio-resources in the context of these new developments. The Convention on Biological Diversity (CBD) led to development of Biological Diversity Act (BDA) 2002 in India which governs access to all genetic resources. The International Treaty on Plant Genetic Resources for Food and Agriculture (ITPGRFA) is another legally binding treaty which has implications to access of 64 crops and forage species under Multilateral System (MS). This exchange would be under the conditions of Standard Material Transfer Agreement (SMTA). As a national obligation for the TRIPS Agreement of WTO, a new legislations, the Protection of Plant Varieties and Farmers Right Act, 2001 and Geographical Indications of Goods (Registration and Protection) Act, 1999 were enacted and suitable amendments made in other exiting IPR legislations, which have a bearing on the product, processes and technologies developed utilizing bioresources.

With these developments, following activities related to access and management of genetic resources need to be considered and understood by all scientists and holders of genetic resources.

Sharing Bioresources with Foreign Agencies

The BDA was enacted in 2002 and its rules enforced in 2004. These regulate access to genetic resources of Indian origin to others. Sharing of bioresources with foreign agencies now requires prior approval of National Biodiversity Authority (NBA) based at Chennai (NBA, 2007). As per section 3 of the Act, all access of genetic resources to non-Indians including multinational companies based in India is to be done with prior approval of the NBA. However, exchange under any collaborative research projects between Government of India and other countries is exempted (section 5) from such approval, but such projects must be approved by at least one Government Agency/Department in India. Therefore, MoUs/or work plans with private agencies /seed companies etc. need approval of the NBA. There has to be a single window system in the country for such type of activities. Some research organizations are singing MoUs for sharing of Indian genetic resources with foreign agencies without prior approval / consent of NBA/DARE, which is now mandatory. Awareness on this issue is very essential and country's sovereign right on its bioresources should be honoured.

Germplasm Explorations and Collection

It is mandatory to inform the State Biodiversity Boards before taking up any survey/collection work about the bioresources as per Section 7 of the BDA, 2002. It has been observed that scientists/breeders of ICAR institutions and SAUs undertaking germplasm explorations independently without consulting NBPGR. This can result in duplication of efforts and areas already explored are revisited and material is duplicated. Such material is often not assigned in National Identity (IC) numbers. The status of past explorations needs to be verified from NBPGR before planning exploration missions from specific areas. If possible, NBPGR scientists need to be involved in any explorations from unexplored areas/trait specific explorations.

Feedback Information about Germplasm Utilization

The NBPGR, New Delhi, is sending material within India and receiving material from other countries under the conditions of Material Transfer Agreements (MTA)/SMTA as agreed under the ITPGRFA. It is important to understand the implications of MTAs or SMTA. Therefore, transfer of material to any third party without the consent/approval of the NBPGR/ICAR should not be allowed. Feedback information, its establishment, maintenance and effective use is essential in ensuring availability of the germplasm in national system for use by other interested breeders/researchers and avoid repeated introductions.

National Identity for Using Germplasm and Publication of its Results

There are many examples of elite Indian germplasm without any National Identity (IC) number being sent to other countries where it was released and commercialized without any benefit to the country. Some chickpea varieties released in Australia are of Indian origin. 'Nap Hal' wheat transferred to USA many years ago was exploited for developing other varieties and its patents filed. India had to file opposition for grant of such patents. Use of National Identity from a centralized database of NBPGR for each germplasm accession should be mandatory and it should be indicated in all publications and future correspondence to avoid naming the same accession with different identities. This will create 'prior art' and help stake claim of national sovereignity and benefit sharing.

Germplasm Registration

During the process of germplasm utilization for breeding, research and experimentation and developing superior genotypes for specific or multiple combinations of traits, many elite lines are developed which give good performance for a specific or a combination of traits but do not show yield superiority over the exiting varieties so as to get released. In addition, raw germplasm identified or selected by scientists (land races/traditional varieties) with one or more outstanding traits such as resistance to disease(s), high nutritional quality, bold grain etc. or with traits of academic or scientific importance (mutants) are valuable genetic resources, but do not qualify to get notified because of poor cultivation performance. With the objectives of giving credit to those scientists who have developed or identified such promising experimental material (including parents of inbred lines) or promising germplasm, and to facilitate flow of germplasm among scientists working in crop improvement programmes, the ICAR initiated germplasm registration mechanism in 1996 to register such promising germplasm. The information about registered germplasm is being published in the *Indian Journal of Plant Genetic Resources, Indian Journal of Genetics and Plant Breeding* and several *Newsletters* such as *ICAR Newsletter, National Bureau of Plant Genetic Resources Newsletter* etc. with the objective of dissemination of information among scientists. Thus, information on germplasm is put in public domain, which has become a mandatory requirement to safeguard the national resources with respect to IPR. The ICAR constituted Plant Germplasm Registration Committee (PGRC) during March 1996 with Deputy Director-General (Crop Science) as the Chairman. So far, 595 germplasm lines have been registered by this committee at NBPGR. Similar registration process has been initiated in case of animal, fish and microbial genetic resources. Applications for plant germplasm registration are required to be submitted at NBPGR through the Institute Germplasm Identification Committee (IGIC). The recommendations of the IGIC should be based on parameters of uniqueness and stability of expression in germplasm. Registration of potentially valuable germplasm should be encouraged on priority by all crop based institutions.

Registration of Extant/New Varieties

In India, agricultural research including development of new plant varieties has largely been the concern of government and public sector institutions. Earlier, India did not have any legislation to protect the plant varieties and in fact no immediate need was felt. However, after India became signatory to the Trade Related Aspects of Intellectual Property Rights Agreement (TRIPs) in 1994, such a legislation was necessitated. Article 27.3 (b) of this agreement requires the member countries to provide for protection of plant varieties either by a patent or by an effective *sui- generis* system or by any combination thereof. The existing Indian Patent Act, 1970 of India excluded agricultural and horticultural methods of production from patentability. The *sui-generis* system for protection of plant varieties was developed integrating the rights of breeders, farmers and village communities, and taking care of the concerns for equitable sharing of benefits. It offers flexibility with regard to protection, compared to other similar legislations existing or being formulated in different countries. The Act covers all categories of plants, except microorganisms. Under the PPVFR Act, any persons claiming to be the breeder of the variety, successor or assignee of the breeder or any farmer, group of farmers, university or public funded agricultural institution claiming to be the breeder, can register any of the four types of varieties, namely new varieties, extant varieties, essentially derived varieties, and farmers' varieties.

According to the PPVFR Act, an extant variety is a broad category covering: (i) varieties available in India that are notified under section 5 of the Seeds Act, 1996; (ii) farmers' varieties; (iii) varieties of common knowledge; or (iv) any other variety that is in public domain (Brahmi *et al.,* 2004). Applications for registration of extant varieties are being accepted to grant protection under PPVFR Act for 12 notified crops, viz. wheat, rice, maize, sorghum, pearlmillet, pigeonpea, lentil, chickpea, mungbean, urdbean, pea and kidney bean in first phase. The registration of extant varieties will be done for a period of three years after notification of the genera. All breeders should take urgent action to get registration of extant/new varieties released at central or state level at the earliest to ensure ownership and benefits sharing on commercialization. The application forms are available at the website of PPVFRA (www.plantauthority.gov.in). The ICAR has entrusted NBPGR with the filing of applications of extant-notified varieties released at central level in first phase, followed by new varieties which have been notified during last one year.

INTERNATIONAL TREATY ON PLANT GENETIC RESOURCES FOR FOOD AND AGRICULTURE

The ITPGRFA became effective from 29 June 2004 after ratification by 40 countries. Presently, the Treaty has 115 Contracting Parties, including India. The objectives of the Treaty are 'the conservation and sustainable use of plant genetic resources for food and agriculture and the fair and equitable sharing of benefits arising out of their use, in harmony with the Convention on Biological Diversity' for sustainable agriculture and food security. Each contracting party of the Treaty has to ensure the conformity of its laws, regulations and programmes with the provisions of the Treaty. Each contracting party shall promote an integrated approach to the exploration, conservation and sustainable use of PGRFA and shall in particular undertake some of the activities like: **(i)** survey and inventory of PGRFA, **(ii)** promote the collection of PGRFA and relevant associated information on those PGR that are under threat or are of potential use, **(iii)** promote or support, as appropriate, farmers and local communities' efforts to manage and conserve on-farm their Plant Genetic Resources for Food and Agriculture, **(iv)** promote *in-situ* conservation of wild crop relatives and wild plants for food production, **(v)** co-operate to promote the development of an efficient and sustainable system of *ex-situ* conservation, giving due attention to the need for adequate documentation, characterization, regeneration and evaluation and **(vi)** take steps to minimize or if possible eliminate threats to PGRFA.

Multilateral System of Access and Benefit Sharing

In exercise of sovereign rights of nations over their PGRFA, the Treaty envisages the establishment of a multilateral system (MS) covering a list of 35 food and 29 forage crops given in Annex 1 of the Treaty (Table 2.1) selected on the criteria of food security and inter-dependence. It includes all PGRFA of the listed crops that are under the management and control of the contracting parties and in public domain. The MS shall also include the PGRFA as listed, and held in the *ex-situ* collections of the International Agricultural Research Centres of the Consultative Group on International Agricultural Research (CGIARs). India being signatory to the International Treaty should form effective policy for realizing optimum benefit from the provisions of the Treaty. Germplasm available with Indian system should be short listed by the concerned crop coordinators in their respective crop. Efforts to designate germplasm should be made with discretion, keeping in view the trade significance and other uses of crop plants, besides the food value, since material to be exchanged under the Treaty is solely for the purpose of food and feed, and excludes from its perview the non-food or pharmaceutical / industrial use of the crops covered under the Treaty.

All countries are interdependent for their PGR requirements and cannot acquire and conserve resources to satisfy all their needs. There is a need to collaborate at local, regional and international levels for acquisition and conservation of germplasm. More emphasis needs to be given to crop and trait-specific explorations. New tools of geographical information system and remote sensing needs to be deployed to supplement the existing ground data in PGR programmes to exploit agro-biodiversity, particularly in difficult/inaccessible areas. Collection and conservation of wild and related species of crop plants that posses many useful genes for resistance/ tolerance for biotic and abiotic stresses needs to be stressed.

Table 2.1. Crops covered under the multilateral system

Crop	*Food crops*	*Observations*
Breadfruit	*Artocarpus*	Breadfruit only
Asparagus	*Asparagus*	
Oat	*Avena*	
Beet	*Beta*	
Brassica complex	*Brassica etc.*	Genera included are: *Brassica, Armoracia, Barbarea, Camelina, Crambe, Diplotaxis, Eruca, Isatis, Lepidium, Raphanobrassica, Raphanus, Rorippa*, and *Sinapis*. This comprises oilseed and vegetable crops such as cabbage, rapeseed, mustard, cress, rocket, radish, and turnip. The species *Lepidium meyenii* (maca) is excluded
Pigeonpea	*Cajanus*	
Chickpea	*Cicer*	
Citrus	*Citrus*	Genera *Poncirus* and *Fortunella* are included as rootstock
Coconut	*Cocos*	
Major aroids	*Colocasia, Xanthosoma*	Major aroids include taro, cocoyam, dasheen and tannia
Carrot	*Daucus*	
Yams	*Dioscorea*	
Fingermillet	*Eleusine*	
Strawberry	*Fragaria*	
Sunflower	*Helianthus*	
Barley	*Hordeum*	
Sweet potato	*Ipomoea*	
Grass pea	*Lathyrus*	
Lentil	*Lens*	
Apple	*Malus*	
Cassava	*Manihot*	*Manihot esculenta* only
Banana/plantain	*Musa*	Except *Musa textilis*
Rice	*Oryza*	
Pearlmillet	*Pennisetum*	
Beans	*Phaseolus*	Except *Phaseolus polyanthus*
Pea	*Pisum*	
Rye	*Secale*	
Potato	*Solanum*	Section tuberosa included, except *Solanum phureja*
Eggplant	*Solanum*	Section melongena included
Sorghum	*Sorghum*	
Triticale	*Triticosecale*	
Wheat	*Triticum etc.*	Including *Agropyron, Elymus* and *Secale*
Faba bean/vetch	*Vicia*	
Cowpea *etc.*	*Vigna*	
Maize	*Zea*	Excluding *Zea perennis, Zea diploperennis* and *Zea luxurians*

Contd.

Genera	Species
	Legume forages
Astragalus	*chinensis, cicer, arenarius*
Canavalia	*Ensiformis*
Coronilla	*Varia*
Hedysarum	*Coronarium*
Lathyrus	*cicera, ciliolatus, hirsutus, ochrus, odoratus, sativus*
Lespedeza	*cuneata, striata, stipulacea*
Lotus	*corniculatus, subbiflorus, uliginosus*
Lupinus	*albus, angustifolius, luteus*
Medicago	*arborea, falcata, sativa, scutellata, rigidula, truncatula*
Melilotus	*albus, officinalis*
Onobrychis	*Viciifolia*
Ornithopus	*Sativus*
Prosopis	*affinis, alba, chilensis, nigra, pallida*
Pueraria	*Phaseoloides*
Trifolium	*alexandrinum, alpestre, ambiguum, angustifolium, arvense, agrocicerum, hybridum, incarnatum, pratense, repens, resupinatum, rueppellianum, semipilosum, subterraneum, vesiculosum*
	Forage grasses
†*Andropogon*	*Gayanus*
Agropyron	*cristatum, desertorum*
Agrostis	*stolonifera, tenuis*
Alopecurus	*Pratensis*
Arrhenatherum	*Elatius*
Dactylis	*Glomerata*
Festuca	*arundinacea, gigantea, heterophylla, ovina, pratensis, rubra*
Lolium	*hybridum, multiflorum, perenne, rigidum, temulentum*
Phalaris	*aquatica, arundinacea*
Phleum	*Pretense*
Poa	*alpina, annua, pratensis*
Tripsacum	*Laxum*
	Other forages
Atriplex	*halimus, nummularia*
Salsola	*vermiculata*

There is a need to adopt complementary conservation strategies involving both *in-situ* and *ex-situ* approaches, especially for horticultural crops. For *in-situ* conservation, due attention is required to be given to genetically rich hot spots including tribal belts and to strengthen and expand the network of germplasm conservation by including all the stakeholders, including the communities. The core collection concept is likely to result in a more structured and efficient approach to identify limited sets of diverse germplasm and utilize the same more effectively. Pre-breeding is needed to incorporate new kind of pest resistance, to bring in new levels of productivity and stability of performance, and to provide quality traits for food and feed products.

Registration of genetic stocks and elite germplasm needs to be encouraged to promote germplasm exchange and effective utilization. In the changing global scenario under the new intellectual property rights (IPR) regime, modalities for benefit sharing by both private and public sectors will also have to be worked out. Further, models will have to be developed for sharing of benefits with farmers and communities as well. Awareness generation among the people at various levels (policy makers, scientists, administrators, farmers etc.) about the value of PGR wealth and its protection is essential.

There is an urgent need to increase the interaction between plant breeders and PGR workers. The interface among different stakeholders is likely to bring out new useful PGR management alternatives. The loss of every gene, species, ecosystem and ultimately bioresources will obviously limit our options for future. Without an intelligent and judicious use of bioresources it will be difficult to achieve sustainable advances in agricultural and industrial production. With the new IPR regime, the thrust on capitalizing our wealth of plant genetic resources has increased, which need to be utilized to the benefit of the country on the whole and to the farming population in particular.

References

Arora, R.K. 1991. Plant diversity in Indian gene centre. *In: Plant Genetic Resources – Conservation and Management.* Poroda, R.S. and Arora, R.K. (Eds) IPGRI, Regional Office for South Asia, New Delhi, India.

Arora, R.K. and Pandey, A. 1996. *Wild Edible Plants of India: Diversity Conservation and Use.* National Bureau of Plant Genetic Resources, New Delhi, India.

Dhillon, B.S. and Saxena, Sanjeev 2003. India Chapter. *In: Plant Genetic Resources in SAARC Countries,* SAIC publication, SAARC Agricultural Information Centre, Bangladesh.

FAO, Rome 2002. The International Treaty on Plant Genetic Resources for Food and Agriculture, Food and Agriculture Organization, Rome, Italy.

Gautam, P.L. 2004. Trends in Plant genetic resources management. *In: Plant Genetic Resources Management.* Dhillon, B.S., Tyagi, R.K., Lal, A. and Saxena, S. (Eds). Narosa Publishing House, New Delhi.

Myers, N., Mittermeier, R.A., Mittermeier, C.G., GAB da Fonseca and Kent, J. 2000. Biodiversity hot spots for conservation priorities. *Nature* **403**:853-58.

Nayer, M.P. 1980. Endemism and pattern of distribution of endemic genera (angiosperm) in India. *J. Econ. Tax. Bot.* **1**:99-110.

NBA, Chennai. 2007. The Biological Diversity Act, 2002 and Biological Diversity Rules, 2004, National Biodiversity Authority Chennai, India, 57 pp.

NBPGR, New Delhi. 2007. Division of Plant Exploration and Collection. *In: Annual Report,* National Bureau of Plant Genetic Resources 2006-2007, NBPGR, Pusa Campus, New Delhi, India, 18-23.

Pratibha, Brahmi, Saxena, Sanjeev and Dhillon, B.S. 2004. The Protection of Plant Varieties and Farmersí Right Act of India. *Current Science* **86**:392-98.

Vavilov, N.I. 1951. Phyto-geographical basis of plant breeding. *In: Selected Writings of NI Vavilov and translated by KS Chester the Origin, Variation, Immunity and Breeding of Cultivated Plants, Chronica Botanica* **13:**364. Waltham Mass., USA.

3

Screening of Banana to Biotic Stresses

M.A. HASAN, R. RAY CHOWDHURY, S. SARKAR AND K.K. MANDAL

BANANA is attacked by several diseases (black sigatoka, Fusarium wilt, bunchy top virus) and pests (burrowing nematode,weevil-borer *etc.*). In breeding for resistance to biotic stresses, basic requirement is availability of an efficient screening technique, which should clearly distinguish resistant genotypes from susceptible ones. To have an environment-friendly cultivar that has minimal pesticide application, a cultivar resistant to diseases and pests may be more important. Therefore, screening of banana cultivars of different genomic groups for tolerance/resistance to pests and diseases was done.

The experiment was conducted at Horticultural Research Station, Bidhan Chandra Krishi Viswavidyalaya, Mondouri, Nadia,West Bengal, to evaluate 39 banana cultivars of different genomic groups for tolerance/resistance to pests and diseases during 2003-2006.The response of cultivars to biotic stress, viz. leaf and fruit scarring beetle, pseudostem weevil, nematode (*Meloidogyne,Rotylenchus* and *Tylenchus*), Sigatoka, Panama wilt and bunchy top were observed.The intensity of infestation was recorded following the code used for evaluation in IPGRI/INIBAP descriptor (1996). The intensity of infestation was scored by coding: 0 = inapplicable, 1 = very low or no visible sign of susceptibility, (CR = completely resistant), 3 = low (HR = highly resistant), 5 = intermediate (MR = moderately resistant), and 7 = high (S = susceptible). Multivariate analysis of biotic parameters were made followed by Dillon and Goldstein (1994) by nearest neighbourhood method of hierarchical cluster analysis of squared Euclidean distance matrix and correlation distance matrix on the basis of characters measured which were either cardinal or ordinal in nature.

Proximity Matrix

The results clearly indicate that Kanchkela 2, Kanchkela 3, Kanchkela 4, Kanchkela 5 as well as Kanchkela 7 and Kanchkela 6, Manua and Madhubash and Bichkela 2 and Bichkela1 were not similar or alike in relation to biotic stress. Similarly, Kanchkela 6 and Pantharaj, Kanthali and Kanthali clone and Giant Governor and Robusta were also close to each other with respect to biotic stress (Table 3.1). Agniswar and Amritsagar were also very close to each other in relation to biotic stress. Among cooking bananas, Behulla remained far away from others. Apart from them, Agniswar and Amritsagar, Martaman and Malbhog, Chini Champa and Kanthali Champa were also very close to each other.

Average Linkage between Group and within Group

Different methods of hierarchical cluster analysis based upon squared Euclidean distance matrix of biotic stress parameters measured and ordinal in nature, was followed to cluster homogeneous cultivars on the basis of distance or similarity with regard to biotic stress parameters under study. The data of ordinal measurement were transformed to range 0 to 1 standardization prior to calculation of distance or proximity matrix. Number of cluster formed and cluster member were displayed in Table 3.1.

Average linkage between group and within group clustering technique on squared Euclidean distance matrix methods 7 clusters were formed. Cluster 1 consisted of Kanthali, Kanthali clone, Kalibhog, German Kanthali, Bagda Kanthali, Krishna Kanthali, Manua, Madhubash, Kanchakelal 1, Kanchakelal 2, Kanchakelal 3, Kanchakelal 4, Kanchakelal 5, Behulla, Pantharaj, Kanchkela 6, Kanchkela 7, Bichkela 1, Bichkela 2, Hill banana, Atteakole were similar in biotic stress parameters. Rampalmartaman, Kanaibanshi, Sabri, Kanthali Champa, Chini Champa were homogeneous in characters in formed second cluster group. Martaman, Doodsagar, Champa, Martaman clone, Chinia were nearest to each other by biotic stress variables to make third cluster. Jahaji, Kabuli, Robusta and Giant Governor formed forth cluster group similarly Agniswar and Amritsagar formed fifth cluster. Cluster 6 and 7 having Malbhog and Kalibow were distinct to formed separate cluster (Table 3.1).

Table 3.1. Details of clusters for evaluating biotic in stress banana cultivars by using average linkage between group and within group clustering methods on squared Euclidean distance matrix

No. of cluster	*Cluster members(allowed distance co-efficient=0.87 and 0.452)*
1.	Kanthali, Kanthali clone, Kalibhog, German Kanthali, Bagda Kanthali, Krishna Kanthali, Manua Madhubash, Kanchakelal1, Kanchakelal2, Kanchakelal3, Kanchakelal4, Kanchakelal 5, Behulla, Pantharaj Kanchkela6, Kanchkela7, Bichkela1, Bichkela2, Hill banana, Atteakole
2.	Rampalmartaman, Kanaibanshi, Sabri, Kanthali Champa, Chini Champa
3.	Martaman, Doodsagar, Champa, Martaman clone, Chinia
4.	Jahaji, Kabuli, Robusta, Giant Governor
5.	Agniswar, Amritsagar
6.	Malbhog
7.	Kalibow

Single Linkage and Complete Linkage Clustering

Single linkage and complete linkage technique of clustering when applied on squared Euclidean distance matrix, it is to found that total 6 clusters are formed in both the methods (Table 3.2). The cultivars like Atteakole, Bichkela 1, Kanchkela 6 and Kanchkela 7 are nearest to each other to form first cluster by considering the biotic stress variables. Martaman, Martaman clone, Kanthali Champa and Chini Champa are similar in biotic stress variables to make second cluster. Cluster 3 is formed by Agniswar, Amritsagar and Jahaji. The forth cluster consists of Kabuli, Robusta and Giant Governor, as they are closest to each other. Malbhog and Kalibow are distinct in biotic stress variables to form fifth and sixth cluster respectively.

Table 3.2. Details of clusters for evaluating biotic stress in banana cultivars by using single linkage and complete linkage clustering methods on squared Euclidean distance matrix.

No. of cluster	*Cluster members(allowed distance co-efficient=0.295 and 1.746)*
1.	Kanthali, Kanthali clone, Kalibhog, German Kanthali, Bagda Kanthali, Krishna Kanthali, Manua, Madhubash, Kanchakelal 1, Kanchakelal 2, Kanchakelal 3, Kanchakelal 4, Kanchakelal 5, Behulla, Pantharaj, Kanchkela 6, Kanchkela 7, Bichkela 1, Bichkela 2, Hill Banana, Atteakole
2.	Rampalmartaman, Kanaibanshi, Sabri, Kanthali Champa, Chini Champa, Martaman, Doodsagar, Champa, Martaman clone, Chinia
3.	Kabuli, Robusta, Giant Governor
4.	Agniswar, Amritsagar, Jahaji
5.	Malbhog
6.	Kalibow

Principal Component Analysis

The factors having Eigen values more than one is considered and thus three factors are extracted which explained 78.5% of total, accounting for variation of study (Tables 3.3, 3.4 and 3.5). Comparing factor loadings of first factor it was found that biotic stresses like Sigatoka, leaf scarring beetle, bunchy top and *Rotylenchulus*, had higher positive loading in contrast to highest negative loading due to weevil. Hence, Amritsagar, Kabuli, Jahaji, Robusta and Giant Governor are susceptible to these positively loaded stress and in contrast Bichkela 2, Kalibhog, Atteakole, Bichkela 1 and Pantharaj were prone to weevil. Additional stress due to panama wilt, *Meloidogyne incognita* clearly affected Malbhog, Martaman, Martaman clone, Sabri, Champa, Chinia, Manua as Madhubash performance as visible due to factor loading and factor score due to second factor. Anitha *et al.*, (1996) also reported variability in resistance of varieties to different genomic groups.

Table 3.3. Eigen value and accounted for variance based on correlation matrix input of biotic stress of banana cultivars

Factor	*Eigen value*	*Variance explained (%)*	*Cumulative variance (%)*
1	3.372	42.151	42.151
2	1.658	20.726	62.877
3	1.249	15.615	78.492

Table 3.4. Component loading based on correlation matrix input for biotic stress of banana cultivars

Character	*Factor 1*	*Factor 2*	*Factor 3*
Sigatoka	0.892	-0.272	-0.013
Panama wilt	0.424	0.831	-0.195
Bunchy top	0.691	-0.627	0.039
Meloidogyne incognita	-0.047	0.614	0.515
Tylenchulus	0.542	-0.115	0.464
Rotylenchulus	-0.207	-0.019	-0.806
Pseudostem weevil	-0.899	-0.195	0.238
Leaf and fruit scarring beetle	0.878	0.268	-0.155

Table 3.5. Factor-wise distribution of performing cultivars for biotic stress due to positively and negatively loaded dominant characters

Factor	*Dominant characters with positive loading*	*Dominant characters with negative loading*	*Positively loaded cultivars*	*Negatively loaded cultivars*	*Variance explained (%)*
1.	Sigatoka, leaf scarring beetle, bunchy top and *Tylenchulus*	Weevil	Amritsagar, Kabuli, Robusta, Giant Governor and Jahaji	Pantharaj, Kalibhog, Atteakole, Bichkela 1, Bichkela 2	3.372
2.	*Meloidogyne* and Panama wilt, leaf and fruit scarring beetle	Bunchy top	Malbhog, Martaman, Martaman clone, Doodsagar, Champa	Agniswar, Amritsagar, Kabuli, Robusta, Giant Governor and Jahaji	1.658
	Rotylenchulus		Agniswar, Amritsagar, Malbhog	1.249	

Third factor is negatively but heavily loaded by *Rotylenchulus* and this stress could only affect the performance of Agniswar, Amritsagar and Malbhog. It is evident from the scatter diagram between factor 1 vs factor 2 (Fig.3.1) that the characters under factor 1, i.e. Sigatoka, leaf and fruit scarring beetle, bunchy top and *Rotylenchulus* were positively and heavily loaded on Robusta followed by Giant Governor and Kabuli. In contrast, highly positive loading of *Meloidogyne*, Panama wilt and leaf and fruit scarring beetle was in the order in banana Champa > Chinia > Malbhog > Martaman > Martaman clone > Doodsagar (Fig.3.1). The cultivars in order of Robusta > Giant Governor > Kabuli > Jahaji and minimum in Atteakole. This result finds support to those of Rachael *et al.* (2004).

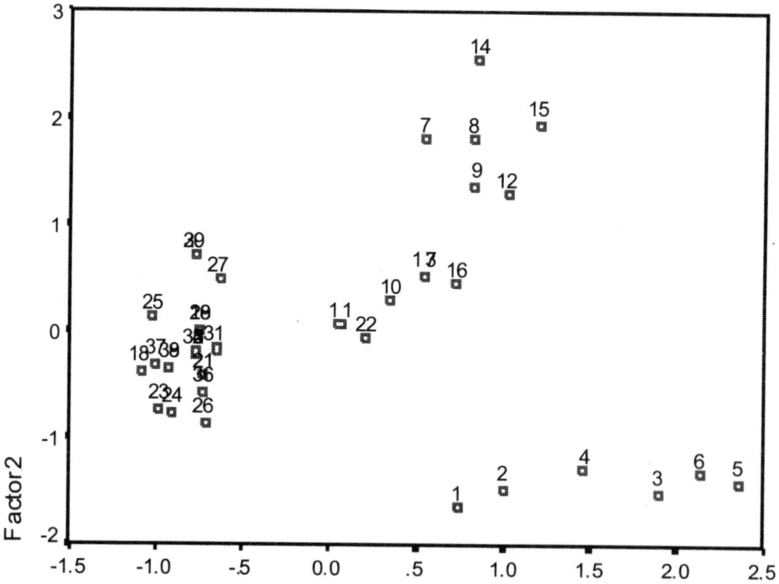

Fig.3.1. Scatter diagram of factor scores between factor 1 vs factor 2

Multidimensional Scaling (MDS) Group Plots for Biotic Stress

While studying the disease aspects of 39 cultivars MDS (multidimensional scaling) result came after 4 iterations under default maximum iterations, convergence and minimum. The S-stress conditions and resulted stress as 0.103 and R^2 as 97%. With the help of stimulus co-ordinated so produced the group plot of cultivars was drawn for such disease study and included in Fig. 3.2. It is found that similar cultivars positioned together, if similarity, did exist in multidimensional data set for multiple characters but converted in two dimensions for better understanding of underlying distance among cultivars. It is concluded that cultivars susceptible to Sigatoka, leaf and fruit scarring beetle, bunchy top and *Tylenchulus* are Amritsagar, Kabuli, Robusta and Giant Governor, while tolerant cultivars are Pantharaj, Kalibhog, Atteakole, Bichkela1 and Bichkela 2. While *Meloidogyne,* Panama wilt and leaf scarring beetle susceptible cultivars are Malbhog, Martaman, Martaman clone, Doodsagar and Sabri and weevil tolerant cultivars are Amritsagar, Kabuli, Jahaji, Robusta and Giant Governor.

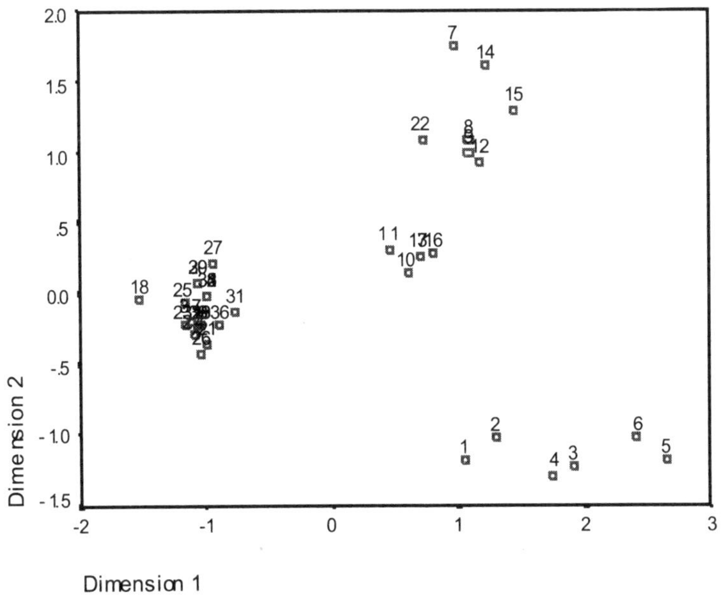

Fig. 3.2. Multidimentional scaling group plots of all variables for biotic stress variables

The numbers in diagram denote name of cultivars in following order: 1, Agniswar; 2, Amritsagar; 3, Kabuli; 4, Jahaji; 5, Robusta; 6, Giant Governor; 7, Malbhog; 8, Martaman; 9, Martaman clone; 10, Rampalmartaman; 11, Kanaibansi; 12, Doodsagar; 13, Sabri; 14, Champa; 15, Chinia; 16, Kanthali champa; 17, Chini Champa; 19, Bichkela 1; 20, Bichkela 2; 21,Hill banana; 22 Kalibow; 23, Kanthali; 24, Kanthali clone; 25, Kalibogh; 26, German Kanthali; 27, Bagda Kanthali; 28, Krishna Kanthali; 29, Manua; 30, Madhubash; 31, Kanchkela 1; 32, Kanchkela, 2; 33, Kanchkela3; 34, Kanchkela 4; 35, Kanchkela 5; 36, Behulla; 37, Pantharaj; 38, Kanchkela 6; 39, Kanchkela 7.

SUMMARY

An experiment was conducted at Horticultural Research Station, Bidhan Chandra Krishi Viswasvidyalaya, Mondouri, Nadia, West Bengal, to evaluate 39 banana cultivars of different genomic groups for tolerance/resistance to pest and diseases during 2003-2006. The response of cultivars to biotic stress, viz. leaf and fruit scarring beetle, pseudostem weevil, nematode (*Meloidogyne, Rotylenchulus* and *Tylenchulus*), Sigatoka, Panama wilt and bunchy top were observed. The intensity of infestation was recorded following the code used for evaluation in IPGRI/INIBAP descriptor (1996). The intensity of infestation was scored by coding 0 = inapplicable, 1 = very low or no visible sign of susceptibility (CR : completely resistant), 3 = low (HR : highly resistant), 5 = intermediate (MR : moderately resistant) and 7 = high (S : susceptible). Multivariate analysis of biotic parameters were made following nearest neighbourhood method of hierarchical cluster analysis of squared Euclidean distance matrix and correlation distance matrix on the basis of characters measured which are either cardinal or ordinal in nature. According to proximity matrix, Kanchkela 2(ABB), Kanchkela 3(ABB), Kanchkela 4(ABB) and Kanchkela 5(ABB) as well as Kanchkela 6 (ABB) and Kanchkcla 7(ABB), Manua (ABB)and Madhubas(ABB) and Bichkela 1(BB) and Bichkela 2 (BB) are similar or alike in response of cultivars to biotic stresses. Likewise Agniswar (AAA) and Amritsagar(AAA) are also very close to each other in relation to biotic stress. Similarly, Kanchkela 6 and Pantharaj (ABB), Kanthali(ABB) and Kanthali clone(ABB) and Giant Governor (AAA) and Robusta (AAA) are also close to each other with respect to biotic stresses. Among cooking bananas, Behula (ABB) remained far away from other cooking banana. Apart from cooking banana Agniswar and Amritsagar, Martaman (AAB) and Malbhog (AAB), Chini Champa (ABB) and Kanthali Champa (ABB) are also very closer to each other. In principal component analysis, the factors having Eigen values more than one are considered and then the three factors are extracted which explained 78.5% of the total accounted for variation of this study. It is found that biotic stresses like Sigatoka, leaf scarring beetle, bunchy top and *Rotylenchulus* had higher positive loading in contrast to highest negative loading due to weevil. Similar to cluster analysis and principal components analysis, the cultivers are grouped together as expected in multidimensional scaling group plot.

References

Anitha, N., Rajamany, L. and Radhakrishnan, T.C.1996. Reaction of banana clones against the major biotic stresses. *Planter* **72**:315-21.

Dillon, R.W. and Goldstein, M. 1984. *Multivariate Analysis: Methods and Applications*, pp. 23-52. John Wiley and Sons.

Rachael, C., Konar, A. and Bandopadhyay, B. 2004 Screening of some banana cultivars against leaf and fruit scarring beetle (*Nodostoma siridipenne* Motsch) in West Bengal. *Proceedings of National Seminar on Present Scenario and Future Strategies,* held at Bidhan Chandra Krishi Viswasvidyalaya, Mondouri, Nadia, West Bengal, India during 11-13 June 2004.

4

Screening of Banana Hybrids for Nematode Resistance

P.S. KAVITHA, T.N. BALAMOHAN, K. POORNIMA AND E.I. JONATHAN

PLANT parasitic nematodes are major constraints to sustainable *Musa* production. Among various nematodes affecting banana and plantain, spiral nematode (*Helicotylenchus multicinctus*), burrowing nematode (*Radopholus similis*), root-knot nematode (*Meloidogyne incognita*) and root lesion nematode (*Pratylenchus coffeae*) are more destructive, causing on an average annual yield losses of 20% worldwide (Sasser and Freckman, 1987). The losses due to infestation by nematodes in banana and plantain vary depending on extent of infection and ability of cultivars to resist or tolerate infestation. Crop rotation and nematicides can be used to control nematodes (Gowen and Queneherve, 1990) but crop rotation is impracticable in areas where banana is grown continuously and the price of nematicides is often prohibitive for small farmers, apart from its extreme toxicity to environment. Evidence on availability of nematode resistance and tolerance sources in *Musa* genepool, has motivated the studies on breeding of certain *Musa* cultivars/ hybrids for nematode resistance. Therefore, hybrids were screened for resistance to *Pratylenchus coffeae* under field condition.

SCREENING OF HYBRIDS FOR NEMATODE RESISTANCE

The phase I evaluation (seed to sucker) was carried out with 24 hybrids. The phase II evaluation (sucker to sucker) was carried out for 19 hybrids already developed. The hybrids were evaluated for their breeding potential and yield besides their resistance to nematodes in a randomised block design replicated thrice. The field was infected predominantly with *Pratylenchus coffeae* occurring at a population density of more than 1 nematode/g of soil. Before planting, nematode population was estimated and then planting was taken up in the sick plot. The population was checked at different stages of crop for screening experiment.

Each hybrid was screened in field in a randomised block design replicated thrice (for phase II hybrids three replications but for phase I hybrids only one since it is single seed segregant). The hybrids were evaluated along with reference cultivars. The susceptible check cultivar used for screening *Pratylenchus coffeae* was Rasthali (AAA), while resistant reference cultivar used for screening was Pisang Lilin (AA). Infestor rows of Rasthali were planted in among hybrids and also along borders to serve as cultures of nematodes.

Root and corm damage assessment for screening studies was done as per the INIBAP technical guidelines (Carlier *et al.*, 2003). All the roots from a standard size excavation of 20cm × 20cm × 20cm extending outward from corm of plant were collected. These roots were divided into 2 categories as dead roots and functional roots. From functional roots, 5 roots were randomly selected and their lengths were reduced to 10 cm. Then the roots were cut length-wise. Scoring was carried out with one half of each of the 5 roots for per cent of root cortex showing necrosis. The maximum root necrosis per root half can be 20% giving a maximum root necrosis of 100% for 5 root halves together. Necrosis of individual roots was recorded. The sum is total root necrosis of sample.

Corms were cleaned to remove the soil with water. The outward half of the corm was selected for damage assessment after removing roots. Number of root bases on selected outward half of corm was counted. Root bases were often in groups of three to four. Nematode related damage appears as blackish purple lesions around the root bases. The number of small lesions and large lesions were counted. Lesions smaller in diameter than the root bases were considered as small lesions, while lesions equal or larger in diameter than root bases were considered as large lesions.

The Results

There were significant differences in ability of hybrids and parents to produce more number of functional roots. Resistant/tolerant hybrids produced thick and more number of healthy roots to overcome nematode infection. Good root development potential favours resistance (Fogain, 1995).

Based on intensity of lesion on roots and corms, they were assessed for their levels of resistance. The total root necrosis ranged from 3.00 to 44.00% in H-04-23 and H-04-02 respectively. The number of dead roots varied from 0.00 to 6.00. The hybrids, H-04-05, H-04-07, H-04-12, H-04-21, H-04-22 and H-04-23, had no dead roots at all. The maximum number of functional roots was observed in H-04-24 (45.00) and minimum number of roots in H-04-07 (6.00). The root lesion index and corm grade ranged from 1 to 5 and 0 to 4 respectively. Among hybrids, nine hybrids had registered minimum root lesion index, while seven maximum root lesion index. Among hybrids, H-04-22 and H-04-23 had registered lowest corm grade. Out of all 24 hybrids, 7 were found to be resistant, 8 tolerant, 6 susceptible and 3 highly susceptible.Under phase I (seed to sucker), field screening of 24 hybrids for *P. coffeae* revealed that H-04-04, H-04-05, H-04-06, H-04-12, H-04-21, H-04-22 and H-04-23 were resistant with a corm grade of 0-1 and root lesion index of 1. Under phase II evaluation, total root necrosis was highest in H-03-09 (63%) and lowest in NPH-02-01 (5%) among hybrids, while Red Banana recorded maximum total root necrosis of 31% and Anaikomban minimum of 5% among parents. The maximum dead root of 61.53% was registered H-03-04, however, H-03-15, H-03-17, H-03-19 and NPH-02-01 were found to have all roots intact. The root lesion index ranged from 1 to 5 and corm grade from 0 to 4 among hybrids and parents.

The hybrids, H-03-05, H-03-11, H-03-13, H-03-14, H-03-15, H-03-17, H-03-19 and NPH-02-01 recorded lowest root lesion index of 1, while NPH-02-01 recorded lowest corm grade of

0 among hybrids. Hybrids, H-04-01, H-04-03, H-04-07, H-04-09, H-04-11, H-04-16, H-04-19 and H-04-24, were found to be tolerant to this nematode with a corm grade of 1-2 and root lesion index of 1-2. The same was confirmed under pot studies under artificially inoculated conditions of 2 nematodes per gram of soil (Table 4.1).

Table 4.1. Root and corm damage assessment in banana hybrids (phase I) under field conditions caused by *P. coffeae*

Hybrid	*Total RN (%)*	*DE (%)*	*Root lesion index*	*Corm grade*	*Reaction status*
H-04-01	11	6.45	1	2	T
H-04-02	44	14.29	5	4	HS
H-04-03	12	4.17	2	2	T
H-04-04	10	5.41	1	1	R
H-04-05	10	0.00	1	1	R
H-04-06	9	2.86	1	1	R
H-04-07	13	0.00	2	2	T
H-04-08	21	25.00	3	4	S
H-04-09	16	5.88	2	1	T
H-04-10	40	23.08	5	4	HS
H-04-11	10	5.56	1	2	T
H-04-12	9	0.00	1	1	R
H-04-13	42	12.50	5	4	HS
H-04-14	40	13.64	5	3	S
H-04-15	37	14.29	4	3	S
H-04-16	15	6.67	2	2	T
H-04-17	32	12.20	5	4	S
H-04-18	35	12.50	5	4	S
H-04-19	15	5.88	2	2	T
H-04-20	32	11.11	5	3	S
H-04-21	8	0.00	1	1	R
H-04-22	4	0.00	1	0	R
H-04-23	3	0.00	1	0	R
H-04-24	15	4.26	2	1	T
Cultivators					
Pisang Lilin	5	0.00	1	1	R
Rasthali	44	39.02	5	4	HS

R, resistant; T, tolerant; S, susceptible; HS, highly susceptible; RN, Root necrosis

All the resistant/tolerant hybrids were found to possess more number of functional roots, root spread, root thickness and fresh root weight compared to susceptible ones. This is an important character considered by breeders in selecting plant material for breeding nematode resistance. Vigorous roots and more root number can allow field tolerance of plants to nematode infection to a certain extent (Gowen, 1993). Resistant/tolerant hybrids produced thick and more number of healthy roots to overcome nematode infection. Good root development potential favours resistance (Fogain, 1995).

Under field screening, resistant hybrids had more number of functional roots and thus with reduced percentage of dead roots. Root necrosis was also comparatively lesser than other hybrids, resulting in minimum root lesion index and corm grade. However, assessment of root predicted considerable number of nematode population in roots, which might be due to inherent ability of hybrids to overcome the invasion of nematodes. The increase in number of functional roots compared to their parents was attributed to heterotic phenomenon manifested in these hybrids due to geographical as well as genetical divergence.

It is concluded that hybrids, H-04-06, H-04-05 and H-04-24, are potential parents that can be used for further breeding programme due to their high male and female fertility along with resistance to nematodes. Further, H-04-06, NPH-02-01, H-03-19, H-03-17 and H-03-05 with better bunch weight besides possessing resistance to nematodes have to be studied further for exploitation.

SUMMARY

The resistant/tolerant hybrids were found to possess better root characters such as more number of functional roots, more root spread, root thickness and fresh root weight compared to susceptible hybrids. Under phase I (seed to sucker), field screening of 24 hybrids for *Pratylenchus coffeae* revealed that hybrids, H-04-04, H-04-05, H-04-06, H-04-12, H-04-21, H-04-22 and H-04-23, were resistant with a corm grade of 0-1 and root lesion index of 1. Under phase II (sucker to sucker), field screening for *Pratylenchus coffeae* revealed, that H-03-05, H-03-10, H-03-11, H-03-13, H-03-14, H-03-15, H-03-17, H-03-19 and NPH-02-01 were resistant with a corm grade of 1-2 and root lesion percentage of 1-2, while H-03-06 was tolerant with a corm grade of 2 and root percentage of 2.

References

Carlier, J., De Waele, D. and Escalant, J.V. 2003. Global evaluation of *Musa* germplasm for resistance to *Fusarium* wilt, *Mycosphaerella* leaf spot diseases and nematodes. Vegine, A. and Pig, C. (Eds). INIBAP Technical guidelines 7. The International Network for the Improvement of Banana and Plantain, Montpellier, France.

Fogain, R. 1995. Study of resistance of *Musa* spp. to nematodes (*Radopholus similis* and *Pratylenchus goodeyi*). Ph.D. Thesis, University of Reading, Berkshire, U.K.

Gowen, S.R. and Queneherve, P. 1990. Nematode parasites of bananas, plantains and abaca *In: Plant Parasitic Nematodes of Subtropical and Tropical Agriculture*, pp. 431-60.

Gowen, S.R. 1993. Possible approaches for developing nematode resistance in bananas and plantains. *In: Breeding Banana and Plantain for Resistance to Diseases* and *Pests.* Ganry, J. (Ed.), **INIBAP,** Montpellier, France, pp. 123-28.

Sasser, J.N. and Freckman, D.W. 1987. A world perspective on nematology: the role of the society. *In: Vistas on Nematology.* Veech, J.A. and Dickson, D.W. (Eds). Society of Nematologists , Inc, Hyattsville, USA, pp. 7-14.

5

Development of Transgenic Banana for Resistance to Diseases

SUKHADA MOHANDAS

BANANA is most important fruit crop of the world. Most cultivated land races of banana are parthenocarpic and sterile triploids, hence classical breeding is a difficult task. Genetic transformation provides an opportunity for single genes or gene combinations to be transferred directly into the desired variety. This allows the variety undergoing transformation to retain all its original characteristics, with simple addition of the desired trait. Molecular breeding plays an important role in crop improvement and, in combination with conventional breeding, will provide the tools with which new improved varieties will be developed at a faster pace in future. It is already a reality, and the technology to transform and express transgenes in banana is already available. Genetic transformation of banana has been accomplished by using techniques such as particle bombardment of embryonic cell suspensions or shoot tips (Navak *et al.,* 1989; Sagi *et al.*, 1995), Agrobacterium-mediated transformation, and introduction of DNA into regenerable cell suspension-derived protoplasts by electroporation (Sagi *et al.*, 2001) or by Agrobacterium-mediated transformation of apical meristem after micro-wounding using gene gun (May *et al.*, 1995).

TRANSGENIC BANANAS

Resistance for Fungal and Bactetial Diseases

Banana is susceptible to a number of serious and debilitating diseases caused by bacterial, viral and fungal pathogens. Of these, fungi are usually regarded as the most destructive. Two of the most important fungal diseases affecting banana are black and yellow Sigatoka diseases (caused by *Mycosphaerella fijiensis* and *M. musicola*, respectively) and Panama disease or Fusarium wilt (caused by *Fusarium oxysporum* f.sp. *cubense*) (Robinson, 1996). Both are known to be toxin-producing fungi that clearly induce large necrotic patches upon infection. Transgenic bananas expressing genes with antifungal activity have been developed at Katholieke Universiteit Leuven (KUL), Belgium (Sagi *et al.*, 2001). Broad range of genes encoding different classes of proteins with activity against phytopathogenic fungi like chitinases (rcc 2 and rcg 3), glucanases, antimicropeptides, defensins have been identified and cloned and a large number of transgenic plants have been generated using particle bombardment and *Agrobacterium*-mediated transformation of embryogenic cell suspension cultures. Molecular characterization of these plants and of their micropropagated progeny has been carried out. Selected lines are being tested for resistance to *Mycosphaerella fijiensis* (black Sigatoka) and *Fusarium oxysporum* (Panama disease) under controlled and/or field conditions and are awaiting clearance from biosafety committee.

East African Highland banana, Gros Michel (GM), a banana cultivar has been transformed with two rice chitinase genes (rcc 2 and rcg 3) (Jacon *et al.*, 1997). MSI-99, a synthetic substitution analogue of magainin, an antimicrobial peptide, has been used by BARC, Bombay, to impart disease resistance in transgenic banana Rasthali (AAB) (Ganapathi, 2001). This peptide inhibited the growth and spore germination of *Fusarium oxysporum* f.sp. *cubense* at 16 μg/ml and transgenics developed were found to be resistant to *Fusarium oxysporum* f.sp. *cubense* under controlled conditions. Scientists at Centre for Tropical Crops and Biocommodities, QUT, Brisbane, have investigated four genotypes of banana to identify R genes with potential to confer resistance to Foc. Five families of R genes were identified and members of each of these gene families are present in all four genomes tested (although there is slight DNA sequence variation) (www.ihbi.qut.com). In addition, RGA2 was shown to be transcribed in three Fusarium-resistant *Musa acuminata* subsp. *malaccensis* progeny and not in three Fusarium-susceptible progeny. The RT-PCR was used to compare R gene expression between susceptible or resistant plants. In addition, this group is working on a cell death suppressor genes –(antiapoptoptic gene) and a Fusarium resistance gene isolated from resistant banana cultivar with different modes of action to be used in combination and evaluate the combined impact on fungal resistance.

Hundreds of transgenic lines developed are currently undergoing molecular characterization and fungal bioassays are underway to test for resistance to Foc. Banana cells transformed with a nucleic acid construct which includes: (i) a first polynucleotide encoding endochitinase, (ii) a second polynucleotide encoding stilbene synthase and (iii) a third polynucleotide encoding superoxide dismutase gave 7 lines of banana resistant to Sigatoka (Jane Vishnevetsky US patent 20060288446). At IIHR, Bangalore, work on genetic engineering of banana Neypoovan and Rasthali which with antimicrobial peptide gene for Fusarium resistance is in progress. Antimicrobial peptide gene cloned from onion seeds and AMP construct prepared in pCAMBIA background has been used to transform Rasthali and Neypoovan cultivars using the method of May *et al.* (1994) modified by Tripati *et al.* (2005). To start with transformation procedure was standardized with GFP gene. Thereafter, transformants generated with AMP gene and glucanase-chitinase gene is under evaluation. Transformation of Rasthali banana with AMP gene has also been standardized and putative transformants have been generated.

Xin-Wu PEI *et al.* (2005) introduced human lysozyme (HL) gene which inhibited *F.oxysporum* f.sp. *cubense in-vitro,* driven by a constitutive cauliflower mosaic virus 35S promoter into banana via Agrobacterium-mediated transformation. After confirming the integration of gene they screened the plants for Fusarium resistance in pots and under field conditions, and short listed two plants highly resistant to Foc R4. They concluded that Agrobacterium-mediated transformation, with assistance of particle bombardment, is a powerful approach for banana transformation and that a transgenic HL gene can induce resistance of the crop to Foc R4 in field.

The plant ferredoxin like protein *(pflp)* gene and hypersensitivity response assisting protein *(hrap)* gene isolated from sweet pepper are being used to develop transgenic banana resistance to banana Xanthomonas wilt (BXW). These genes code for novel plant proteins that can intensify the harpinPSS-mediated hypersensitive response. These proteins function via iron depletion, antimicrobial action and harpin triggered HR enhancing. The transgenic banana plants expressing transgene (*pflp* or *hrap*) is being tested by artificial inoculation (*in-vitro* screening) to evaluate for BXW resistance, in controlled laboratory conditions and further in containment facilities.

Resistance for Viral Diseases

Banana bunchy top babu virus, streak badnavirus, bract mosaic potyvirus, cucumber mosaic cucumovirus are viruses infecting banana. For RNA viruses, coat protein and replicase which prevents multiplication and RNA mediated resistance which has led to post transcriptional gene silencing- RNAi technology is the current approach for virus resistance. In RNAi technology, plant generates 20-22 nucleotide molecules which are complementary to target RNA. These molecules hybridise with target RNA and form dsRNA which is degraded by plant encoded dsRNase. Waterhouse *et al.* (2001) have developed series of vectors in which target RNA sequence (derived from target virus is inserted as DNA) as an inverted repeat with the repeats separated by an intron. Once transformed into plants and transcribed, a novel dsRNA is formed and this is rapidly recognised by the plant as an aberrant RNA and is silenced. Plant is pre-immunised for invasion by homologous virus and is effectively immune to infection by that virus.

There is no expression of recombinant protein in transformants. It is highly sequence specific. But plant viruses exist as a continuum of strains which mostly differ from one another in their genome sequence. This can be partially overcome by selecting an integrated sequence after surveying a large number of isolates of BBTV which is a ssDNA virus. Coat protein mediated resistance is found to be not so effective. Expressing a defective (mutated or antisense) replicase protein which binds to the target viral DNA but lacks the function of replication (no DNA replication) is attempted. Transgenic banana resistant to BBTV has been developed by John Hu's group at Hawai using mutated replicase gene and using gene silencing technology. The plants are under evaluation. QUT is adopting a strategy of constitutive expression of mutated Reps and virus activated cell death suppressor gene.

Resistance for Insect Pests

Banana weevil is a serious problem facing banana industry. Proteinase inhibitors are expressed naturally as plant defense mechanism against insect attack. They operate by disrupting protein digestion in insect mid gut via inhibiting proteinases. Transformation with two proteinases a combination of cysteine and serine proteinases is considered as an advanced strategy to combat the weevil problem. Vegetative insecticidal proteins (VIPs) can be used against weevil. The VIP1 and VIP2 are contained in culture supernatants of *Bacillus cereus* (Warren, 1997), while VIP3 can be found in supernatants of *B. thuringiensis*. They function by causing gut paralysis followed by lysis of the gut epithelium cells, completely arresting gut function and leading to the death of the insect (Duck and Evola, 1997).

Alpha-amylase inhibitors (AI) and chitinase enzymes might also have a future potential for weevil control. Alpha-amylase inhibitors operate by inhibiting the enzyme alpha-amylase, which breaks down starch to glucose in insect gut (Le Berre-Anton *et al.*, 1997; Morton *et al.*, 2000). They are divided into two types, AI-1 and AI-2, isolated from common and wild beans (*Phaseolus vulgaris*), respectively. Ishimoto *et al.* (1996) produced transgenic adzuki beans with enhanced resistance to bean bruchids, which are Coleopteran insects. Alpha-amylase inhibitors may be of interest for banana weevil control in GM banana.

Resistance for Nematodes

The *Radopholus similis* is a major problem. It is a burrowing, migratory and endoparasite. The Meloidogynae or root knot nematode is sedentary endoparasite. Protylenchus a migratory endo/ectoparasite, Helicotylenchus, a spiral nematode migratory ectoparasitic and sedentary parasite and *Hoplolaimus pararobustus* are major nematodes infecting banana. Prteinase inhibitors of plant origin like Trypsin, Serine, Cysteine, which inhibit nematode development and reduce fecundity of females in Meloidogynae have been used to control nematodes (Kiggundu *et al.*, 2003). The cystatin suppresses the nematode's ability to grow, lay eggs and build to population levels that damage crops. As cystatins are part of the human diet (present in cereal seeds, eggs etc.), they have no effect on human digestion or health. Transgenic banana plants containing and expressing different cystatin genes were challenged with *Meloidogyne incognita,* exhibited levels of resistance. This was the first demonstration worldwide of transgenic East African Highland Bananas (EHAB). Cystatin genes were also introduced into another type of AAA bananas (Cavendish) leading to levels of resistance to *Radopholus similis*. The R gene stacking cultivars with partial resistance to nematode when transformed with OclD86 gave enhanced full resistance. Co-delivery of OclD86 and CpTI as a fusion protein linked by a short peptide established an additive effect. Lectins- concanavalin A suppresses *M. incognita*. The Bt proteins have effects on saprophagous nematodes (Marroquin *et al.*, 2000). Short peptides disrupting chemoreception showing prevention and invasion of plant roots by nematodes have been developed (Bing Liu *et al.*, 2005).

Delayed Ripening

Bananas for extended shelf-life have been developed by suppressing production of ethylene by different private companies and are at various stages of testing (www.frost.com/prod/). Producing vaccine against Hepatitis B in plants is one of the major areas of research (Atkinson, 2004). The BARC group, Bombay, has succeeded in producing a Hepatitis B antigen in bananas (Sunil Kumar *et al.,* 2005).

Vaccine Production

Banana can be consumed to get vaccinated against Hepatitis B. Embryonic cell suspensions of banana Rasthali have been transformed with "s" gene of Hepatitis B surface antigen (HbsAg) using Agrobacterium-mediated transformation. Expression levels of antigen in plants grown under *in-vitro* conditions as well as greenhouse hardened plants were estimated by ELISA. Expression of HbsAG in fruits was confirmed by Rt-PCR. The team is at present aiming to enhance the expression levels by using abundant pulp protein promoters and by incorporating certain regulatory sequences.

Promoters

There is a need to develop promoters for expressing transgenes differentially under specific conditions or being organ specific. The KUL group is employing T-DNA tagging method with promoterless codon-optimized luciferase (*luc*) gene to identify promoters directly from banana genome. They are generating large populations (tens of thousands) of independent transgenic cell colonies via highly efficient *Agrobacterium*-mediated transformation of embryogenic cell suspensions. Combined with an ultrasensitive imaging system this allows high-through put screening for luciferase (LUC) activation. Screening for putative constitutive promoters results in a gene activation frequency of 1-2.5%. In the sample containing a few hundred transformed cell colonies more than five colonies with LUC activation by a putative promoter can be identified in the LUC image. The use of *luc* gene also allows a combined screening for different inducing factors.

References

Atkinson, H. 2004. Strategies for resistance to nematodes in www. Promusa.org/Research/ gen_trans_strategies_nematode.

Bing Liu, Jennifer K. Hibbard, Peter Edward Urwin, Howard J. Atkinson, 2005. The production of synthetic chemodisruptive peptides in planta disrupts the establishment of cystnematodes. *Plant Biotechnology Journal* **3**(5):487-96.

Duck, N. Evola, S. 1997. Use of transgenes to increase host plant resistance to insects opportunities and challenges, *In: Advances in Insect Control — the Role of Transgenic Plants.* Carozi, N., Koziel, M. (Eds). Taylor and Francis, London, pp. 45-56.

Ishimoto, M. Sato, T., Chrispeels, M.J., Kitamura, K. 1996. Bruchid resistance of transgenic azuki bean expressing seed á – amylase inhibitor of common bean. *Entomol. Exp. Appl.* **79:**309-15.

Jacon, G., Remy, S., and Sági, L. 1997. Development of improved East African Highland banana. In Genetic transformation for the improvement of Bananas. *Annual Report 1997*, Katholieke Universiteit Leuven. http://www.biw.kuleuven.be/english.

Kiggundu, A., Michael Pillay, Altus Biljoen, Clifford Gold, Wilberforce Tushemereirwe, Karl Kunert (2003) Enhancing banana weevil (*Cosmopolitus sordidus*) resistance by genetic modification: A perspective. *African Journal of Biotechnology* **2**(12):563-69.

Leena Tripathi, 2003. Genetic engineering for improvement of Musa production in Africa. *African Journal of Biotechnology*, **2**(12):503-508.

May, G.D., R Afza, HS Mason, A Wiecko, FJ Novak, C.J. Bio/Technology, 1995 - nature.com... Generation of Transgenic Banana (Musa acuminata) Plants via Agrobacterium-Mediated Transformation. *Bio Technology* **13:**486-92.

Novak, F.J., Afza, R., Van Duren, M., Perea-Dallos, M., Conger, B.V. and Xiaolang, T.1989. Somatic embryogenesis and plant regeneration in suspension cultures of dessert (AA and AAA) and cooking (ABB) bananas (*Musa* spp.). *Bio Technology* **7**:154-59.

Robinson 1996. Agrobacterium mediated transformation of embryogenic cell suspensions of banana cultivar Rasthali AAB). *Plant Cell Reports.* **20:**157-62.

Sagi, L., Panis, B., De Smet, K., Remy, S., Swennen, R. and Cammue, B.P.A.1995. Genetic transformation of banana and plantain (*Musa* spp.) via particle bombardment. *Bio Technology* **13**:481-85.

Sagi, S.Remy, J.B.P.Hernandez, B.P.A. Cammue and Swennen 2001. *Transgenics Banana in Biotechnology in Agriculture and Forestry*, Vol 57. Transgenic Crops II. Springer-Verlag Berlin Heidelber.

Sunil Kumar, G.B., T.R. Ganapathi, C.J.Revathi, L.Srinivas, V.A.Bapat. 2005. Expression of hepatitis B surface antigen in transgenic banana plants. *Planta* **222:**484-93.

Xin-Wu Pei, Shi-Kai Chen, Rui-Ming Wen, Shang Ye, Jia-Qin Huang, Yong-Qiang Zhang, Bing-Shan Wang, Zhi-Xing Wang, Shi-Rong Jia 2005. Creation of Transgenic Bananas Expressing Human Lysozyme Gene for Panama Wilt Resistance. *Journal of Integrative Plant Biology* **47**(8):971-77.

LeBerre-Anton, V., Bompard-Gilles, C., Payan, F., Rouge, P., 1997. Characterization and functional properties of the alpha-amylase inhibitor (alpha A-1) from kidney bean (Phaseolus vulgaris) seeds. *Biochim. Biophys. Acta.* **1343:**31-40.

Morton, R.L., Schoroeder, H.E., Bateman, K.S., Chrispeels, M.J., Armstrong, E., Higgins, T.J.V., 2000. Bean ò-amylase inhibitor-I in transgenic peas (Pisum sativum) provided complete protection from pea weevil (Bruchus piorum) under field conditions. *Proceeding of National Academy Science,* **97:**3820-825.

Winterer, J., 2002. The mixed success of proteinase inhibitors to combat insect pests in transgenic crops. *Ag. Biotech. Net.* **4:**1-7.

Warren, G.W., 1997. Vegetative insecticidal proteins: novel proteins for control†of corn pests. *In: Advances in Insect Control: The Role of Transgenic Plants.* Carozzi, Koziel, M. (Ed.) Taylor and Francis,London, pp. 109-21.

6

Advances in R&D in Transgenic Banana

A. Das, S.N. Gupta and B. Bandyopadhyay

There are diverse germplasms of banana traditionally cultivated in different regions of India having remarkable genetic differences. In spite of their overall genomic grouping like AAA, AAB and ABB, variations still exist within the same group (Singh and Chadha, 1993). Traditional procedure to characterize banana plants by morphological descriptors has many limitations. Different communities refer to the local cultivars by different names and lack of clear clone identity in the crop has resulted in unnecessary duplication with regard to collection, conservation and research (Onguso *et al.*, 2004). To alleviate this problem, application of modern fingerprinting technology through protein and/or DNA markers have been recommended for accurate selection of banana clones (Robinson, 1996). Randomly Amplified Polymorphic DNA (RAPD) markers are usually preferred for the initiation of this kind of work, as the technique is simple, versatile, relatively inexpensive and able to detect minute differences (Williams *et al.*, 1990).

The AICRP on Tropical Fruits at BCKV centre has assembled a diverse collection of banana germplasm from different regions of the country. Among these collections a larger part contains different indigenous clones or habitats of banana from eastern and north-eastern region of India. Wide phenotypic variations among these cultivars have been noticed for fruit quality, plant types, bearing habit, disease, pest infestation etc. The extent of genetic variation among 24 of such selected clones and the primary establishment of their genetic identity on the basis of RAPD markers have been established (Das *et al.*, 2004).

In further investigation, three selections locally named as Baishchhara, Bagda and Kanthali Clone-1 were found promising for commercial cultivation in different agro-ecological situations. All these three germplasms belong to genomic group ABB. However, Baishchhara is a cooking type, while Bagda and Kanthali Clone-1 are dessert type. The phenotypic observation and records predicted that these cultivars originated from mother clone Kanthali. To establish genetic identity of these clones, their SDS-PAGE protein profiles and DNA banding pattern with 10 RAPD markers were studied and compared with three other standard ABB genotypes. The results of analysis are presented in this chapter.

ADVANCES IN R&D IN TRANSGENIC BANANA

Plant Material

Banana genotypes belonging to ABB genomic group, maintained at BCKV centre, West Bengal, were used for experiment. The genotypes were: 1. Bagda (dessert); 2. Baisha (dessert); 3. Kanthali (dessert); 4. Kanthali Clone1 (dessert); 5. Baishchhara (cooking) and 6. Pantharaj (cooking). Among these genotypes, Baisha (2) and Kanthali (3) were closely related genotypes (Das *et al.*, 2004) and Pantharaj was a standard cooking type banana which was used as reference material. The selected clones, Bagda, Kanthali Clone 1 and Baishchara, were compared with these reference genotypes.

SDS-PAGE Protein Analysis

Total soluble proteins from banana leaves of 6 clones were extracted and estimated as per the procedure of Lowry *et al.* (1951). Sodium dodecyl sulfate poly acrylamide discontinuous gel electrophoresis (SDS-PAGE) was performed following the procedure described by Laemmli (1970). Each well in the gel was loaded with equal volume (25 µl) having equal quantity of protein in solution (300 µg) for each sample. A standard molecular weight protein marker of range 14300-97400 Daltons (Bangalore Genei) was loaded in each gel with test samples. The SDS-PAGE gels were visualized, analyzed and documented by Ultra lum gel documentation system (Model no. EBW 20) with total lab gel analysis software to determine the number, relative mobility and molecular weight of bands.

DNA Extraction

Genomic DNA was extracted from tender leaves of plants using Plant DNA CTAB Extraction Kit (KT-55) of Bangalore Genei Ltd. (Bangalore, India). A very small leaf sample (3 cm in length and approximately 36 mg of weight) was required to yield good quality DNA within a period of 3 - 4 hours. Tender leaf samples were collected, cleaned using sterile double distilled water, cut into small pieces, kept separately in 1.5 ml tubes and stored in -20°C temperature before extraction. The DNA pellet was re-suspended in 20 µl of TE buffer (Tris-HCl 10mM and EDTA 1mM pH 8.0). Quantification of DNA was made by electrophoresis of extracted DNA of an amount of 5-10 µl alongside with the known amount of uncut bacteriophage lambda DNA in an agarose gel (1%). The concentration of extracted DNA was estimated according to the method of Kahangi *et al.* (2002).

PCR Procedure

Amplification was achieved by protocol outlined by Williams *et al.* (1990) with slight modifications. The DNA was amplified with Operon decamer primers selected by preliminary screening for giving higher levels of polymorphism and reproducible fragment patterns. Ingredients of each reaction included template DNA 25-30 ng, dNTPs 200 µM each, *Taq* DNA polymerase 1.5 unit, $MgCl_2$ 2mM, 1x PCR buffer (Bangalore Genei Ltd) and 15 ng of each decamer primer in a total volume of 25 µl. The amplification was performed in a thermocycler (Gene Amp PCR System 9700, Applied Bio Systems). The complete reaction consisted of 45 cycles, each cycle consisting of three steps; denaturation at 92 °C for 30 seconds, annealing at 38 °C for 30 seconds, extension at 72 °C for one minute, with an initial denaturation at 94 °C for 30 seconds and a final extension at 72 °C for five minutes, followed by cooling at 4 °C temperature. Amplification fragments were separated on 1.2% agarose (Sisco Research Laboratory) gels containing ethidium bromide (0.5 μg/ml of agarose) at 60 V for six hours in Tris Borate EDTA buffer.

Data Analysis

Amplified fragments from all the primers, which were reproducible over two amplifications, were scored by Total Lab gel documentation software (Ultra Lum,Inc. USA). The size of fragments (molecular weight in base pair) was estimated by using a 100-bp ladder marker, which was run along with amplified products. Bands were manually scored '1' for presence and '0'for absence and binary data were used for statistical analysis. A genetic dissimilarity matrix was calculated according to Squared Euclidean Distance that estimated all pair-wise differences in amplification products (Sokal and Sneath, 1973) and cluster analysis was performed based on Ward's method using a minimum variance algorithm (Ward, 1963).

TRANSGENIC BANANA

The analysis of SDS-PAGE of leaf proteins of banana clones resulted 9 subunit bands for each. The molecular size of these protein subunits ranged between 32KD to 102 KD. In spite of some variations in intensities of bands, polymorphism in banding pattern was not noticed.

In RAPD analysis, a total of 66 unambiguous, readable and reproducible genomic DNA fragments were produced from selected 10 primers. The number of amplified fragments varied from 4 to 9 with an average of 6.7 fragments per primer (Table 6.1). Out of 67 DNA fragments, 49 were polymorphic. The size of fragments ranged from 200bp to 1400bp. The amplification products obtained by primer OPAT 04 is illustrated in Fig. 6.1. In every PCR reaction, no DNA fragments were found in negative control, while similar banding pattern were amplified in positive control, indicating contamination free PCR ingredients and consistency of protocol.

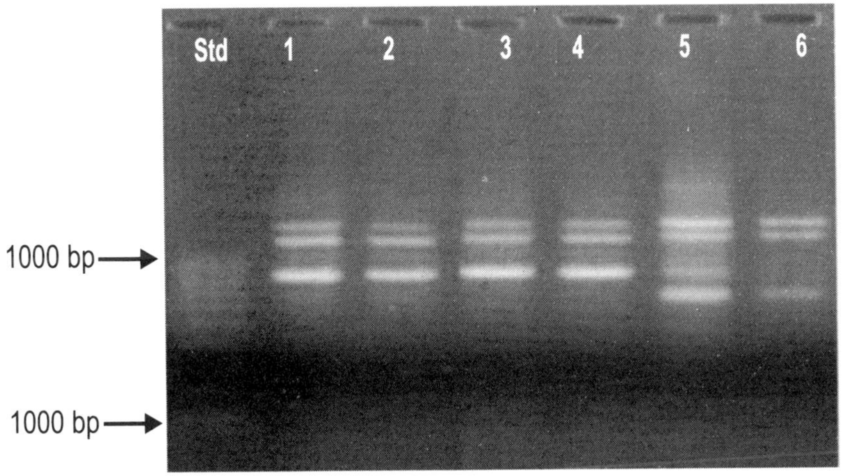

Fig. 6.1. RAPD banding pattern of six banana clones by primer OPAT 04; Lane: std. Standard 100bp ladder DNA; 1, Bagda; 2, Baisha; 3, Kanthali; 4, Kanthali Clone 1; 5, Baishchhara and 6, Pantharaj.

Table 6.1. Synthetic deoxyribonuceotide 10-mer random primers used for RAPD profiling of banana clones

Primer no.	*Nucleotide sequence (5´–3´)*	*No. of amplified fragments*	*Fragment size (kb)*
OPA01	CAGGCCCTTC	8	0.2 – 1.3
OPA07	GAAACGGGTG	5	0.5 – 1.2
OPA08	GTGACGTAGG	5	0.5 – 1.2
OPAT03	GACTGGGAGG	7	0.5 – 1.4
OPAT04	TTGCCTCGCC	7	0.6 – 1.3
OPH04	GGAAGTCGCC	4	0.5 – 1.1
OPH15	AATGGCGCAG	7	0.6 – 2.0
OPM06	CTGGGCAACT	9	0.4 – 1.4
OPM13	GGTGGTCAAG	9	0.4 – 1.3
OPN07	CAGCCCAGAG	5	0.6 – 1.3

Based on estimated proximity matrix scale generated from the data of RAPD, highest genetic dissimilarity (32) was noticed among clones Baischhara (5) and Kanthali clone 1 (4). The closest proximity (10) was noticed among the clones Kanthali (3) and Baisa (2) (Table 6.2). The dendrogram made from these data showed that six banana clones formed two distinct major clusters, 'A' and 'B'. The cluster 'A' was formed by dessert types of banana and in cluster B only two cooking types were present. The distance between two clusters was maximum 100 units. In cluster A, Kanthali (3) and Baisa (2) were in closest proximity and they have nearly 65-70 unit linkage distance from selected clones Bagda and Kanthali Clone-1. The distance among selected clone Baischhara and standard cooking type cultivar Pantharaj was more than 50 units.

Table 6.2. Squared euclidean distance matrix of six banana clones based on RAPD markers

Clone	*Bagda*	*Baisha*	*Kanthali*	*K. Clone 1*	*Baischhara*	*Pantharaj*
Bagda	0					
Baisha	18	0				
Kanthali	20	10	0			
K. clone 1	26	20	18	0		
Baischhara	24	26	26	32	0	
Pantharaj	21	19	17	23	17	0

The result of RAPD analysis of 24 clones of banana was reported (Das *et al.*, 2004). The cultivars Kanthali and Baisa were also included in that study. The RAPD bands generated from 6 clones of banana taken in this investigation were also compared with that of previous analysis. The dendrogram thus generated from altogether 28 clones clearly showed that 6 clones of present investigation were quite distinct from each other. Among them, Bagda, Kanthali clone -1. and Baisa, showed closer relationship with cultivar Kanthali. The Clones, Baischhara and Pantharaj, the cooking types, although formed separate group in this dendrogram had closest proximity with Kanthali than other established cultivars of banana. In banana, indigenous varieties like Kanthali have potentials to create genetic variations. The somatic mutation during long-term and continuous vegetative propagation of such popular cultivars might be possible source of genetic variation and development of new clones (Kahangi *et al.*, 2002). The present investigation through RAPD analysis could clearly predict that three selections of AICRP, Tropical Fruits, BCKV centre, were although basically originated from indigenous variety Kanthali had distinguished genetic identities.

SUMMARY

The eastern region of India is endowed with diverse germplasm resources of banana. Traditional procedure to characterize such clones by morphological descriptors and genomic grouping has many limitations as variation within an established cultivar also exists. From the diverse collection of banana germplasms by AICRP Tropical Fruits, BCKV Centre, three elite types commonly named as Baishchhara, Bagda and Kanthali Clone-1, were selected based on their distinguished field performances. All these three germplasm belong to genomic group ABB, however, Baishchhara is a cooking type and Bagda and Kanthali Clone-1 are dessert type. To establish genetic identity of these clones, their SDS-PAGE protein profiles and DNA banding pattern with 10 RAPD markers were studied and compared with three other standard ABB genotypes. The selected clones showed their distinct fingerprint characters that were more prominent in RAPD profiles. It appeared that three selections had close proximity with common banana Kanthali than other banana genotypes belonging to ABB group.

References

Das, A., Sarkar, J., Mondal, B. and Mishra, D.K. 2004. RAPD profiling of some banana clones collected from different parts of India. *In: Banana Industry- Present Scenario and Future Strategies.* Bandyopadhyay, B., Sarkar, S.K., Mishra, D.K., and Barui, F.K. (Eds). BCKV, West Bengal, pp. 20-23.

Kahangi, M. E., Lawton, M. A. and Kumar, C.A.C.Y. 2002. RAPD profiling of some banana varieties selected by small-scale farmers in Kenya. *J. Hort. Sci. Biotech.* **77**:393-98.

Laemmli, U.K. 1970. Cleavage of the structural protein during the assembly of the head of bacteriophage T-4. *Nature* **227**:680-85.

Lowry, O.H., Rosebrough, N.J., Furr, A.L. and Randall, R.J. 1951. Protein measurement with the Folin phenol reagent. *J. Biol. Chem.* **193:**265-75.

Onguso, J., Kahangi, M., Ndiritu, W.D. and Mizutani, F. 2004. Genetic Characterization of cultivated bananas and plantains in Kenya by RAPD markers. *Sci. Hort.* **99:**9-20.

Robinson, J.C. 1996. *Bananas and Plantains.* CAB International, Wallingford. U.K. 238p.

Singh, H.P. and Chada, K.L. 1993. Genetic resources of banana. *In: Advance in Horticulture,* Vol. I. Chada, K.L. (Ed.) Malhotra Pub. Co. New Delhi, pp. 123-42.

Sokal, R.R. and Sneath, P.H. A. 1973. *Principles of numerical taxonomy.* W.H. Freeman and Co., San Francisco, USA.

Ward, J.H. 1963. Hierarchic grouping to optimize an objective function. *J. Amer. Stat. Asso.* **58:**236-39.

Williams, J.G.K., Kubelik, A.R., Livak, K.J., Rafalaski, J. A. and Tingey, S.V. 1990. DNA polymorphism amplified by arbitrary primers are useful as genetic markers. *Nucl. Acids Res.* **18**:6531-35.

7

Establishment of Embryogenic Cell Suspensions for Development of Transgenic Hill Banana Resistant to Banana Bunchy Top Virus

R. SELVARAJAN, R. RAJMOHAN, V. BALASUBRAMANIAN, T. RAJESH, S. UMA AND M.M. MUSTAFFA

BANANA and plantain are most important fruit crops in the world. Hill banana (Virupakshi, AAB) belonging to subgroup Pome, is commercially cultivated for its best flavour, taste and quality in Sirumalai and lower Pulney hills as a shade crop in coffee plantations or sole crop in Tamil Nadu. This banana is also used for making panchamirtham which is offered to God. This hill banana is devastated by banana bunchy top virus (BBTV) infection and it will be extinct in this region if due care is not taken up in coming years. The disease is present round the year owing to presence of viral inoculum and viruliferous vector in cool hilly climate. Moreover, banana is grown perennially with no irrigation made disease uncontrollable through conventional methods. Supply of virus-free tissue cultured hill banana by the Government of Tamil Nadu, was not successful owing to quick spread in subsequent crops. Therefore, to find out a suitable solution to rejuvenate the hill banana using modern scientific approaches like biotechnological interventions was necessary.

Genetic engineering and other biotechnological applications can be utilized for hill banana improvement and are based on reliable plant regeneration protocols. Developing embryogenic cell suspensions is a prerequisite for genetic manipulation of banana such as developing transgenics. The productivity of somatic embryos depends on quality of embryogenic cultures. Moreover, it can be affected by medium composition and environmental conditions. Main setbacks in formation of banana somatic embryogenic callus are contamination and phenolic compound accumulation in the media. It has been successful only in a few banana cultivars such as Bluggoe, ABB (Novak *et al.*, 1989; Dhed'a *et al.*, 1991; Panis *et al.*, 1993); Grande Nain, AAA (Novak *et al.*, 1989; Escalant *et al.*, 1994; Co^te *et al.*, 1996; Becker *et al.*, 2000;

Escalant *et al.*, 1994; Co^te *et al.*, 1996; Grapin *et al.*, 1996) Mysore, Rasthali (AAB) (Escalant *et al.*, 1994; Ganapathi *et al.*, 2001). Phenolic exudation in media is higher in hill banana than other cultivars. Therefore, to develop efficient embryogenic cell suspensions of hill banana Virupakshi (AAB) to evolve BBTV resistant transgenic plants was essential.

ESTABLISHMENT OF EMBRYONIC CELL SUSPENSION FOE TRANSGENIC BANANA

Virus-free indexed hill banana plantation was established at NRC for Banana during 2005-06, as the source of continuous supply of male buds for establishing embryogenic calli and cell suspensions. The male buds were retested for presence of BBTV by PCR as described by Selvarajan *et al.* (2007) before using for callus induction. The male inflorescences were reduced to 3 cm tips by removing enveloping bract and floral initials were washed with sterile distilled water followed by 70% ethanol for 15 minutes. Finally, tips were reduced to 1 cm in length under sterile conditions. The very young enveloping bracts contained in inflorescence tips were removed individually and primordia of male flower buds were excised and inoculated in culture medium. Media contains basal MS salts supplemented with growth regulators and anti-oxidants such as charcoal, ascorbic acid and sodium citrate. The MA media described by Ma (1991) has been used which consists of MS macro, micronutrients and MA vitamins supplemented with Picloram (10 μM), 5.7 μM IAA, 5.4 μM NAA, 87 mM sucrose and instead of agar, phytagel-2.5g used as solidifying agent. The MA media without any antioxidant (M_1), MA with 2mg/litre (M_2) and 2mg ascorbic acid charcoal (M_3) were used. The immature flower positions 16 to 8 were placed at 135° angle in media plate. Inoculated male flower buds were kept under dark at 24 ± 1 °C. The development of embryogenic calli was observed once in 15 days. All inoculates were periodically examined for phenolic secretion, blackening of media and development of friable embryogenic calli.

The Results

The hill bananas were indexed for BBTV and suckers were planted in field and the plants were reconfirmed to be free of virus by PCR method. Male buds which were positive for BBTV were discarded. Three treatments, viz. M_1: MA media alone, M_2: MA medium incorporated with charcoal (2mg/litre) and M_3 was MA medium with ascorbic acid (2mg/litre) were used for callus induction. Totally, 247 male flower buds were used. In these 82 buds for M_1, 89 buds for M_2 and 76 buds for M_3 were used. Male flower initials inoculated in M_1 has showed blackening within 3 days (Fig.7.2) and did not show any response of growth and structural variation till third week. After fourth week, the explant became curled and in fifth week it started bulging and became circular with slight blackening around the explants. It took approximately 7-8 months for callus formation. In M_1, out of 892 explants used, 38.3% formed callus and the rest became blackened and dried. In M_2, flower buds gave response on the third day itself. After second week, slight blackening appeared around the buds. Explant became circular and bulged in third week. Within fourth month, the explants turned into callus. Out of 1,008 flower primordia, 524 became callus and callus formation was 51.9%. In M3, ascorbic acid incorporated MA medium, explant started responding in first week itself. Curling and bulging of explants was started in fourth week onwards. Out of 912, 383 explants became callus and the percentage of callus formation was 41.9. (Table 7.1). Blackening area increased with gradually in all treatments. But charcoal contained media blackening was slow and it was the reason the callus got induced in medium within 4 months. But in M_3, explants started blackening at a higher rate than media containing charcoal. Most of the callus obtained was of compact calli which had non-embryogenic cells. Some of the calli were yellowish. Mixed types of callusing were also observed. Only 24 friable, transparent embryogenic calli were obtained from all the three treatments (Fig.7.1). Embryogenic cells (Fig.7.3) obtained were frequently transferred into fresh suspension medium and multinucleate, dividing cells were obtained more in M_2 than other treatments.

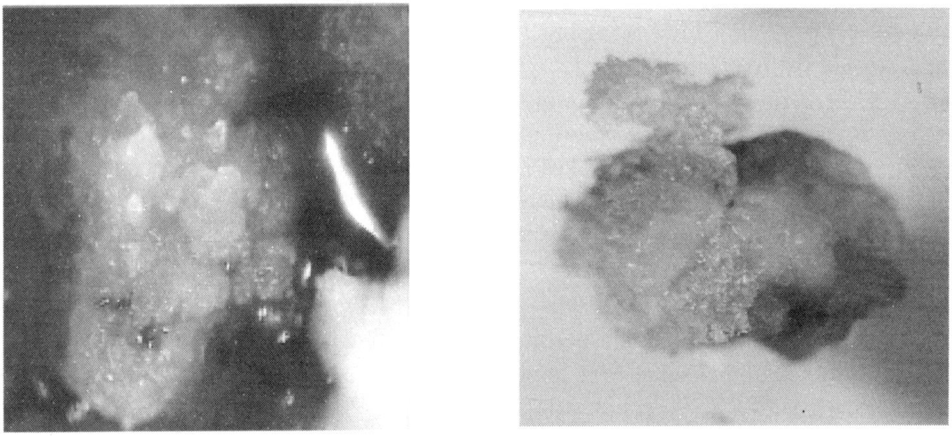

Fig. 7.1. Embryogenic calli of hill banana (Virupakshi, AAB)

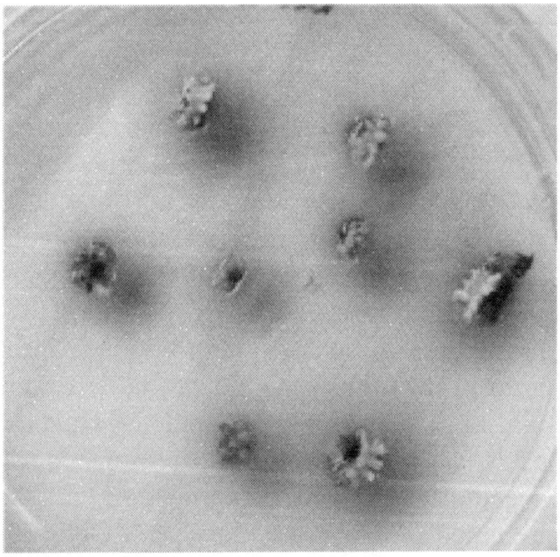

Fig. 7.2. Blackening of medium around the explant in treatment M_1 due to phenolic exudation

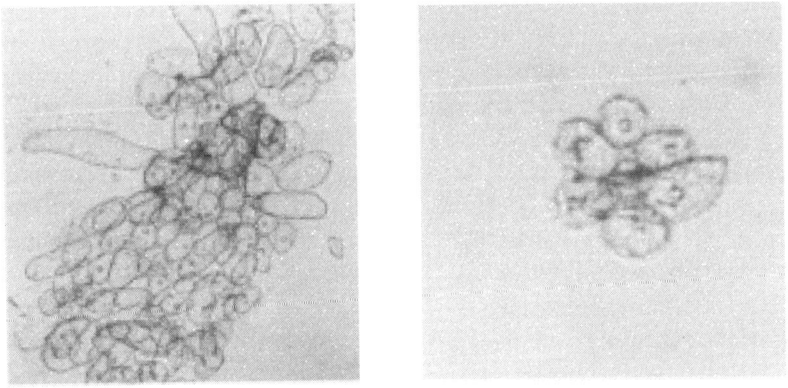

Fig. 7.3. Blackening of medium around the explant in treatment M_1 due to phenolic exudation

Table 7.1. Emerging callus from male flower bud in antioxidant treatment media

Treatment	*Total number of plates*	*Total number of flower buds*	*No. of explants became blackish*	*No. of explants turned into callus*	*Blackening (%)*	*Callus obtained (%)*
M_1	82	892	550	342	61.7	38.3
M_2 (charcoal)	89	1008	484	524	48.1	51.9
M_3 (ascorbic acid)	76	912	524	383	58.1	41.9

We have standardized a protocol for developing hill banana ECS in short time using immature male flowers. Presence of antioxidants in media inhibited oxidation of phenolics and helped in explant growth and morphogenesis *in-vitro* (Constantin *et al.*, 1977). Antioxidant such as charcoal adsorb toxic phenolic compound present in media and reduced blackening around the explants. As a medium supplement, activated charcoal has been reported to stimulate embryogenesis (Kohlenback and Wernicke, 1978). The activated charcoal reduced brown/black pigments and other known colourless toxic compounds than other antioxidants. All inoculates should be periodically transferred to appropriate medium to avoid drying and blackening of explants. Periodically transferred explants produced embryogenic callus quickly than those kept for a longer time in the same medium. These embryogenic calluses were transferred to suspension media for maturation of embryogenic cells. Embryogenic cells have been be used as explants for genetic transformation (Ganapathi *et al.*, 2001).

SUMMARY

Hill banana (Virupakshi, AAB), is commercially cultivated in Sirumalai and lower Pulney hills of Tamil Nadu. This highly valued, socio - economically important hill banana is devastated by Banana bunchy top virus (BBTV) infection and has been almost extinct in this region. Genetic engineering and biotechnological tools becoming used to develop transgenic resistance to plant viruses. The PRSV resistant transgenic papaya is a well-known success story. Establishment of embryogenic cell suspension is prerequisite to develop transgenic banana. The immature male flower buds were used for developing ECS. Main problem in formation of banana somatic embryogenic callus was accumulation of phenolic compound in the media. In this study, different antioxidant such as charcoal, ascorbic acid and sodium citrate were used. Charcoal was found to be better in minimizing oxidation. Use of semisolid media also reduced the oxidation of phenol. Different combinations of media were tried for early induction of embryogenic calli for hill banana. Basic MS media supplemented with 5.7μM IAA, 5.4μM NAA and 87mM sucrose and 18 μM 2.4D or Picloram (10 μM)with different antioxidant such as ascorbic acid (2mg/litre) or charcoal (2mg/litre) was used. Charcoal containing media showed reduced level of phenolic accumulation than control media (without antioxidant) and ascorbic acid media.Picloram gave significantly more success rate in callus induction than other combinations. The viral indexed male buds used for development of somatic embryogenic calli. The immature flower positions at 16-8 were placed at 135° angle to absorb sufficient amount of nutrients. All inoculates transferred twice in a month to avoid drying of explants. Curling of immature male flowers started within a month of initiation. After 4-6 month proembryos developed.

References

Becker, D.K., Dugdale, B., Smith, M.K., Harding, R.M. and Dale, J.L. 2000. Genetic transformation of Cavendish banana (*Musa* spp. AAAgroup) cv. Grand Naine via microprojectile bombardment. *Plant Cell Rep.* **19**:229-34.

Constantin, M.J., Henke, R.R. and Mansur, M.A. 1977. Effect of activated charcoal on callus growth and shoot organogenesis in tobacco. *In vitro* **13**:293-6.

Co^te, FX, Domergue, R., Monmarson, S., Schwendiman, J., Teisson, C. and Escalant, J.V. 1996. Embryogenic cell suspensions from male flower of *Musa* AAA cv. Grand Naine. *Physiol. Plant.,* **97**:285-90.

Dhed'a, D., Dumortier, F., Panis, B., Vuylsteke, D. and De Langhe, E. 1991. Plant regeneration in cell suspension cultures of cooking banana 'Bluggoe' cultivar (*Musa* spp. ABB group). *Fruits,* **46**:125-35.

Escalant, J.V., Teisson, C., and Co^te, F.X. 1994. Amplified somatic embryogenesis from male flowers of triploid banana and plantain cultivar (*Musa* spp.). *In-Vitro Cell Dev. Biol.,* **30**:181-86.

Ganapathi, T.R., Higgs, N.S., Balint-Kurti, P.J., Arntzen, C.J., May, G.D. and Van Eck, J.M. 2001. *Agrobacterium* mediated transformation of embryogenic cell suspensions of the banana cultivar Rasthali (AAB). *Plant Cell Rep.,* **20**:157-62.

Grapin, A., Schwendiman, J., and Teisson, C. 1996. Somatic embryogenesis in plantain banana. *In-Vitro Cell Dev. Biol.* **32**:66-71.

Kohlenback and wernicke. 1978. Investigation on inhibitory effort of agar and function of active carbon in anther culture. *Z. Pflanzen Physiol.,* **86**:463-64.

Ma, S.S.1991. Somatic embryogenesis and plant regeneration from cell suspension culture of banana. *Proceedings of Symp Tissue Culture Horticulture Crops* pp. 181-88. National Taiwan University, Taipei, Taiwan.

Novak, F.J., Afza, R., Van Duren, M., Parea-Dallos, M., Conger, B.V. and Xiaolang, T. 1989. Somatic embryogenesis and plant regeneration in suspension cultures of dessert (AA and AAA) andcooking (ABB) bananas (*Musa* spp.). *Biol/Techology* **7**:154-59.

Panis, B., Wauwe, A.V. and Swennen, R. 1993. Plant regeneration through direct somatic embryogenesis from protoplast of banana (*Musa* spp.). *Plant Cell Rep.,* **12**:403-07.

Selvarajan, R., Balasubramanian, V., Dayakar, S., Uma, S., Ahlawa, Y.S. and Sathiamoorthy, S. 2007. Molecular diagnosis of BBTV (Indian isolate) coat protein gene by polymerase chain reaction. *In: Banana: Technological Advancements,* Singh, H.P and Uma, S. (Eds). Association for the Improvement in Production and Utilization of Banana (AIPUB), National Research Centre for Banana, Trichy, India, pp 431-34.

8

Tissue Culture in Banana with Low-Cost Resource Utilization

A. Das and S.N. Gupta

Plant tissue culture is identified as the most potential technology for developing countries for production of high quality and disease-free planting material. It is also expected that tissue culture based industries could generate rural employment, particularly for women (FAO, 1999). However, it is experienced that there are difficulties in selling tissue culture products because of its high unit production cost than conventional system of production (Savangikar, 2004). In India and other developing countries, banana plantlets are the first largescale products of commercial tissue culture industries. In India, out of 90 commercial micropropagation units, 32 are closed down mainly due to high cost of production. This constraint of using full potential of tissue culture has now been addressed for the invention of reliable and cost-effective tissue culture methods without compromising on quality (Savangikar, 2004). The cost of production in tissue culture is mainly based on input costs of chemicals, media, energy, labour and capital. In India, although labour is comparatively cheaper, cost of chemicals and electricity is high that can share nearly 40% of unit cost of products. For example, agar is an essential substance in all types of tissue culture as gelling agent, requires at concentration of 8-10g/litre and costs @ Rs 3,500-4,000/kg.

Sugar is another important constituent of the media as carbon source, which requires at the concentration of 30-40g/litre and laboratory grade costs @ Rs 300-400/kg. If alternatively isabgol (market cost @ Rs 300/kg) can be used in place of agar and table sugar (market cost @ Rs 20/kg) in place of laboratory grade sucrose, unit cost of production can be drastically minimized. High quantities of these constituents are also required in rooting media for promotion of sufficient rooting in multiple shoots so that they can survive well in hardening. Instead of using gelling agents like agar, scope is there to use some natural plant residues like sugarcane bagasse etc. that not only reduce the cost but also might promote better root growth than in agar media (George, 1993). Therefore, to find out better substitutes of components of both multiplication and root initiation media, particularly for distilled water, laboratory grade sucrose and agar is of utmost importance. Variation in response of banana explants on media replaced with different substitutes is recorded and presented in this chapter.

LOW-COST RESOURCE UTILIZATION

Banana Giant Governor (AAA) was taken as the plant material for the experiment. The procedure standardized for micropropagation by earlier works was followed (Das *et al.*, 1998; Das *et al.*, 2004). For the trial of multiple shoot formation of explants, in normal MS medium (Murashige and Skoog, 1962) substitutions of three major components, e.g. distilled water, sucrose and agar, were made in different ways that formed altogether eight treatments as :1, standard: MS medium + BAP (3mg/litre) without any substitution following Das *et al.* (2004); 2, substitution of distilled water by tap water; 3, substitution of agar (8g/litre) by isabgol (35 g/ litre); 4, substitution of laboratory grade sucrose (30g/litre) by table sugar (30g/litre); 5, substitution of both distilled water by tap water and agar by isabgol (35 g/litre); 6, substitution of both distilled water by tap water and laboratory grade sugar by table sugar (30g/litre); 7, substitution of both laboratory grade sugar by table sugar and agar by isabgol; 8, substitution of distilled water by tap water, laboratory grade sugar by table sugar and agar by isabgol. The procedure of excision of apical buds, sterilization and also inoculation was followed as described by Das *et al.* (1998 and 2004). From each apical bud 4 vertical sections were made and each section was used as the single explant.

For the initiation of roots of multiple shoots, basic medium used was MS (half strength) containing IBA (1mg/litre) and charcoal (100mg/litre) (Das *et al.*, 2004). Six treatments were created replacing agar in media by sawdust (400g/litre), groundnut husk (500g/litre), rice husk (500g/litre), coir pith (400g/litre), coconut fibre (100g/litre) and isabgol (30g/litre). In all treatments in addition to these substitutions, boiled tap water and table sugar were also used as the substitutes of distilled water and laboratory grade sucrose respectively. After root initiation, the plants were transferred to hardening medium containing soil: sand: organic manure mixed in a 1:1:1 ratio in small nursery pots. The numerical data of different shoot and root growth parameters generated from the above experiments were recorded for each treatment and analyzed by complete randomized design with the help of Agris 7.01 software.

THE RESULTS

In experiment first, problem arose in trials where fresh tap water was used for preparing the media instead of double distilled water. The problem was high rate of precipitation in solution prior to adding agar or gelling agent for preparation of medium. The precipitation was noticed after adding the gelling agents and formation of medium in the culture containers. The chemical analysis of water from water-testing laboratory showed that tap water contained high amount of bicarbonate (0.279mg/ litre), calcium (71.34mg/ litre), magnesium (81.17mg/ litre) and chloride (8.125mg/ litre) in comparison with distilled water where these elements were present in very trace amount. It was apprehended that presence of high amount of bicarbonate in tap water reacting with added macro and microelements in solution and creating sedimentation problems. Boiling the required amount of tap water for half-an-hour before use as solvent and then allowing it for sedimentation for several hours could solve problem. After this experience was gained, in all treatments with tap water, prior to use, this procedure was followed. The sedimentation problem during solution preparation and also in media was not noticed further.

The time required for multiple shoots initiation was more or less same in all the treatments (Table 8.1). Within a period of 30-32 days after inoculation initiation of multiple shoots were noticed following the normal process of organogenetic development. About 5-7 multiple shoots were normally observed in each culture 45 after days inoculation (Table 8.1). The number of multiple shoots generated in each treatment at this stage although had no significant variations, highest number of multiple shoots was observed in treatment where only substitution was made for agar by isabgol in medium. This was followed by the media where substitution was done only for distilled water by tap water and also where substitutions were made for both distilled water and agar by tap water and isabgol respectively (Table 8.1). In all the treatments, where table sugar was used in place of laboratory grade sucrose, some initial vitrification symptoms were noticed at early stage of development of multiple shoots that was found to be recovered at later stage. Use of table sugar in combination with other substitutes resulted in lesser number of multiplc shoots in cultures than other treatments (Table 8.1). The multiple shoots from all media at their 60 days growth as shown were transferred to root initiation media (Fig. 8.1).

Table 8.1. Multiple shoot initiation for banana explants on MS media substituted by different treatments

Treatment (substitution)	*After 30 days in culture*		
	Days required for root initiation	*No. of roots/plantlet*	*Average root length (cm)*
Sawdust	12.50	3.42	0.48
Groundnut husk	12.67	3.12	0.62
Rice husk	12.34	3.42	0.84
Coir pith	11.83	4.00	1.94
Coconut fiber	6.50	5.25	5.40
Isabgol	11.34	4.67	5.30
Control	6.34	3.92	5.60
CD (5%)	0.771	1.133	2.110
SEm (±)	0.268	0.373	0.983

In root initiation, plantlets developed in medium containing coconut fibre had in all respect comparable performance as in normal control medium. Initiation of root was within 6-7 days in this medium and plantlets could develop 5-6 numbers of roots of an average 5.4 cm length

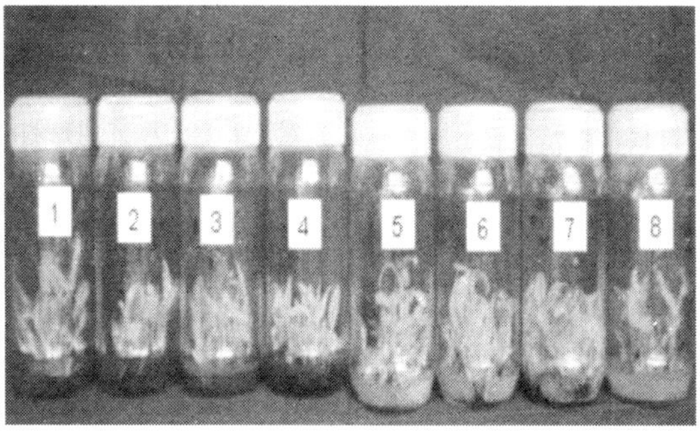

Fig.8.1. Multiple shoots development of banana in MS media substituted with: 1 tap water + table sugar + isabgol; 2 tap water + isabgol; 3 table sugar + isabgol; 4 isabgol; 5 table sugar + tap water; 6 tap water; 7 table sugar; 8 normal at 60 days after inoculation.

within 30 days culture in this rooting medium (Table 8.2, Fig 8.2). In all other media, e.g. sawdust, groundnut husk, rice husk, coir pith and isabgol initiation of root was noticed after 11-12 days of transfer. Among these treatments, better root growth in terms of number of roots and root length were observed in medium where isabgol was used. In rest of the media, growth rate both in terms of roots and shoots were slow (Table 8.2). After one month in rooting media the plantlets were transferred for hardening. The plantlets generated from coconut fibre rooting medium showed best growth in terms of number of leaves, shoot length and also survival percentage (Table 8.3). This was followed by plantlets generated from isabgol rooting medium. The growth of plantlets in media substituted with groundnut husk, rice husk and coir pith was moderate. Their survival in hardening was 50-62%. The plantlets rooted in sawdust medium had very poor performances in growth and their survival rate in hardening was also very poor.

Table 8.2. Root growth for banana plantlets on MS rooting media substituted by different treatments

Treatment (Substitution)	*Initiation of multiple shoots (days after inoculation)*	*Average number of multiple shoots (45 days after inoculation)*
Normal	31.67	5.517
Tap water	31.34	5.750
Isabgol	32.50	5.958
Table sugar	31.50	5.708
Tap water + isabgol	32.50	5.750
Tap water + table sugar	31.16	5.125
Table sugar + isabgol	32.34	4.750
Tap water + table sugar + isabgol	32.34	5.375
CD 5%	0.602	NS
SEm (±)	0.211	1.447

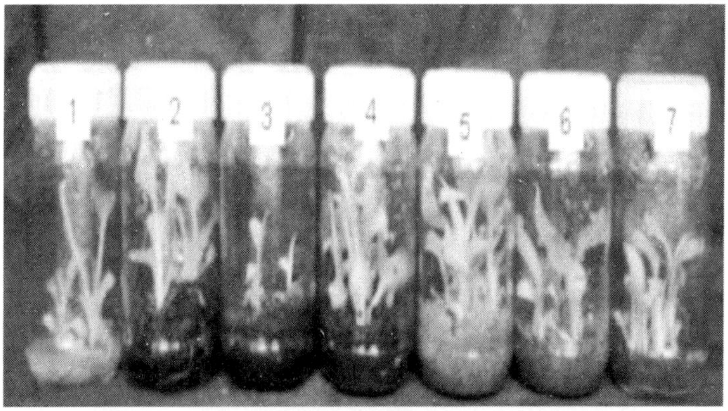

Fig.8.2. Development of roots in multiple shoots of banana in MS media substituted with: 1 normal; 2 coconut fibre; 3 sawdust; 4 coir pith; 5 rice husk; 6 ground nut husk; 7 isabgol.

Table 8.3. Number of leaves, shoot length of plantlets developed in different rooting media and their survival at hardening

Treatment(substitution)	After 30 days in culture		
	No. of leaves	Shoot length (cm)	Survival at hardening (%)
Sawdust	3.40	2.10	33.4
Groundnut husk	3.64	2.82	62.5
Rice husk	4.46	4.58	62.5
Coir pith	4.34	3.47	50.0
Coconut fiber	5.17	5.30	84.0
Isabgol	4.74	5.16	84.0
Control	3.78	5.12	77.0
CD (5%)	0.654	0.448	-
SEm (±)	0215	0.154	-

Thus, in tissue culture, tap water could be used. However, some problems of sedimentation might occur due to presence of high bicarbonate content which needs to be solved by boiling water. In tissue culture, substitution of distilled water by ordinary tap water was also found possible (Ganapati *et al.*, 1995). The use of table sugar in multiplication medium showed the initial symptoms of vitrification that was recovered at the later stage of development. In micropropagation at initial stage of development, vitrification might occur due to too much water available and low agar concentration (Peirik, 1987). For replacing high cost agar for tissue culture, isabgol derived from seeds of *Plantago ovata,* might be a good alternative due to its polysaccharide and colloidal nature as reported earlier by Babbar and Jain (1998). The efficiency of gelling agent was noticed for both in multiple shoot initiation and rooting. The root initiation instead of using any gelling agent and simple use of coconut fibre was more effective both in terms of plantlet growth as well as cost reduction. For root initiation, aeration of tissues and of roots was vital that could be provided by a porous substrate better than it would be agar or static liquid (George, 1993). Better rooting observed in this experiment in coconut fibre, might be due to better aeration of tissues that resulted in healthy plantlets.

SUMMARY

The invention of reliable and cost-effective tissue culture methods without compromising on quality has been addressed to use full potential of this technology. The experiment was conducted with this outlook on banana tissue culture where in different combinations, distilled water was substituted by normal tap water; laboratory grade sucrose was substituted by table sugar and agar was substituted by isabgol in common MS multiplication medium. In root initiating MS medium, in addition to the above substitutions, sawdust, groundnut husk, rice husk, coir pith and coconut fibre were used in place of agar. The time required for initiation of multiple shoots was more or less similar in each treatment. Number of multiple shoots was found higher in medium where only isabgol substituted agar followed by the medium where both isabgol and tap water substituted agar and distilled water respectively. When laboratory grade sucrose was substituted by table sugar, initial vitrification was noticed. Among the root promoting media, initiation of roots was found earlier when coconut fibre was used. Root development was better in both the media containing coconut fibre and isabgol. Plantlets raised in these two media also responded well in hardening.

References

Babbar, S.B. and Jain, N. 1998. Isabgol as an alternative gelling agent in plant tissue culture media. *Plant Cell Rep.,* **17:**318-22.

Das, A., Pal, A.K. and Chaudhuri, S. 1998. Banana tissue culture-variation in response of four genotypes. *The Hort. J.,* **11:**13-20.

Das A., Gupta, S.C., Sarkar, J., Mondal, B., Barui, F.K. and Sarkar, S.K. 2004. Differential response in *in-vitro* micropropagation of available germplasms of banana *In: Banana Industry- Present Scenario and Future Strategies.* Bandyopadhyay, B., Sarkar, S.K., Mishra, D.K. and Barui, F.K. (Eds), pp. 33-35.

FAO, Rome. 1999. Biotechnology: Committee on Agriculture, 15th Session, 25-29 January 1999, Food and Agriculture Organization of the United Nations, Rome, Italy.

Ganapathi, T.R., Mohan, J.S.S., Suprasanna, P., Bapat, V. A. and Rao, P.S. 1995. A low cost strategy for *in vitro* propagation of banana. *Curr. Sci.,* **68**(6):646-50.

George, E.F. 1993. *Plant Propagation by Tissue Culture,* part 1. The Technology, *Exegetics Limited,* pp.337-56.

Murashige, T. and Skoog, F. 1962. A revised medium for rapid growth and bioassays with tobacco tissue cultures. *Physiol. Plant.,* **15**:473-97.

Pierik, R.L.M. 1987. *In vitro* culture of high plants. *Martinus Nijhoff Publishers,* pp. 50.

Savangikar, V. A. 2004. Role of low cost options in tissue culture. *IAEA TECDOC-*2004(1384*),* pp.11-15.

9

Overcoming Occurrence of Variants in Tissue-Cultured Banana

V. Mache Gowda, T. Shiddlingeswara, G.C. Raghavendraprasad and P. Narayana Swamy

Propagation of Cavendish group of bananas is commercially exploited in India in recent years. *In-vitro* propagated plants having many advantages, viz. regular and large-scale availability; elite clones can be multiplied in quick succession, disease and pest free, earliness, uniform blooming and higher yields. But, occurrence of variants is a limiting factor in the acceptance of *in-vitro* plantlets for large-scale cultivation. Therefore, to elucidate the causes and remedies for occurrence of variants in tissue culture Cavendish group of bananas is quite essential (Scowcroft, 1984).

OCCURRENCE OF VARIANTS IN BANANA

A field survey of tissue culture banana grown gardens and visit to banana tissue culture laboratories around Bangalore as well as a series of laboratory studies, viz. type of explant, BAP concentration in media, number of generations in relation to occurrence of variants as well as identification of variants by RAPD markers and GA treatment technique to cull out variants were undertaken.

Tissue culture banana grown gardens survey indicated that, field variants observed were of three types, i.e. mixtures of genotypes (5-10%), mutants (5-8%) and viruses infection (2-5%). This was mainly because of care not taken at the time of explants collection, improper labeling and handling more than one variety in the laboratory at a time.

Combined laboratory studies involving two Cavendish group genotypes (Robusta and Dwarf Cavendish), cytokinin (BAP) concentration in basal MS media and sub-culturing up to tenth subculture reveals that, growth parameters (height, leaf length, leaf breath and leaf index) were significantly maximum up to eight subculture cycle and BAP concentration up to 2.50 mg/litre at deflasking stage as well as 40 days after deflasking in Robusta and Dwarf Cavendish (Tables 9.1 and 9.2). This clearly shows that beyond eighth subculture cycle and more than 2.50 mg/litre BAP in basal MS media induces dwarf plantlets. Dwarf Cavendish plantlets recorded lower growth than Robusta plantlets.

HEALTHY MOTHER PLANTS SELECTION

↓ ← 1. Apply RAPD to confirm the mutants and virus occurrence in mother plants.

SELECTION OF EX-PLANT

↓ ← 2. Avoid small buds/bulbils

TISSUE CULTURE LABORATORY

↓ ← 3. Restrict the subculture level upto eight generation and cytokinin level in the media not more than 3ppm

DE-FLASKING

↓ ← 4. Sort out rosette appearance plantlets and treat the plant with 100 ppm GA_3

PRIMARY NURSERY

↓ ← 5. 40 days after deflasking doubtful plants having leaf area index <1.5. should be discarded

SECONDARY NURSERY

↓ ← 6. Discard dwarf, leathery leaved plantlets. Random sampling of plantlets and subject for RAPD for mutants and viruses identification

RELEASING PLANTLETS FOR PLANTING

Interventions to be followed in banana tissue culture protected to avoid the occurrence of mutants and virus in plantlets

Table 9.1. Effect of subculture cycles, BAP concentration in media and duration on tissue culture plantlets of Robusta banana

Growth parameter	*Leaf length (cm)*				*Leaf breadth (cm)*				*Leaf index (cm)*				*Plantlet height (cm)*			
BAP concentration	*C1*	*C2*	*C3*	*C4*	*C1*	*C2*	*C3*	*C4*	*C1*	*C2*	*C3*	*C4*	*C1*	*C2*	*C3*	*C4*
Subculture							Deflasking stage									
SC-6	5.9	5.9	5.1	5.2	2.1	2.0	2.1	1.7	2.8	2.9	2.4	2.9	5.5	5.4	5.8	4.8
SC-7	7.9	7.6	6.7	7.4	2.8	2.5	2.1	2.5	2.8	2.9	3.1	2.9	6.6	6.7	4.8	6.1
SC-8	6.3	5.1	5.8	2.6	2.2	1.9	2.0	1.3	2.8	2.7	2.8	2.0	6.4	4.6	4.0	2.3
SC-9	3.2	3.2	3.4	1.7	1.2	1.4	1.3	0.9	2.6	2.3	2.6	1.8	2.2	2.4	2.4	1.4
SC-10	2.6	2.3	1.9	1.7	1.0	1.0	0.9	1.0	2.3	2.3	2.1	1.3	1.9	2.0	1.6	1.4
SEm ±		0.89				0.41				0.10					0.87	
CD (5%)		0.66				0.3				0.31					0.64	
Subculture							40 days after deflasking									
SC-6	14.4	14.5	15.3	14.1	6.8	6.6	6.6	5.3	2.1	2.2	2.3	2.6	13.2	13.3	13.4	12.7
SC-7	16.5	15.6	13.3	12.7	8.7	7.5	5.7	5.8	2	2.1	2.3	2.2	14.1	13.5	10.8	8.3
SC-8	18.7	18.0	18.2	14.3	8.3	7.4	7.8	6.6	2.2	2.4	2.3	2.1	15.0	15.9	15.0	10.9
SC-9	13.1	13.5	11.7	11.6	6.3	6.2	5.0	5.4	2.2	2.2	2.3	1.9	8.9	10.2	6.9	7.6
SC-10	7.3	6.8	3.1	3.3	2.9	2.8	1.5	1.6	2.4	2.3	1.9	1.3	4.3	4.6	2.5	3.1
SEm ±		2.44				1.41				0.07					2.34	
CD (5 %)		1.81				1.04				0.2					1.73	

NOTE: subculture cycles: SC-6, SC-7, SC-8, SC-9, SC-10
BAP concentration: C_1, -1.25mg/litre; C_2, 2.50mg/ litre; C_3,5.00mg/ litre; C_4, 7.5mg/ litre

Table 9.2. Effect of subculture cycles, BAP concentration in media and duration on growth parameters of tissue cultured plantlets of Dwarf Cavendish banana plantlets

Growth parameter	*Leaf length (cm)*				*Leaf breadth (cm)*				*Leaf index (cm)*				*Plantlet height (cm)*			
BAP concentration	*C1*	*C2*	*C3*	*C4*	*C1*	*C2*	*C3*	*C4*	*C1*	*C2*	*C3*	*C4*	*C1*	*C2*	*C3*	*C4*
Subculture							Deflasking stage									
SC-6	4.4	5.1	5.3	4.8	1.7	1.8	2.2	1.9	2.1	2.2	2.3	2.6	4.0	4.5	5.1	4.0
SC-7	5.5	6.9	5.8	5.7	2.4	2.3	2.2	2.1	1.9	2.1	2.3	2.2	4.6	5.7	4.8	4.8
SC-8	5.2	5.1	6.5	5.1	2.1	2.0	2.8	2.1	2.1	2.4	2.3	2.0	4.2	4.4	4.5	3.4
SC-9	2.9	3.1	3.2	2.9	1.2	1.4	1.3	1.4	1.9	2.1	2.3	1.9	2.3	2.3	3.8	3.0
SC-10	2.2	2.6	2.1	1.6	0.9	1.1	0.9	0.8	2.3	2.3	1.9	1.3	1.5	1.8	1.7	1.3
SEm ±		0.89				0.41				0.17				0.87		
CD (5%)		0.66				0.3				0.53				0.64		
Subculture							40 days after deflasking									
SC-6	12.4	11.5	11.2	11.3	6.0	5.2	5.1	5.5	2.6	2.8	2.4	2.5	10.0	8.9	8.3	8.7
SC-7	13.6	14.9	14.3	12	6.5	6.9	4.9	6.6	2.4	2.9	2.6	2.7	9.2	10.6	10.0	8.2
SC-8	13.1	13.4	13.9	14.3	6.6	5.5	6.8	7.5	2.4	2.5	2.4	2.5	10.0	9.9	11.9	11.6
SC-9	9.7	11.5	10.5	9.1	4.7	6.0	4.8	4.4	2.4	2.1	2.4	1.8	6.0	7.9	6.1	6.4
SC-10	6.5	7.1	5.9	2.7	2.7	3.2	2.3	1.4	2.4	2.2	2.3	1.3	3.6	4.4	2.9	1.9
SEm ±		2.44				1.41				0.17				2.34		
CD (5%)		1.81				1.04				0.53				1.73		

NOTE: subculture cycles: SC-6, SC-7, SC-8, SC-9, SC-10.
BAP concentration: C_1, 1.25mg/litre; C_2, 2.50mg/ litre; C_3, 5.00mg/ litre; C_4, 7.5mg/litre

Table 9.3. Effect of exogenous applications of GA_3 (100 ppm) on growth of tissue cultured banana plantlets (45 days after planting)

Growth parameter	*Leaf length (cm)*				*Leaf breadth (cm)*				*Leaf index (cm)*				*Plantlet height (cm)*			
BAP concentration	*C1*	*C2*	*C3*	*C4*	*C1*	*C2*	*C3*	*C4*	*C1*	*C2*	*C3*	*C4*	*C1*	*C2*	*C3*	*C4*
Subculture							Deflasking stage									
SC-6	8.5	8.5	10.2	8.9	4.7	4.5	4.5	3.6	1.8	1.9	2.2	2.4	7.7	7.8	7.6	7.9
SC-7	8.6	8.0	6.6	5.3	5.6	5.0	3.5	3.4	1.5	1.6	1.9	1.5	7.5	6.6	5.9	5.5
SC-8	12.2	12.9	12.5	11.6	6.0	5.5	5.6	5.4	2.0	2.3	2.3	2.1	8.5	11.4	11	8.5
SC-9	9.9	10.2	8.4	9.9	5.0	4.9	3.6	4.4	2.0	2	2.3	2.2	6.5	7.9	4.5	6.0
SC-10	4.7	4.5	1.2	1.4	1.8	1.8	0.9	1.1	2.6	2.5	1.3	1.2	2.5	2.5	1.1	1.6
SEm ±		0.22				0.21				0.08				0.20		
CD at (5%)		0.67				0.64				0.27				0.61		
Subculture Dwarf Cavendish																
SC-6	8.1	6.4	6.1	6.5	4.4	3.9	3.0	3.5	1.8	1.6	2.0	1.9	5.8	4.5	3.4	4.8
SC-7	8.1	8.0	8.4	6.7	4.2	4.5	2.6	4.8	1.9	1.8	3.3	1.4	4.5	4.7	5.1	3.3
SC-8	7.5	8.5	7.4	9.2	4.5	3.6	4.0	5.3	1.6	2.2	1.9	1.7	5.8	5.5	7.5	8.2
SC-9	6.9	8.4	7.2	6.2	3.5	4.6	3.4	3.3	1.9	1.8	2.1	1.9	3.7	5.5	2.3	3.4
SC-10	4.3	4.4	3.7	1.1	1.8	2.0	1.7	0.9	2.4	2.2	2.1	1.1	2.0	2.5	1.1	0.6
SEm ±		0.21				0.22				0.07				0.13		
CD (5 %)		0.63				0.66				0.22				0.39		

NOTE: subculture cycles: SC-6, SC-7, SC-8, SC-9, SC-10

BAP concentration: C_1,1.25mg/litre; C_2,2.50mg/ litre; C_3,5.00mg/litre; C_4-7.5mg/litre

To identify dwarf variants in nursery stage, experiment was designed to treat the plantlets in 100 ppm GA_3 immediately after deflasking for 5 seconds and followed by planting in hardening media. The results indicated (Table 9.3) that plantlets treated with (100ppm) GA_3 recorded more growth up to eighth subculture plantlets and lower BAP concentration treatments. In ninth and tenth subculture and more than 2.5mg/litre BAP concentration in basal media raised plantlets growth was not influenced due to GA_3 treatment. This indicated that low growth (dwarfness) of plantlets was not influenced even by GA_3 treatments. These findings are in confirmation with those of Damasco *et al.*(1993) in Cavendish banana. The molecular characterization of tissue culture banana plantlets as suggested by somaclonal (Shiddlingeswara *et al.*,2004) variants , bunchy top virus (Xie and Ha, 1995) and streak virus (Lockhart, 1995) proved good tools for identification of variants at nursery stage before planting (*see* Box).

SUMMARY

Field survey of tissue-cultured banana grown gardens reveled that variants observed were of three types, i.e. mixtures, mutants and viruses infection.In order to overcome the occurrence of variants in banana tissue culture some measures are recommended. They are: (i) tissue culture laboratories should take due care in labeling and preventing mixing of varieties/lots, (ii) taking explants only from sword suckers,(iii) cytokinine (BAP) concentration in MS media not to exceed more than 2.50ppm,(iv) sub-culturing should be restricted to eighth generation, (v) subject plantlets to gibberellic acid response test to callout the somaclonal variants at deflasking stage,(vi) before releasing of plantlets for planting, random testing for purity and viruses has to be checked by adopting molecular characterization by RAPD.

References

Damasco, O.P., Godwin, I.D., Smith, M.K. and Adkins, S.W., 1993, Variation in micropropagated bananas and plantains: Literature survey. *Fruits* **48**:15-22.

Lockhart, B.E.L. 1995, Development of detection methods for banana streak virus (BSV). INIBAP. *Annual Report,* 1994. International Network for the Improvement of Banana and Plantain, Montpellier, France.

Scowcroft, W.R., 1984, Genetic variability in tissue culture: impact on germplasm conservation and utilization. A *Technical Report,* IBPGR, Rome.

Shiddlingeswara, T., Nachegowda, V., Narayanaswamy, P. and Krishnamurthy, K.B. 2004, Confirmation of filed identified somaclonal variants in tissue cultured Cavendish banana using RAPD markers. *Proceedings of National Seminar on Banana Industry Present Scenario and Future Strategies,* pp. 17-21.

Xie, W.S and Hu, J.S. 1995. Molecular cloning, sequence analysis, and detection of banana bunchy top virus in Hawaii. *Phytopathologhy* **85:**339-47.

10

In-Vitro Selection of Salt Tolerant and its Commercial Application

Daksha Bhatt and Prashant

BANANA is major fruit crop in Gujarat. There are four serious problems associated with production of banana. They are mainly diseases caused due to root-knot nematodes, *Erwinia* rot and thrips. The last but not the least and overlooked so far, is the deterioration of fruit quality and quantity due to increase in salinity of irrigation water, leading to reduced productivity. Over 1,000 acres of land only in the delta region of River Mahi in central Gujarat where banana was traditionally cultivated, is now unsuitable for traditional banana cultivation. The salinity has increased from 700 to 1400- 2000 ppm. Banana cultivation in severely affected soils is totally abandoned, whereas in regions with 1200-1400 TDS, banana is still cultivated but the yield and quality of fruits have been severely affected. Therefore, *in-vitro* selection methods to generate banana clones, which can tolerate high salt content in soil and give quality yield comparable to those grown in normal soils, are needed.

IN-VITRO SELECTION METHODS

A 5 acre of tissue cultured banana plantation of Grand Naine in the area with high salt level of irrigation water (2000 ppm) was selected for survey of sport of salt toelarant plant. Most of the plants in field had stunted growth with severe leaf necrosis along margins (Figs 10.1a and b). The size of fruits was also much smaller resulting in poor yield compared to normal plants (Fig. 10.2).

Suckers were collected from one unique, very rare plant that was showing salt resistance. Their fruit morphology as well as their yield were comparable to those growing with irrigation water of 500-600 ppm. Shoot cultures were initiated from such suckers using standard protocol followed in the laboratory. They are designated as salt resistant cultures (=SR-M1 cultures). Along with this, shoot cultures were also initiated from plants growing in regular irrigation water with 500-600 ppm TDS and were designated as unselected control cultures (UN Cultures). Shoot cultures were grown on MS medium (2) supplemented with 12 μM benzyl-amino purine and 160 mg/litre adenine sulfate (medium T-1). In next subculture, shoots were transferred to a

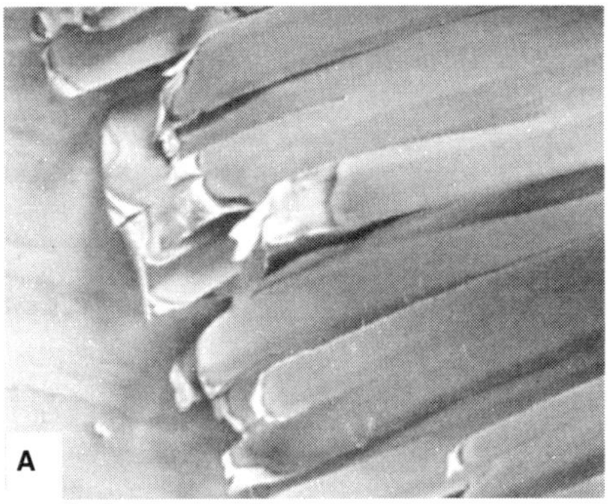

Figs. 10.1a. Effect of salinity on leaf necrosis.

Fig. 10.1b. Stunted growth of banana plants.

Fig. 10.2. Effect of salinity on quality of fruits of Grand Naine: Fruits harvested from plants grown with normal irrigation water of 700ppm (*left*) and fruits of the same variety irrigated with 1400 ppm TDS water (*right*).

medium with similar composition but with increased salt concentration having total EC of 7 mSi prepared using water having EC around 2.2 mSi (T-2 medium). During three subsequent subcultures, EC was gradually raised to 7.64 (T-3), 8.23 (T-4), and 8.82 mSi (T-5). Thereafter, cultures were maintained on media of maximum salt level (T-5) until the time they were ready for primary hardening.

Shoot elongation was achieved in medium devoid of growth regulators. Rooting was initiated in shoots by addition of 15 μM NAA to medium. Plantlets 4-5 cm long with 3-4 leaves and well-developed roots were kept for primary hardening in portrays filled with cocopeat. They were allowed to grow in polytunnel in greenhouse for four weeks. After proper root development, plantlets were transplanted in polybags containing a mixture of soil:farmyard manure (2:1) and were grown for 45 days. Fully hardened plants were about 30-45 cm tall. They were planted in the same area from where the original selection of mother plants was done. These plants were designated as M_1 generation. Along with M_{-1} plants that were regenerated from SR-MI cultures, plants regenerated from UN cultures of T_{-1} medium were also planted for comparison. Recurrent selection from M-1 plants was done having high yield (above 35 kg bunch/plant) and good quality of fruits.New cultures were initiated in the same manner as described above and they are designated as SR-M2 cultures. The micropropagation procedure used was similar to above except that after culture initiation on T-1 media, they were multiplied for one round on T-2 media and then directly multiplied on T-5 media for subsequent 7 cycles. Regenerated plants from these cultures were denoted as M-2 generation. Control plants (UN) were also propagated. During both seasons, data on culture growth, rate of multiplication, and culture quality were collected for both SR and UN cultures.

THE RESULTS

The control cultures (UN cultures) on standard multiplication media (T-1) had a rate of multiplication between 2.7 and 3.2 up to 7 cycles (Table 10.1). None of the cultures showed any phenolic exudation in medium and development of pseudostem and leaves was normal (Fig. 10.3).

Table 10.1. Growth of control cultures and selected culture on various media

Selected cultures ñSR-M-1					*Control cultures -UN*	
	Medium	*No.of cultures inoculated*	*No. of shoots obtained*	*Multiplication rate*	*Medium used*	*Average multiplication rate*
S.C.Cycle						
1	T- 1	70	150	2.14	T- 1	2. 00
2	T- 2	100	270	2.70	T- 1	2.70
3	T- 3	250	550	2.20	T- 1	2.80
4	T- 4	550	1120	2.00	T- 1	3.20
5	T- 5	1000	2230	2.23	T- 1	2.80

In contrast to the control, cultures raised from selected clones (SR-M1 cultures) had a lower rate of multiplication. The rate of multiplication remained more or less uniform even though total salt content and EC of culture media gradually increased. The cultures initially showed phenolic exudation in medium and poor pseudostem growth. However, after second cycle, cultures growth was stabilized and phenolic exudation was also reduced. The comparison of multiplying cultures of UN-control and selected salt tolerant SR -M1 plants on T-1 medium.

In all, over 5,000 M-1 plants were finally generated and planted during first year. Plants fruited at usual time and gave an average of 25 kg bunch weight, which was higher to control plants, and showed an average of 18 kg bunch weight. From this M1 cycle, superior selections of mother plants having bunch weight of 35 kg/plant were selected for micropropagation during the second year.

Fig. 10.3. Multiplying cultures from control plants (*left*) and selected clone (*right*) on T-1 media.

It was observed that after the recurrent selection, multiplication rate of SR-M2 cultures on T-5 media was comparatively higher than that of UN (control) plants on T-1 media in all cycles (Table 10.2). This finding was different from that observed in M1 cycle where SR-M1 cultures showed lower multiplication rate on regular media. This may be due to adaptation of both plants as well as culture to selection pressure at field and culture levels. Moreover, no phenolic exudation was observed in cultures in M2- cycle.

Table 10.2. Growth and multiplication rate of SR-M2 and UN cultures

SR-M2 cultures			*UN cultures*		
Cultures inoculated	*Cultures produced*	*Rate of multiplication*	*Cultures inoculated*	*Cultures produced*	*Rate of multiplication*
60	100	1.68	60	85	1.42
60	99	1.65	60	88	1.46
60	96	1.60	60	94	1.56
60	97	1.62	60	102	2.14

Improved growth of plants regenerated from SR-M2 cultures in M-2 cycle and grown in saline soil was also reflected in field. When the growth of normal plants and the selected ones was compared, it was observed that plants were better in vegetative growth characteristics (data not shown). Both types of plants fruited simultaneously, but the quantitative and qualitative parameter of yield and fruit quality was much superior in SR plants - selected clones compared to the control (Table 10.3 and Fig. 10.4).

Table 10.3. Comparison of growth parameters of plants in M-2 cycle

Growth parameter(mean of 300 plants)	Control (UN) plants	Selected (SR) plants
Length of bunch	88 cm	92 cm
Bunch weight (kg)	26.3 kg	31.2 kg
Fruit length	19.5 cm	21.08 cm
Fruit girth	11.5 cm	12.44 cm
No. of hands/bunch	11.8	12

It is clear that *in-vitro* selection can be used effectively to improve salt tolerance in banana. The selected clones after micropropagation behaved true-to-type to parental characters of banana Grand Naine. The somaclonal variation generated by *in-vitro* is stable. There are reports in the literature on selection of salt resistant somaclones in many species. However, there have been only two reports on record for research on salt resistance in banana. Ulisses *et al.* (2002) reported *in-vitro* selection of Nanicao banana using 80 and 100 mM sodium chloride. Selected plants were planted in soil with salinity levels having EC of 0.2, 7.0 and 14 mSi. Plants did not survive on 14 m Si soils. Alterations in isoenzyme bands characteristics were indicated due to somaclonal variation. However, no data on field performance is available. In Egypt, Ismail *et al.* (2005) reported the transgenic Grand Naine banana conferring salt resistance when grown in saline soils had better survival and improved morphological characters such as pseudostem growth and leaf area. They also studied peroxidase isoenzyme patterns and attributed the changes in number and intensity of bands to somaclonal variation. However, data on field performance is not available.

Control bunch — Salt resistant (SR) bunch

Fig. 10.4. Banana bunch harvested from UN- control plants as well as from selected SR-M2 clones.

Table 10.4. Financial loss to Banana growing farmer due to high TDS water

Parameter	Cost (Rs)
Organic manure	3,000
Chemical fertilizer potash and phosphatic fertilizer	1,500
Reduction in yield (25%)	10,000
Total loss	14,500

Hence this is first report showing field performance of salt resistant plants produced *in-vitro* by recurrent selection technique and evaluated on a commercial scale. The results obtained during both subsequent generations have encouraged us to scale up the production of this clone and plant it in saline salt regions in various parts of Gujarat. The salinity problem results into great financial loss to farmers (Table 10.4). With use of plants from selected clones, this loss is totally reduced bringing an overall increase in productivity and income to farmers.

SUMMARY

Banana requires relatively sweet water less than 700 ppm TDS for its growth. In Gujarat, large area of more than 1,000 acres around the delta of Mahi river was traditionally been used for cultivation of banana. But since last 10 years the ingress of Arabian Sea in this area has increased the TDS of groundwater to about 1,400-2,000 TDS. As a result, banana cultivation in the area above 1600 TDS has totally been discontinued, whereas in area with 1200-1400 TDS, banana being still cultivated but the yield and quality of fruits has been severely affected. We used combination of natural selection and *in-vitro* recurrent selection techniques to generate banana plants, which can grow very well in areas with 1400-1600 TDS. The plants have been commercially evaluated for three years in this area and they have been a great success.

References

Ismail, I.A., Salama, M.I., Hamid, N.A. and Sadik, A.S. 2005. Production of transgenic banana plants conferring tolerance to salt stress. *Annals of Agricultural Science (Cairo)* **50**:263-79.

Murashige, T. and Skoog, F. 1962. A revised medium for rapid growth and bioassays with tobacco tissue cultures. *Physiol. Plant.* **15**:473-97.

Ulisses, C., Camara, T. R., Willadino, L., de Albuquerque, C., Martins, L., de Freitas, N. 2002. Isoenzyme characterization of the clones of banana nanic„o subjected to salinity *Rev. bras. eng. agrìc. ambient.* **6** (2).

11

Karyomophology in Natural Banana Hybrids

A. Rekha and S.C. Hiremath

Banana an ancient fruit could be assumed that it has evolved with the civilization (Krishnamurthi and Seshadri 1958). Edible bananas are placed under *M.paradisica* L. for plantain and *M.sapientum* L. for banana, which was later considered to be hybrids of two main species *M.acuminata* and *M.balbisiana* (Dodds and Simmonds 1948). Banana and plantain cultivars have been evolved from Eumusa group of species. Almost all cultivars have been derived from *M.acuminata* (donor of 'A' genome) and *M.balbisiana* (donor of 'B' genome) (Simmonds 1962). Edibility of mature fruits of diploid *M.acuminata* (AA) came as a result of two mutation events, female sterility and parthenocarpy (Simmonds 1962). Triploid AAA cultivars were derived from these diploids, which could be due to crosses between edible diploids and wild *M.acuminata* sub-species, resulting in a wide range of AAA phenotypes.

These triploids are most vigorous, have larger fruits, thus replacing the AA diploids of Asia. The diploid and triploid *acuminata* cultivars were taken by man to the native areas of *balbisiana* which resulted in natural hybridization and formation of hybrid progeny with the genomes: AB, AAB, ABB. It is thought that subsequent dispersal of edible bananas out of Asia was brought about by man. Secondary diversification within the groups of cultivated bananas are the results of somatic mutations. The meiotic studies among 17 South Indian banana varieties by Agrawal (1983 a,b, 1987,1988a,b) revealed the cytological abnormalities. He has reported that none of them showed restitution nuclei formation and sterility was due to genetic reasons. Meiotic studies made among male sterile triploids exhibited no chromosomal irregularities in vegetatively propagated crop like banana and structural changes are subjected to selective elimination by sexual process and hence the nuclei is always in unbalanced state. Chromosome counts on 20 Bihar banana cultivars were reported by Roy and Sharma (1951). Valsalakumari and Sivaraman Nair (1990) reported in 98 cultivars. Somatic chromosomes of 53 *Musa* land races and hybrids were reported by Osuji *et al.* (1996) and 16 *Musa* species and land races by Dolezel *et al.* (1998). Chromosome counts and ploidy levels were determined for 60 *Musa* accesssions, plantains including somaclonal variants by Osuji (1995).

KARYOMORPHOLOGY IN NATURAL BANANA HYBRID

AB Group

They are of purely south Indian origin. Ney Poovan (Kerala and Tamil Nadu), Elakki Bale (Karnataka) are backyard bananas in Malabar Coast, which are cultivated in shade of arecanut palms. It has pink tinge along petiole margins. They are medium-sized plants with slender yellowish-green pseudostems. Fruits have very thin, papery peel and white firm flesh, which is very sweet and preferred for its quality.

AAB Sub-group

Pome Sub-group

Virupakshi, also known as hill banana, is tall vigorous. The cultivar is commercially important in Tamil Nadu. The fruits are angular with broad base, having thick leathery fibrous skin, with yellowish sweet flavour. Kullan and Rajapuri are semi-dwarf mutants that bear small sub-horizontal slightly curved fruits with persistent male flowers.

Mysore Sub-group

Poovan, Mysore, Champa and Palayankodan are other names of this cultivar. They are extremely vigorous with cylindrical tightly packed bunches, have short plump bottle necked fruits, which ripen to attractive yellow colour with acid sweet pulp.

Plantain Sub-group

Nendran, also known as Ethakkai in Kerela, usually has male axis with persistent bracts. Its fruits are either eaten raw or cooked. Pulp is hard that becomes softer after cooking. It is also used as vegetable and is popular fruit in coastal Karnataka and Kerela. It has wide uses in making chips and baby foods. Fruits are longer, thicker than other bananas. Bunch is loose not compact. Velethan is a type from Travancore area has a distinct pink pseudostem, petiole and midrib.

Silk Sub-group

Principal cultivar is Rasthali which is also called as Rasabale, Malbhog, Amruthpani and Poovan. Because of its high susceptibility to Panama wilt disease, its cultivation has drastically reduced since past 40 years. It is not very vigorous compared to Mysore and has smaller bunches of pale green fruits. The ripe fruits are thin-skinned and fall down easily from the bunch. The flesh is creamy-white, soft-textured, very tasty and sweet with a distinct flavour.

ABB Group

The cultivars of group ABB are generally vigorous. The starchy fruits are mainly used for cooking purpose easily distinguishable by widely spaced hands of large, angular, spreading, starchy fruits. Peduncle is conspicuously long and the bunch is pendulous. The cultivar Klue teparad was earlier classified as natural tetraploid (ABBB) by Simmonds based on the morphological characteristics. The latest studies on its chromosome number and DNA content revealed the triploid nature of the cultivar. Hence it was grouped under ABB group by Jenny *et al.* (1997). The plants are tall with a height of 3.00-3.10 m. The pseudostems and leaves are green. The pseudostesms are large with a 22.0-24.0 cm diameter. Monthan is a popular culinary banana, which is somewhat drought tolerant. Muthia also a cooking banana from North-East

Hilly region. Karpooravally has heavier bunches of comparatively smaller fruits with thin skin. It has a tall statured plants with light green stem, known as Pisang Awak.

Cytological Methods

The root apices of all accession's are collected from healthy suckers planted in pots. The root tips are washed thoroughly to remove soil/dust particles and pre-treated with 0.003M 8-Hydroxyquinone for two hours at 14-16°C. The pre-treated root tips were fixed in Carnoy's-II fixative, viz. 6:3:1 of absolute alcohol : glacial acetic acid: chloroform. The fixed materials are transferred to 70% alcohol after 24 hours of fixation and stored.

The stored root tips are rinsed in distilled water and hydrolysed in a water bath with 1 N HCl at 60° C for 5 minutes. The hydrolysed root tips are transferred to Lilee (1951) or Schiff's reagent after rinsing in distilled water, and stored in dark for one-and-a-half to 2 hours. Schiff's reagent stains the actively dividing meristematic tissue to deep magenta colour. The stained tips are squashed with a drop of 1% aceto-carmine, after removing the root cap and observed under microscope. The slides are scanned for well-spread metaphse chromosomes; such cells are selected for karyomorphological studies.

The cells are photographed and are enlarged to suitable size. The chromosomes are very small in size and sticky in nature needing hard tapping to obtain better chromosome spreads. Hence it was difficult to obtain good picture of well-spread plates. Increase in hydrolysis time result in poor staining and moderate pictures. The chromosomes are measured manually from camera lucida drawings. The measurements are recorded as millimeter's (mm) and converted to micrometers (μm), using conversion factor. Levan *et al.* (1964) system is followed for description of karyotypes and their classification, which is done based on the arm ratio (long arm/short arm).

Levan *et al.,* 1964 system was followed for description of karyotypes and their classification, which is done based on the arm ratio (long arm/short arm). The chromosomes are categorized as follows: SAT-m (chromosomes with median centromere having satellite), SAT-sm (chromosomes with sub-median centromere having a satellite), SAT-st (chrmpsomes with sub-terminal centromere having a satellite), m (chromosomes with median centromere), sm (chromosomes with sub-median centromere) and St (chromosomes with sub-terminal centromere).

THE RESULTS

Cultivated AB Genomic Group

The chromosome number in all three collections was determined as 2n=22 chromosomes. The karyotypic formula in Elakki Bale was 2n=2x=2SAT-m+10m+10sm. It consisted of three types of chromosomes. There was one pair of chromosome with satellite and median centromere. Five pairs of chromosomes had median centromere and five pairs of chromosomes with sub-median centromere. The chromosomes were 1.05-1.89 μm in length and absolute length was 16.82 μm. The karyotypic formula of cultivar Ney Poovan is 2n=2x=2SAT-m+18m+2sm. It consisted of three types of chromosomes. One pair of chromosome with median centromere and satellite, which measured 0.42 μm. There were 9 pairs of chromosomes with median centromere. There was also a pair of chromosome with sub-median centromere. The length of chromosomes varied from 1.02 to 2.33 μm and absolute length is 17.47 μm.

The karyotypic formula of banana Rasa Kadali is 2n=2x=2SAT-m+16m+4sm. There are three types of chromosomes. One pair of chromosome with median centromere had satellite of 0.42 μm in length. There are 8 pairs of chromosomes with median centromere, and two pairs of chromosomes with sub-median centromere. The chromosome lengths were 1.30-2.14 μm, and absolute chromosome length was 19.83 μm.

The comparative karyomorphological data of three cultivars are given in Table 11.1. The cultivars have similar satellite lengths. There is variation in composition of chromosomes. The absolute length and chromosome lengths also varied. Elakki Bale had lowest absolute length of 16.82 μm and Rasa Kadali had highest absolute length of 19.84 μm. The chromosomes are paired based on their short arm and long arm lengths as A and B chromosomes could not be differentiated.

Table 11.1. Karyomorphological data among three cultivars of AB group

Accession	*Karyotypic formula*	*SAT-Size (mm)*	*Chromosoms length (mm)*	*Absolute length (mm)*
Elakki Bale	2SAT-m+10m+10sm	0.42	1.05-1.89	16.87
Ney Poovan	2SAT-m+18m+2sm	0.42	1.02-2.33	17.47
Rasakadali	2SAT-m+16m+4sm	0.42	1.30-2.14	19.84

Group AAB

The cultivars of group AAB are divided into four sub-groups based on their fruit types.

(a) Sub-group 'Pome'

The karyotypic formula of cultivar Rajapuri is 2n=3x=3SAT-m+28m+2sm. It consisted of three types of chromosomes. There are three chromosomes with median centromere and satellite measuring 0.42 μm in length. It has 28 chromosomes with median centromere. The chromosome length varied from 1.21 to 2.52 μm. The absolute length of chromosomes is 39.61 μm.

The karyotypic formula of Kullan is 2n=3x=2SAT-sm+1SAT-m+26m+4sm. It consisted of four types of chromosomes. There are two chromosomes with sub-median centromere and satellite on its short arm measuring 0.42 μm in length. There is also one chromosome with median centromere and satellite measuring 0.42 μm in length. It has 26 chromosomes with median centromere and four chromosomes with sub-median centromere. The average chromosome lengths varies from 1.26 to 2.47μm, and absolute length is 37.80 μm.

The karyotypic formula of Virupakshi is 2n=3x=3SAT-m+27m+3sm. It consisted of three types of chromosomes. There are three chromosomes with median centromere and satellite measuring 0.42 μm. It has 27 chromosomes with median centromere and 3 chromosomes with sub-median centromoere. The range of chromosome length is 1.00-2.58 μm, the absolute length being 37.28 μm.

The karyotypic formula of Ladan Small is 2n=3x=3SAT-m+23m+6sm+1 st. It consisted of four types of chromosomes. There are three chromosomes with medium centromere and satellite measuring 0.42 μm. There are 23 chromosomes with median centromere and six chromosomes with sub-terminal chromosome. The range of chromosome lengths is 43.78 μm.

The karyotypic formula of Kunnan is 2n=3x=2SAT-m+23m+8sm. It consisted of three types of chromosomes. There were two chromosomes with sub-median centromere. The range of chromosome length is 1.06-2.10 m. The absolute length is 37.02 μm.

The comparative karyomorphological aspects of cultivars of AAB group, sub-group 'Pome' are given in Table 11.2. Variations are observed in composition of chromosomes. The cultivar Ladan small has a chromosome with sub-terminal centromere. All the cultivars has similar satellites. The cultivar Kullan had a sub-median chromosome with satellite. There were variations in the absolute length of chromosomes. The cultivar Ladan small had highest absolute length of 43.78 µm and Kunnan had lowest absolute length of 37.02 µm.

Table 11.2. Karyomorphological data among five cultivars of AAB group/sub-group Pome

Accession	*Karyotypic Formula*	*SAT-Size (mm)*	*Chromosoms length (mm)*	*Absolute length (mm)*
Rajapuri	3SAT-m+28m+2sm	0.42	1.21-2.52	39.61
Kullan	3SAT-sm+1SAT-m+26m+4sm	0.42	1.26-2.47	37.80
Virupakshi	3SAT-m+27m+3sm	0.42	1.00-2.58	37.28
Ladan small	3SAT-m+23m+6sm+1st	0.42	1.36-2.93	43.98
Kunnan	3SAT-m+23m+8sm	0.42	1.06-2.10	37.02

(b) Sub-group 'Mysore'

The karyotypic formula in Palayan Kodan is 2n=3x=2SAT-m+29m+1sm+1 st. There were four types of chromosomes. Two chromosomes had median centromere and satellite, which measure 0.42 µm. There are 29 chromosomes with median centromere and 1 chromosome with sub-median centromere. There is also 1 chromosome with sub-terminal centromere. The chromosome length is between 1.21-2.52 µm. The absolute length is 38.91 µm. The somatic chromosome measurements are as per Table 11.3.

Table 11.3. Karyomorphological data of cultivar Palayan Kodan AAB sub-group Mysore.

Accession	*Karyotypic formula*	*SAT-Size (mm)*	*Chromosome length (mm)*	*Absolute length (mm)*
Palayan kodan	2SAT-m+29m+1sm+1st	0.42	1.21-2.52	38.91

(c) Sub-group 'Plantains'

The karyotypic formula of Nendran is 2n=3x=2SAT-m+28m+3sm. There are three types of chromosomes. It has two chromosomes with median centromere and satellite of 0.42 µm in length. There are 28 chromosomes with median centromere and three chromosomes with sub-median centromere. The range of chromosome length is 1.21-2.21 µm. The absolute length is 35.38 µm.

The karyotypic formula of Velethan is 2n=3x=2SAT-m+28m+2sm+1 st. It consisted of four types of chromosomes. There are two chromosomes with median centromere and satellite of 0.42 µm length. It has 28 chromosomes with median centromere and two chromosomes with sub-median centromere. There is one chromosome with sub-terminal centromere. The length of chromosomes is 1.47-2.73 µm, absolute length is 42.87 µm.

The comparative karyomorphological data on two cultivars of Plantain sub-group are given in Table 11.4. Structural composition of karyotypes varied. Velethan has one chromosome with sub-terminal centromere. Both the cultivars are identical in other aspects though there are morphological differences. The chromosome length and absolute length are higher in Velethan. Nendran has lower absolute length of 35.88 μm.

Table 11.4. Karyomorphological data among two cultivars of group AAB sub group plantains

Accession	*Karyotypic formula*	*SAT-Size (mm)*	*Chromosoms length (mm)*	*Absolute length (mm)*
Nendran	2SAT-m+28m+3sm	0.42	1.21-2.21	35.88
Velethan	2SAT-m+28m+2sm+1st	0.42	1.47-2.93	43.07

(d) Sub-group 'Silk'

The karyotipic formula of Rasthali is 2n=3x=2SATm+25m+6sm. It has three types of chromosomes. There are two chromosomes with median centromere and satellite of 0.42 μm length. Also there are 25 chromosomes with median centromere. It had 6 chromosomes with sub-median centromere. The range of chromosome length was between 1.11 to 2.20 μm. The absolute length was 36.94 μm.

The karyotypic formula of Rasabale a collection from Shimoga district, is 2n=3x=3SAT-m+27m+3sm. It consisted of three types of chromosomes. Three chromosomes have satellite, which measure 0.42 μm and median centromere. There were 27 chromosomes also with median centromere , and three chromosomes have sub-median centromere. The lengths of chromosomes is 1.26-2.73 μm. The absolute length is 39.45 μm.

The karyotypic formula of Nanjangud Rasabale is 2n=3x=2SAT-sm+1SAT-m+20m+10sm. It consisted of four types of chromosomes. There are three chromosomes with satellite of length 0.42 μm of which two had sub-median centromere and one is with median centromere. There are 20 chromosomes with median centromere and 10 chromosomes with sub-median centromere. The length of chromosome is 1.16-3.15 μm, absolute length being 48.48 μm.

The comparative karyomorphological data of three cultivars of sub-group 'Silk' are given in Table 11.5. Differences are observed in karyotypes. Nanjangud Rasabale has wide range of chromosomes. The longest chromosome was almost three times that of the smallest. It also has highest absolute length of 48.48 μm. Rasthali has lowest chromosome length of 36.94 μm.

Table 11.5. Karyomorphological data among Three cultivars of group AAB sub-group Silk

Accession	*Karyotypic formula*	*SAT-Size (mm)*	*Chromosoms length (mm)*	*Absolute length, (mm)*
Rasthali	2SAT-m+25m+6sm	0.42	1.11-2.20	36.94
Rasabale	3SAT-m+27m+3sm	0.42	1.26-2.73	39.45
Nanjangud rasabale	2SAT-m+27m+3sm	0.42	1.16-3.15	48.48

ABB Genomic Group

The karyotypic formula of Klue Teparad is 2n=3x=2SAT-m+23m+8sm. There are three types of chromosomes. Two chromosomes have median centromere and satellite measuring 0.42 μm in length. It has 23 chromosomes with median centromere and 8 chromosomes had sub-median centromere. The length of chromosome is 1.00-2.52 μm.

The karyotypic formula of Cuba is 2n=3x=2SAT-m+27m+4sm. It has three types of chromosomes. There are two chromosomes with satellite, which measured 0.42 μm, and median centromere. The 27 chromosomes have median centromere and 4 chromosomes had sub-median centromere. The range of chromosome length is between 1.16-2.42 μm, the absolute length being 40.08 μm.

The karyotypic formula of Muthia is 2n=3x=2SAT-3+28m+3sm. It had three types of chromosomes. Two chromosomes have median centromere and satellite measuring 0.42 μm. There are 28 chromosomes with median centromere and 3 chromosomes have sub-median centromere. The chromosome length is 1.00-2.94 m, the absolute length being 39.56 μm.

The karyotypic formula of Nadu was 2n=3x=2SAT-m+31m. It has two types of chromosomes. There are two chromosomes with satellite and median centromere the satellite measure 0.42 μm. in length. The remaining 31 chromosomes have median centromere. The chromosome length varies from 1.00 to 2.26 μm.

The cultivar Karpooravally (1) had karyotypic formula as 2n=3x=2SAT-m+25m+6sm. It has three types of chromosomes. Two chromosomes have satellite measuring 0.42 μm. in length and median centromere. There are 25 chromosomes with median centromere and 6 chromosomes with sub-median centromere. The length of chromosomes is 0.94-2.31 μm the absolute length being 35.49 μm.

The somatic chromosome measurements of Karpooravally (2) are given in Table 11.5. The karyotypic formula of cultivar is 2n=3x=2SAT-m+31m. There are two types of chromosomes. It have two chromosomes with median centromere and satellite measuring 0.42 μm in length. The other 31 chromosomes have median centromere. The chromosome lengths is 1.26-2.31 μm the absolute length being 37.01 μm.

The karyotypic formula of Monthan is 2n=3x=2SAT-m+26m+5sm. There are three types of chromosomes. Two chromosomes have median centromere and satellite, which measure 0.42μm in length. There are 26 chromosomes with median centromere. The other five chromosomes have sub-median centromere. The range of chromosomes length is 1.10-2.84 μm. The absolute length is 44.02 μm.

The comparative karyomorphological data of group ABB are as per Table 11.6. There are variations in karyotypes, but Nadu and Karpooravally (2) had similar karyotypic formula and almost same absolute length. All the cultivars have the same length of satellite. The absolute length varies among cultivars. Monthan has highest absolute length of 44.02 μm while Karpooravally (1) lowest absolute length of 35.49 μm.

Table 11.6. Karyomorphological data among seven cultivars of group ABB.

Acc. Name	Karyotypic formula	SAT-Size (mm)	Chromosoms length (mm)	Absolute length(mm)
Kule teparad	2SAT-m+23m+8sm	0.42	1.00-2.52	36.20
Cuba	2SAT-m+27m+4sm	0.42	1.16-2.42	40.08
Muthia	2SAT-m+28m+3sm	0.42	1.00-2.94	39.56
Nadu	2SAT-m+31m	0.42	1.00-2.26	37.19
Karpooravally(1)	2SAT-m+25m+6sm	0.42	0.94-2.31	35.49
Karpooravally(2)	2SAT-m+31m	0.42	1.26-2.31	37.01
Monthan	2SAT-m+26m+5sm	0.42	1.10-2.84	44.02

Natural hybridization between *M.acuminata* (AA) × *M.balbisiana* (BB), polyploidy, parthenocarpy and sterility coupled with vegetative propagation have led to diversification of major cultivated groups of banana/plantain. The chromosomes are small as also mentioned by Osuji (1995). Karyotypes of three AB group cultivars studied are not homogenous and show karyotypic heteromorphism. In these diploid AB cultivars (2n=2x=22) 11 chromosomes of A and B genome cannot be recognized by chromosome morphology.

Karyotypes in 11 triploid cultivars of AAB genomic group (2n=2x=33) are analyzed. Generally karyotypes are asymmetrical and show strong karyotypic heteromorphism.

Among cultivars of ABB, Nadu and Karpooravally (1) showed symmetrical karyotypes, while asymmetrical karyotypes are found in Klue teparad, Cuba, Muthia, Monthan and Karpooravally (2) cultivars. Thus, ABB cultivars show an evolutionary trend from symmetry to asymmetry in karyotypes. Asymmetric karyotype may be due to chromosomal structural changes like deletions, inversions or translocations which are said to be common in *Musa* (Oritz, 2000).

These studies have helped in understanding the karyotypic structure of Indian banana varieties, which are said to be natural hybrids of *M. acuminata* and *M. balbisiana* and evolved as AB, AAB and ABB genomic group cultivars.

SUMMARY

Banana cultivars have evolved from two main species, *M. acuminata* and *M.balbisiana*, which belong to the section Eumusa. It is necessary to understand its relationships and homology in its natural hybrids. The existing natural hybrids are AB genomic group cultivars, which are of south Indian origin. The fruits are known for its special taste and quality. They are male and female sterile. The other important groups are AAB with four subgroups, *viz.* plantain on important group with dual purpose fruits which also used in processing industry, Silk group with Rasthali as preferred cultivar, Pome group consisting of Hill banana and Mysore subgroup having high yields also known as fill basket. Another important set of cultivated bananas belong to ABB genomic group consisting of cooking bananas.Three cultivars of diploid AB group, 11 cultivars of AAB group, viz. 5 from subgroup Pome, one cultivar of Mysore subgroup, three cultivars of subgroup Silk and two cultivars of Plantain subgroup were included in this study. Seven cultivars of ABB group were studied for their Karyomorphological status. The studies confirmed the ploidy status of most of the cultivars. There are variations in the karyotypes, in terms of absolute length of chromosomes. However, A and B chromosomes could not be differentiated based on their morphological appearance. It has helped in solving some confusions like, cultivar Klue teparad was earlier classified as natural tetraploid (ABBB) by Simmonds based on plant morphological characteristics but was found to be a triploid cultivar. The latest studies on its chromosome number and DNA content revealed the triploid nature of cultivar. The present study also confirms the triploid nature of cultivar Klue teparad. The cultivar Kunnan, a collection from Kerala, which is placed in AB group, is found to be triploid in nature with 2n=33 chromosomes.

References

Agrawal, P.K. 1983. Cytogenetical Investigations in Musaceae. I. Meiotic Studies in South Indian bananas. *Cytologia* **48:**847-52.

Agrawal, P.K. 1987. Cytogenetical Investigations in Musaceae. II. Meiotic studies in eight male sterile triploid banana varieties of India. *Cytologia* **52:**451-54.

Dodds, K.S. and Simmonds, N.W. 1948. Genetical and cytological studies of *Musa*. IX. The origin of an edible diploid and the significance of interspecific hybridization in the banana complex, an addendum on the nomenclature of edible banana. By E.E.Cheesman. *Journal of Genetics* **48:**285-96.

Dolezel, J., Dolezelova, M. Roux, N. and Vanden Hourve, I. 1998. A novel method to prepare slides for high-recolution chromosome studies in *Musa* Spp. *Infomusa* **7** (1): 3-4.

Krishnamurthi, S and Seshadri, V.S. 1958. Origin and evolution of cultivated bananas. *Indian J Hort.* **15:**135-45.

Levan, A.K., Fredga and Sandberg, A.A. 1964. Nomenclature for centromeric positions on chromosomes. *Heriditas* **52:**201-220.

Lilee, R.D. 1951. To prepare Schiffís reagent. *Stain Technology* **26:**163.

Oritz. R, 2000. Understanding the *Musa* genome: An update. *Proceedings of International Symposium on Bananas and Plantains for Africa. Acta Horticulture* **540:**157-63.

Osuji, J, 1995. Morphological, anatomical and cytogenetic characterization of the genus *Musa*. L (Musaceae). Ph D. Thesis, University of Port Harcourt, Nigeria.

Osuji, J.O., Okoli, B.E. and Oritz, R. 1996. An improved procedure for mitotic studies of the eumusa section of the genus *Musa* L. (*Musaceae*). *Infomusa* **5**(1):12-14.

Roy, R.S. and Sharma, G. 1951. Chromosome studies of Bihar bananas. *Indian J Genet Pl. Breeding* **11:**211-14.

Shepherd, K., 1999. Cytogenetics of the genus *Musa*. INIBAP Montpellier.

Simmonds N.W. 1962. *Evolution of Bananas.* Longman Publications London.

Valsala Kumari, P.K. and P.C.S. Nair 1990. Cytotaxonomical studies on banana cultivators. *South Indian Horticulture* **38**(4):177-82.

12

Breeding Banana Cultivars and Hybrids for Resistance to Nematodes

P.S. Kavitha, T.N. Balamohan, D. Veeraragavathatham and K. Poornima

BANANA is targeted by many pests and diseases. Among them, nematodes are considered to be the most important pests of banana and plantain, for which remedies are the need of the hour. India is one of the largest banana-producing countries, accounting 15% of the global production. Crop losses caused by nematodes to bananas are very high, with an average annual yield losses of 20% worldwide. Development of nematode resistance with quality and food safety, therefore assumes priority in banana breeding programmes (Simmonds, 1966). In India, the occurrences of few nematodes have been reported, out of which *Pratylenchus coffeae* (lesion nematode) and *Radopholus similis* (burrowing nematode) are destructive in Tamil Nadu. Hence, a screening study was conducted against both of them. The ultimate aim of the breeding programme is to develop secondary triploids *via* tetraploid × diploid crosses as suggested by Rowe (1984).

BREEDING BANANA FOR RESISTANCE TO NEMATODES

The study was taken up to develop new synthetic hybrids resistant to nematodes and also to identify breeding potential of hybrids for further improvement. Commercial triploids like Peykunnan (Pisang Awak, ABB), Poovan (syn. Mysore, AAB), Rasthali (syn. Rasthali, AAB) and Hill Banana, which are grown on an extensive scale in South India, were utilized as female parents. Three potential diploids like Anaikomban, Pisang Lilin and Erachivazhai, which are resistant/tolerant to nematodes and *Fusarium* wilt were used as pollen parents. The evolved hybrids were then subjected to field as well as pot culture experiments for examining their level of resistance/susceptibility to nematodes.

The *Musa* breeding was carried out in two phases, *viz.* production of promising hybrids and testing of elite hybrids for their potential. The banana hybrids evolved were evaluated at two stages, *viz.* preliminary stage and advanced stages. The former involves two types of trials, *viz.,* Early Evaluation Trial (EET) or phase-I evaluation and other is Preliminary Yield Trial (PYT) or phase-II evaluation (Ortiz, 1995). Unlike the plant crop, F_1 seedlings take more time for shooting (16-18 months). The breeders cannot assess yield potential of F_1 in first seedling generation because F_1 cannot express full yield potential and it spends more potential energy for corm development.

The hybrids were evaluated for their breeding potential and yield besides their resistance to nematodes in a randomised block design replicated thrice. For both field and pot studies, hybrids were evaluated along with reference cultivars. In field condition, there was occurrence of *Pratylenchus coffeae* only and hence screening for *Radopholus similis* was done only under pot condition by artificial culturing and inoculation of nematodes.The susceptible check cultivar used for screening both nematodes was Rasthali (AAB, syn. Silk), while resistant reference cultivar used for screening *Pratylenchus coffeae* was Pisang Lilin (AA) and for *Radopholus similis* the resistant reference cultivar used was Yangambi (AAA). The screening studies were done based on root and corm damage assessment as followed in INIBAP technical guidelines (Carlier *et al.,* 2003).

DEVELOPMENT OF RESISTANCE TO NEMATODE

Hybridization involving 6,820 crosses resulted in germination of 53 seedlings with mean of 4.85% germinability. Among phase I banana hybrids, two each diploids, and triploids, five tetraploids and 15 pentaploids were observed. Field screening of 24 hybrids for *Pratylenchus coffeae* revealed that hybrids, H-04-05, H-04-06, H-04-12, H-04-21, H-04-22 and H-04-23, were resistant and H-04-01, H-04-03, H-04-09, H-04-12, H-04-16 and H-04-19 and H-04-24, were tolerant (Table 12.1). Under pot culture evaluation, H-04-05, H-04-06, H-04-12, H-04-22 and H-04-23, were resistant and H-04-21 and H-04-24 tolerant to *Pratylenchus coffeae* (Table 12.2), while H-04-06, H-04-22 and H-04-23 were resistant and H-04-05, H-04-12, H-04-21 and H-04-24 were tolerant to *Radopholus similis* (Table12.3).

Table 12.1. Root and corm damage assessment in banana hybrids (Phase I) under field conditions caused by *P. coffeae*

Hybrid	*Total root necrosis (%)*	*Dead roots*	*Root lesion index*	*Corm grade*	*Reaction status*
H-04-01	11	6.45	1	2	T
H-04-03	12	4.17	2	2	T
H-04-05	10	0.00	1	1	R
H-04-06	9	2.86	1	1	R
H-04-09	16	5.88	2	1	T
H-04-12	9	0.00	1	1	R
H-04-16	15	6.67	2	2	T
H-04-19	15	5.88	2	2	T
H-04-21	8	0.00	1	1	R
H-04-22	4	0.00	1	0	R
H-04-23	3	0.00	1	0	R
Cultivars					
H-04-24	15	4.26	2	1	T
Pisang Lilin	5	0.00	1	1	R
Rasthali	44	39.02	5	4	HS

R, Resistant; T, tolerant; S, susceptible; HS, highly susceptibl

Table 12.2. Root and corm damage assessment in banana hybrids (Phase I) on 90th DAI caused by *P. coffeae* under pot culture

Hybrid	*Total root necrosis (%)*	*Dead roots (%)*	*Root lesion index*	*Corm grade*	*Reaction status*
H-04-05	8	0.00	1	1	R
H-04-06	6	0.00	1	0	R
H-04-12	10	2.70	1	1	R
H-04-21	10	1.75	1	1	T
H-04-22	8	0.00	1	0	R
H-04-23	7	0.00	1	1	R
H-04-24	18	3.57	2	2	T
Cultivar					
Pisang Lilin	9	2.17	1	0	R
Rasthali	58	24.49	5	4	HS

Table 12.3. Root and corm damage assessment in banana hybrids (Phase I) on 90th DAI caused by *R. similis* under pot culture

Hybrid	*Total Root Necrosis (%)*	*Dead roots (%)*	*Root lesion index*	*Corm grade*	*Reaction status*
H-04-05	8	0.00	1	0	T
H-04-06	6	1.79	1	1	R
H-04-12	18	2.56	2	2	T
H-04-21	10	1.82	1	1	T
H-04-22	6	2.78	1	0	R
H-04-23	5	0.00	1	0	R
H-04-24	18	4.92	2	2	T
Cultivars					
YKm#5	7	2.38	1	0	R
Rasthali	48	25.93	5	4	HS

R, Resistant; T, tolerant; S, susceptible; HS, highly susceptible

Field screening of phase II hybrids for nematodes revealed that, H-03-05, H-03-11, H-03-13, H-03-15, H-03-17, H-03-19 and NPH-02-01 were resistant and H-03-06 was tolerant along with better agronomic traits (Table 12.4). Under pot screening experiments, hybrids H-03-05, H-03-17 and H-03-19 were resistant and H-03-11, H-03-13, H-03-15 and NPH-02-01 were tolerant, while H-03-06 was susceptible to *Pratylenchus coffeae* (Table 12.5). On the other hand, H-03-05 and H-03-13, were resistant and H-03-11, H-03-15, H-03-17, H-03-19 and NPH-02-01 were tolerant to *Radopholus similis* (Table 12.6).

It was interesting to note that a good number of hybrids observed to be resistant under field conditions were found to be tolerant under pot culture. This is due to more population of nematodes bombarded in root zone for forceful infection (Dosselaere *et al.,* 2003). In pot, the nematodes are inoculated in 45 days old seedlings, when roots are young and tender. However, under field condition, higher population of nematodes is normally recorded at the time of shooting and harvesting (Carlier *et al.,* 2003). When plants are in shooting stage, they possess larger and longer root system, with increased biochemical and enzyme activities, resulting in more resistant reaction to nematodes. This might be the probable reason in some hybrids showing resistance under field condition, become susceptible when inoculated artificially.

Table 12.4. Root and corm damage assessment in banana hybrids (Phase II) infected by *P. coffeae* under field conditions

Hybrid	Total root necrosis (%)	Dead roots (%)	Root lesion index	Corm grade	Reaction status
H-03-05	9	12.50	1	1	R
H-03-06	18	13.33	2	2	T
H-03-11	10	13.33	1	1	R
H-03-13	15	11.11	1	2	R
H-03-15	21	0.00	2	2	R
H-03-17	8	0.00	2	1	R
H-03-19	8	0.00	1	1	R
NPH-02-01	5	0.00	1	0	R
Cultivar					
Pisang Lilin	5	0	1	0	R
Rasthali	42	37.93	5	4	HS

Table 12.5. Root and corm damage assessment in banana hybrids (Phase II) for *P. coffeae on* 90th DAI under pot culture

Hybrid	*Total root necrosis (%)*	*Dead roots (%)*	*lesion index*	*Corm grade*	*Reaction level*
H-03-05	10	18.18	2	2	R
H-03-06	23	30.77	3	2	S
H-03-11	18	15.38	2	2	T
H-03-13	18	17.24	2	2	T
H-03-15	25	31.82	3	3	T
H-03-17	12	14.08	2	2	R
H-03-19	10	19.35	2	2	R
NPH-02-01	8	10.96	1	1	T
Cultivar					
Pisang Lilin	6	7.40	1	1	R
Rasthali	46	38.46	4	4	HS

Table 12.6. Root and corm damage assessment in banana hybrids (Phase II) for *R. similis* at 90th DAI under pot culture

Hybrid	*Total root necrosis (%)*	*Dead roots (%)*	*Root lesion index*	*Corm grade*	*Reaction status*
H-03-05	8	16.67	1	1	R
H-03-11	13	17.54	2	2	T
H-03-13	10	18.18	1	2	R
H-03-15	20	27.66	2	2	T
H-03-17	6	10.77	1	1	T
H-03-19	10	12.70	1	2	T
NPH-02-01	12	10.00	2	2	T
Cultivar					
YKm#5	5	14.28	1	1	R
Rasthali	40	40.00	5	4	HS

R, resistant; T, tolerant; S, susceptible; HS, highly susceptible

It is concluded that hybrids, H-04-06, H-04-05 and H-04-24, are potential parents that can be used for further breeding programme due to their high male and female fertility along with resistance to nematodes. The hybrids H-04-06, NPH-02-01, H-03-19, H-03-17, and H-03-05, with better bunch weight besides possessing resistance to nematodes have to be studied further for exploitation (Table 12.7). Among hybrids, triploid hybrid H-04-21, with small stature, akin to Rasthali (syn. Silk, AAB), tolerant to nematode and wilt needs further evaluation, since cultivar Rasthali a predominant cultivar grown for its fruit taste and aroma and area of cultivation is shrinking due to its susceptibility to *Fusarium* disease for which nematode is predisposal agent.

Table 12.7. Breeding value of the potential phase I and phase II banana hybrids

Hybrid	*Parents*	*Genome*	*Bunch weight (kg)*	*TSS (%)*	*Male fertility*	*Female fertility*	*Fusarium wilt*	*Nematodes*
NPH-02-01	H201 x Anaikomban	AAB	19.00	21.80	P	P	R	T
H-03-13	Peykunnan x EV	AABB	12.00	21.60	P	P	R	R
H-03-15	H-02-32 x PL	AAB	8.00	20.80	P	P	R	R
H-03-17	Peykunnan x PL	AABB	10.50	21.20	P	P	R	R
H-03-19	Peykunnan x EV	AABB	14.00	22.00	P	P	R	R
H-04-06	H-02-32 x PL	AABB	17.50	24.80	P	P	R	R
H-04-21	H-02-10 x PL	AAB	7.50	26.00	A	A	R	R
H-04-24	Peykunnan OP	AAB	12.00	22.80	P	P	R	T

P, Present; R, resistant; T, tolerant; A, absent; EV, erachivazhai; PL, Pisang Lilin

SUMMARY

Breeding banana with nematode resistance is an alternate strategy for controlling the nematodes, simultaneously ensuring environmental safety. The evaluation was done for hybrids of two phases against *Pratylenchus coffeae* and *Radopholus similis*. The results revealed that hybrids H-04-06, H-04-05 and H-04-24, are potential parents that can be used for further breeding programme due to their high male and female fertility along with resistance to nematodes. The triploid hybrid, H-04-21, akin to Rasthali is tolerant to nematode and needs further evaluation.

References

Carlier, J., De Waele, D. and Escalant, J.V. 2003. Global evaluation of *Musa* germplasm for resistance to *Fusarium* wilt, *Mycosphaerella* leaf spot diseases and nematodes. *Performance Evaluation.* Vegine, A. and Pig, C. (Eds). INIBAP Technical guidelines 7. The International Network for the Improvement of Banana and Plantain, Montpellier, France.

Dosselaere, N., Araya, M. and De Waele, D. 2003. Effect of pot volume on root growth, *Radopholus similis* reproductive potential and its damage on bananas. *Info Musa* **12**(1):17-21.

Ortiz, R. 1995. *Musa* genetics. *In: Bananas and Plantains.* Gowen, S (Ed.), Chapman Hall, London, U.K., pp. 84-109.

Rowe, P.R. 1984. Breeding bananas and plantains. *In: Plant Breeding Review* **2:**135-55.

Simmonds, N.W. 1966. *Bananas.* Longman, London.

13

Genetic Variability and Correlation in Ecotypes of Nendran Banana

P.S. Kavitha, T.N. Balamohan, D. Veeraragavathatham and K. Poornima

Critical assessment of genetic variability existing in that particular crop and extent to which the character to be improved is heritable. Burton (1952) pointed out that calculating the genetic coefficient of variation along with heritability can assess a best picture of the amount of advancement to be expected by selection. Ramanujan and Thirumalachar (1967) suggested that heritability estimate in broad sense will be reliable, if accompanied by a high genetic advance. Johnson *et al.* (1955) and Swarup and Changle (1967) also consider that heritability estimates along with genetic gain are useful and reliable than heritability estimates alone in predicting the selection response. Effectiveness of selection based on phenotypic performance can be more useful and reliable only if selection is based on heritability estimates alonge with genetic gain. Correlation studies will help in predicting growth and yield performance. Association of yield with its component characters is of immense value in selection of superior genotypes. Above all these, knowledge of extent of variability in germplasm is an essential prerequisite in any breeding programme. Therefore, a critical assessment was made on 17 biometrical characters in 12 Nendran banana ecotypes.

VARIABILITY AND CORRELATION IN ECOTYPES OF NENDRAN BANANA

Twelve Nendran ecotypes were planted in a completely randomized block design as per Panse and Sukhatme (1967) (Table.13.1). Suckers of 12 Nendran ectotypes of banana almost uniform in size collected from different parts of Kerala were maintained in the Instructional Farm, College of Agriculture, Vellayani, Thiruvananthapuram. A spacing of 2.0m × 2.0m was adopted. The cultural practices as per the package of practices recommendations (KAU, 1996) were followed. Suckers of Kaliethan, Koonoor Ethan, Quintal Banana and Vellayani Nendran were collected, while those of Padalamurian, Mysore Ethan, Myndoli, Attu Nendran, Chengazhikodan, Changanasseri Nendran, Manjeri Nendran and Zanzibar from Banana Research Station (BRS), Kannara, Thrissur. The observations were recorded on plant height, number of leaves/plant, leaf width, suckers/plant, bunch traits like number of fingers/hand and bunch, bunch weight, hand weight, bunch length and finger characters like, length, girth, weight and volume of finger, ripe fruit weight and pulp weight. Crop duration like days taken for planting to shooting and sugar:acid ratio were also recorded. Biometrical data were collected and statistically analysed following Fischer (1960). From the analysis of variance, phenotypic and genotypic coefficient of variation (PCV and GCV), phenotypic and genotypic variance (PV and GV) were calculated (Burton, 1952). Haeritability estimates (Burton and de Vane, 1953), genetic advance (Allard, 1960) and correlation coefficient by the method of Al-Jibouri *et al.* (1958) were studied.

Table 13.1. Cultivars, banana types, ploidy and genomic composition of Nendran ecotypes

Ecotype	*Type*	*Ploidy*	*Genomic composition*
Vellayani Nendran	Dessert /cooking	3x	AAB
Padalamurian	Dessert /cooking	3x	AAB
Myndoli	Dessert /cooking	3x	AAB
Chengazhikodan	Dessert /cooking	3x	AAB
Attu Nendran	Dessert /cooking	3x	AAB
Kaliethan	Dessert /cooking	3x	AAB
Koonoor Ethan	Dessert /cooking	3x	AAB
Mysore Ethan	Dessert /cooking	3x	AAB
Zanzibar	Dessert /cooking	3x	AAB
Quintal banana	Dessert /cooking	3x	AAB
Changanasseri Nendran	Dessert /cooking	3x	AAB
Manjeri Nendran	Dessert /cooking	3x	AAB

THE RESULTS

The PCV were higher than GCV for all the characters. A significant difference was recorded among ecotypes of Nendran banana for different plant parameters. The results indicated wide variation in general mean for each character. The highest variation was shown by plant height, finger weight, volume of finger, ripe fruit weight and days taken from planting to shooting. The lowest variation was recorded by leaves/plant, suckers/plant, fingers/hand, bunch weight, hand weight, finger length and girth of finger(Table 13.2).

Generally, phenotypic component of variance was higher than the genotypic component of variance and the extent of latter component also showed that they are mostly heritability in nature (Table 13.3). Estimates of different genetic parameters clearly showed that phenotypic variance were higher than genotypic variance. This clearly suggests that environment influenced the expression of all the characters. This is in line with those Nayar *et al.* (1979). The variation between phenotypic coefficient of variation as well as genotypic coefficient of variation is due to that phenotypic variability included genetic variability and effect of environment besides genotypic-environment interaction. Since major part of variability is genetic, the same can be utilised for crop improvement programmes through selection. Sreerangaswamy *et al.* (1980) also obtained similar results in culinary and dessert type bananas. The PCV was 7.64% for leaf width and for fingers/bunch 49.38%. The highest PCV was observed in fingers/bunch (49.38 %), followed by suckers/plant (47.48 %), bunch weight (44.22 %), hand weight (42.63 %), ripe fruit weight (39.48 %), volume of fruit (39.47 %) and pulp weight (39.41 %). The lowest PCV value was observed in leaf width (7.64 %), followed by plant height (12.01%), leaves/

plant (13.60%), days taken for planting to shooting (14.01%) and finger girth (16.39). The GCV is a better tool to understand useful variability, as it is free from environmental components.

Table 13.2. Phenotypic variability in 12 Nendran ecotypes of banana

Character	*Mean ± S.E.*	*Range*	*CD (5%)*
Plant height (cm)	314.71 ± 11.0	387.0 - 264.6	24.3
Leaves/plant	8.96 ± 0.30	11.6 - 7.2	0.70
Leaf width (cm)	69.49 ± 1.58	81.3 - 64.0	3.2
Suckers/plant	10.75 ± 1.40	24.8 - 5.40	3.1
Days taken for planting to shooting	247.0 ± 9.90	303.6 - 174.2	21.8
Bunch weight (kg)	13.658 ± 1.7	30.4 - 6.5	3.7
Fingers/hand	10.451 ± 0.6	14.2 - 7.6	1.4
Fingers/bunch	52.717 ± 7.4	102.8 - 18.4	16.3
Finger weight (g)	254.54± 27.3	507.4 - 180.1	61.2
Bunch length (cm)	59.445 ± 3.4	81.74 - 47.90	7.5
Finger length (cm)	23.737± 1.6	37.3 - 17.8	3.6
Finger girth (cm)	13.612± 0.6	17.6 - 10.3	1.4
Volume of fruit (cc)	248.12 ± 28.2	503.9 - 173.0	62.0
Ripe fruit weight (g)	248.12 ± 26.2	480.2 - 156.5	51.4
Pulp weight (g)	186.44 ± 21.1	362.0 - 115.5	46.5
Sugar: acid ratio	60.706 ± 3.8	79.7 - 31.9	8.40
Hand weight (g)	2.473± 0.3	4.5 - 1.50	0.6

Table 13.3. PCV, GCV, ECV, heritability and genetic advance at 5% of morphological characters of Nendran ecotypes

Character	*GCV (%)*	*PCV (%)*	*ECV (%)*	*Heritability (%)*	*Genetic advance (5 %)*
Plant height (cm)	12.0075	12.1008	1.4996	0.9846	24.5448
Leaves/plant	11.3492	13.6036	7.5002	0.6960	19.5049
Leaf width (cm)	7.2447	7.635	2.41	0.9004	14.1611
Suckers/plant	45.1891	47.4756	14.5559	0.9060	88.6062
Days taken for planting to shooting (days)	13.8466	14.0138	2.1577	0.9763	28.1839
Bunch weight (kg)	42.9321	44.2188	10.5893	0.9427	85.8668
Fingers/hand	20.4671	22.8682	10.2006	0.8010	37.7354
Fingers/bunch	48.4186	49.3847	9.7208	0.9613	97.7909
Finger weight (g)	37.8228	37.9542	3.1555	0.9931	97.7909
Bunch length (cm)	19.6626	20.2343	4.7758	0.9443	39.3606
Finger length (cm)	23.8408	24.3941	5.1665	0.9551	47.9979
Finger girth (cm)	15.4995	16.3974	5.3517	0.8935	30.1805
Volume of fruit (cc)	39.3226	39.4778	3.4964	0.9922	80.6863
Ripe fruit weight (g)	39.3226	39.4778	3.4964	0.9922	80.6863
Pulp weight (g)	39.2395	39.4125	3.6887	0.9912	80.4786
Sugar : acid ratio	21.5886	22.9319	7.7333	0.8863	41.8675
Hand weight (g)	40.2761	42.6343	13.9829	0.8924	78.3796

The GCV helps in comparison and measurement of genetic variability among different characters. The GCV ranged from 7.24% for leaf width to fingers/bunch of 48.41%. The highest GCV was recorded from fingers/bunch, followed by suckers/plant (45.19%), bunch weight (42.93%), hand weight (40.28 %), volume of finger (39.32%), ripe fruit weight (39.32 %), finger weight (37.32%) and pulp weight (39.24%). Rajeevan and Geetha (1982) also supported this study with high estimates of GCV. The lowest GCV value was observed in leaf width, followed by plant height (12.01%), leaves/plant (11.35%), days taken for planting to shooting (13.85 %) and finger girth (15.49 %). Rajeevan and Geetha (1982) observed high PCV and GCV values for bunch weight, number of fingers/bunch, number of hands/bunch, length of finger and weight of finger of 40 banana cultivars. The difference in PCV and GCV and very low estimates of GCV indicate the immense influence of environment of the manifestation of this character. The similar findings were also made by Sreerangaswamy *et al.* (1980). The difference between phenotypic and genotypic coefficient of variation observed with fingers/bunch, bunch weight, finger length, volume of finger, ripe fruit weight and pulp weight suggests that environmental effect in these characters. The estimate of heritability separate the genetic variability from phenotypic variability and indicated the possibility and extent to which improvement can be brought about through proper selection. Heritability in broad sense gives the amount of heritable potion of a character. Characters possessing high heritability can be improved directly through selection as they are less affected by the environment. Along with heritability, correlations, genetic advance and genetic gain are also useful parameters in deciding the favourable response of selection in a population.

The magnitude of heritability indicated effectiveness of selection based on phenotypic performance (Johnson *et al.,* 1955). He also stated that a consideration of heritability estimate and genetic advance jointly would enable one to arrive at a more reliable conclusion. In heritability, all characters showed high heritability, which ranged from 69.60% for leaves/plant to 99.31% for finger weight. The characters like plant height (98.46 %), days taken from flowering to shooting (97.63 %), suckers/plant (90 %), leaves/plant (69.60 %), leaf width (90 %), fingers/ hand (80.10 %), fingers/bunch (96.13 %), bunch weight (94.27 %), hand weight (89.24 %), length of bunch (94.43 %), weight of finger (99.31 %), length of finger (95.51 %), girth of finger (89.35 %), volume of finger (99.22 %), ripe fruit weight (99.22 %), pulp weight (99.12 %) and sugar:acid ratio (88.63 %) had high heritability. Their relatively higher values of heritability imply that large proportion of phenotypic variance was attributable to genotypic variance. The high heritability for fingers/bunch, weight of finger and ripe fruit weight obtained in present studies are in agreement with those of Sreerangaswamy *et al.* (1980). The high heritability was also reported for number of hands (Rajeevan and Geetha, 1982), suckers/plant and volume of finger (Valsalakumari and Nair, 20), bunch length (Rosamma and Namboodiri, 1990) and bunch weight (Uma *et al.*, 2000). Heritability is being demonstrated by Katiyar *et al.* (1974). Ramanujam and Thirumalachar (1967) reported that heritability estimates in broad sense would be reliable if accompanied by a high genetic advance.

There was a wide variation among characters for their genetic advance. Genetic advance varied from 14.16% for leaf width to 97.79% for fingers/bunch and finger weight. The fingers/bunch and weight of finger followed by suckers/plant, bunch weight, ripe fruit weight, volume of finger, pulp weight and hand weight showed higher genetic advance along with a high heritability. This clearly suggests that these characters are mainly of additive types as reported by Johnson *et al.* (1955). The lowest genetic advance obtained for leaf width, leaves/plant, plant height, days taken for flowering to shooting and finger girth. Fingers/bunch with high value of PCV, GCV and heritability coupled with genetic advance indicate that the character is predominantly controlled by additive gene action. This is supported by hypothesis proposed by Panse (1957) suggesting that characters showing high heritability coupled with GA were governed by additive gene effects. High heritability does not necessarily mean a high genetic advance for a particular character (Allard, 1960). Heritability along with genetic advance is more useful than heritability alone in predicting the result and effect of selecting the best individuals (Johnson *et al.*, 1955). Uma *et al.* (2000) reported that plant height with very high value of heritability and moderate value of genetic advance, revealing relatively low influence of environment on this trait of silk ecotypes of banana.

Correlation provides information on nature and extent of association between characters in a population. The component characters always show interrelationships. When selection procedure is applied on a trait, population under selection is improved not only for trait but also for other characters associated with it. This facilitates simultaneous improvement of two or more characters. Therefore, analysis of yield in terms of phenotypic and environmental correlation coefficients of component characters leads help in understanding characters that can form the basis of selection. The phenotypic and genotypic correlations are presented in Tables 13.4 and 13.5. In correlation studies, bunch weight had significant positive phenotypic correlation with fingers/bunch, fingers/hand, finger girth, bunch length, sugar:acid ratio and hand weight. The highly significant phenotypic correlation of bunch weight with fingers/bunch was obtained in present study is in agreement with those of Rosamma and Namboodiri (1990). Significant association of bunch weight with fingers/hand at genotypic level was reported by Krishnan and Shanmugavelu (1983) and Rosamma and Namboodiri (1990). Significant positive association of fingers/bunch with bunch weight at phenotypic and genotypic level is in agreement with those of Krishnan and Shanmugavelu (1983) and Rosamma and Namboodiri (1990). The positive phenotypic and genotypic association with number of fruits/bunch and bunch length obtained is in conformity with the findings of Sunilkumar (1997). Significant phenotypic and genotypic correlation of finger girth with length of finger seen in the present study was also reported earlier by Sunilkumar (1997). Weight of finger showed positive correlation with girth of finger and length of finger supported by findings of Sunilkumar (17).

The association analysis through correlation and heritability estimates and values of genetic advance and genetic gain are incapable of providing a true picture of relative merits or demerits of role of each component to final yield, since an individual component may either have a direct influence in improvement of yield or indirect role through other components in improvement of yield, or both.

SUMMARY

Twelve Nendran ecotypes of banana (*Musa* spp.) were studied for their genetic variability and correlation among various vegetative, fruit and yield traits. The genetic and phenotypic coefficient of variance, heritability, genetic advance and coefficient of correlation were estimated for 17 traits which included plant height, suckers/plant, leaves/plant, leaf width, days taken for planting to shooting, bunch weight, bunch length, hand weight, fingers/bunch, fingers/hand,

Table 13.4. Phenotypic correlation coefficient of variation of 17 traits of 12 Nendran ecotypes

	Plant height	*Leaves/ plant*	*Leaf width*	*Suckers/ plant*	*Planting to shooting*	*Bunch weight*	*Fingers/ hand*	*Fingers/ bunch*	*Finger weight*	*Bunch length*	*Finger length*	*Finger girth*	*Volume of fruit*	*Ripe fruit weight*	*Pulp weight*	*Sugar: acid ratio*	*Hand weight*
Plant height	1.000	0.439	0.534	0.174	0.603	0.601	0.383	0.397	-0.080	0.226	0.105	-0.131	-0.075	-0.075	-.0117	0.406	0.253
Leaves/plant		1.000	0.515	0.600	0.057	-0.097	-0.115	--0.288	0.444	0.463	0.499	0.139	0.441	0.441	0.496	0.730	0.503
Leaf width			1.000	0.682	0.131	0.286	0.331	0.101	0.633	-0.055	0.661	0.519	0.638	0.638-	0.587	0.478	0.742
Suckers/plant				1.000	-0.019	-0.152	-0.040	-0.328	0.758	-0.369	0.753	0.622	0.757	0.757	0.695	0.419	0.707
Planting to shooting					1.000	0.298	0.425	0.502	-0.465	0.269	-0.225	-0.338	-0.449	-0.449	-0.570	0.117	-0.229
Bunch weight						1.000	0.568	0.773	-0.166	0.707	-0.078	0.068	-0.159	-0.159	-0.203	0.020	0.302
Fingers/hand							1.000	0.796	-0.046	0.535	0.061	0.004	-0.027	-0.027	-0.127	-0.032	0.201
Fingers/bunch								1.000	-0.384	0.740	-0.313	-0.183	-0.371	-0.371	-0.449	-0.192	-0.043
Finger weight									1.000	-0.360	0.900	0.793	0.999	0.999	0..983	0.396	0.860
Bunch length										1.000	-0.272	-0.014	-0.354	-0.354	-0.380	-0.167	-0.021
Finger length											1.000	0.770	0.909	0.909	0.890	0.419	0.916
Finger girth												1.000	0.799	0.799	0.802	0.095	0.856
Volume of fruit													1.000	0.909	0.981	0.389	0.867
Ripe fruit weight														1.000	0.981	0.389	0.867
Pulp weight															1.000	0.378	0.830
Sugar:acid ratio																1.000	0.378
Hand weight																	1.000

Table 13.5. Genotypic correlation coefficient of variation in traits of Nendran ecotypes

	Plant height	*Leaves/ plant*	*Leaf width*	*Suckers/ plant*	*Planting to shooting*	*Bunch weight*	*Fingers/ hand*	*Fingers/ bunch*	*Finger weight*	*Bunch length*	*Finger length*	*Finger girth*	*Volume of fruit*	*Ripe fruit weight*	*Pulp weight*	*Sugar: acid ratio*	*Hand weight*
Plant height	1.000	0.513	0.559	0.173	0.621	0.623	0.435	0.406	-0.078	0.230	0.107	-0.134	-0.072	-0.072	-0.117	0.406	0.253
Leaves/plant		1.000	0.631	0.788	0.092	-0.095	-0.143	-0.366	0.546	-0.565	0.600	0.181	0.541	0.541	0.495	0.730	0.503
Leaf width			1.000	0.733	0.149	0.297	0.434	0.125	0.669	-0.050	0.706	0.584	0.679	0.679	0.587	0.478	0.742
Suckers/plant				1.000	-0.008	-0.166	-0.027	-0.339	0.796	-0.411	0.828	0.697	0.798	0.798	0.695	0.419	0.707
Planting to shooting					1.000	0.311	0.468	0.523	-0.472	0.285	-0.235	-0.368	-0.457	-0.457	-0.570	0.117	-0.229
Bunch weight						1.000	0.631	0.792	-0.171	0.755	-0.083	0.069	-0.165	-0.165	-0.203	0.020	0.302
Fingers/ hand							1.000	0.878	-0.044	0.623	0.060	0.020	-0.028	-0.028	-0.127	0.032	0.201
Fingers/bunch								1.000	-0.391	0.780	-0.316	-0.204	-0.378	-0.378	-0.449	-0.192	0.043
Finger weight									1.000	-0.368	0.925	0.844	1.000	1.000	0.983	0.396	0.860
Bunch length										1.000	-0.290	-0.042	-0.363	-0.363	-0.380	-0.167	-0.021
Finger length											1.000	0.811	0.933	0.933	0.890	0.419	0.916
Finger girth												1.000	0.849	0.849	0.802	0.095	0.856
Volume of fruit													1.000	1.000	0.981	0.389	0.867
Ripe fruit weight														1.000	0.981	0.389	0.867
Pulp weight															1.000	0.378	0.830
Sugar:acid ratio																1.000	0.378
Hand weight																	1.000

length, girth, weight and volume of finger, ripe fruit weight, sugar:acid ratio and pulp weight. A remarkable variability was observed among collections for these characters. All the characters showed higher estimates of broad sense heritability, whereas genetic advance was recorded very high in volume of finger, followed by weight of finger, ripe fruit weight, pulp weight and fingers/bunch. The value of high PV, GV, PCV, GCV, heritability and genetic advance makes it a prime character for direct selection. The weight of finger, bunch weight, volume of finger and fingers/bunch showed high genetic advance and high heritability are other important characters which have to be considered for selection of clones.

References

Al-Jibouri, H.A., Miller, P.A. and Robinson, H.F. 1958. Genotypic and environmental variances and covariance in upland cotton cross of inter specific origin. *Agron. J.*, **50:**633-36.

Allard, R.W. 1960. *Principles of Plant Breeding*. John Wiley & Sons Inc., New York, 460 pp.

Burton, C. W. 1952. Quantitative inheritance in grasses. *In: Proc. 6th Int. Grassland Congr.* **1:**277-83.

Burton, G.W. and De Vane, E.V. 1953. Estimating heritability in tall fescue (*Festuca arundinacea*) from replicated clonal materials. *Agron. J.* **45:**478-81.

Fischer, R.A. 1960. *The Design of Experiments*. Hafner Publishing Co. Inc. New York, pp. 50-67.

Johnson, H.W., Robinson, H.F. and Comstock, R.E. 1955. Estimates of genetic and environmental variability in soyabeans. *Agron. J.* **47:**314-18.

Katiyar, R.P., Mishra, B., Singh, S.N. and Chauhan, Y.S. 1974. Genetic variability, heritability and genetic advance of yield and its components in Indian mustard. *Indian J. Agrl. Sci.*, **44**(5):291-93.

KAU, 1996. *Package of Practices Recommendations: Crops 1996.* 11th edn., Directorate of Extension, Kerala Agricultural University, Thrissur, 115pp.

Krishnan, B. M. and Shanmugavelu, K.G. 1983. Correlation studies in banana cv. Robusta. *South Indian Hort.*, **31**(2):110-11.

Nayar, N.K., Lyla, K.R. and Mathew, V. 1979. Estimation of genetic variability in banana (*Musa* spp.) for quantitative traits: 1. the desert types. *Indian J. Agrl. Sci.*, **9:**414-16.

Panse, V.G. 1957. Genetics of quantitative characters in relation to plant breeding. *Indian J. Genet.*, **17:**318-29.

Panse, V.G. and Sukhatme, P.V. 1967. *Statistical Methods for Agricultural Workers*, 2nd edn. Indian Council of Agricultural Research, New Delhi, 108 pp.

Rajeevan, P.K. and Geetha, C.K. 1982. Variability studies in banana. *South Indian Hort.*, **34:**197-200.

Ramanujam, S. and Thirumalachar, D.K. 1967. Genetic variability of certain characters in red pepper (*Capsicum annuum*, L.). *Mysore J. Agric.*, **1:**30-36.

Rosamma, C. A. and Namboodiri, K.M.N. 1990. Genetic analysis of yield in banana, *Agric Res. J. Kerala,* **28:**1-8.

Sreerangasamy, S.R., Sambandamurthi, S. and Murugesan, M. 1980. Genetic analysis in banana. *In: Proceedings of National Seminar on Banana Production Technology,* Muthukrishnan, C.R. and Abdulkhader, J.B.M. (Eds.) Tamil Nadu Agricultural University, Coimbatore, pp. 50-56.

Sunilkumar, K. 1997. Selection of superior types of Kaliethan (*Musa* AAB group) Nendran. M.Sc.(Hort.), thesis, Kerala Agricultural University, Vellanikkara, Thrissur.

Swarup, V and Chaugle, D.S. 1967. Studies on genetic variability in sorgham. 1. Phenotypic variations and its heritable components in some quantitative characters contributing towards yield. *Indian J. Genet.*, **22:**31-36.

Uma, S., Dayarani, M., Singh, H.P., Shyam, B. and Sathiamoorthy S. 2000. Studies on genetic variability in banana silk subgroup (AAB). *Indian J. Hort.*, **57**(2):106-109.

Valsalakumari, P.K. and Nair, P.C.S. 1986. Genetic variability in banana. *Agric. Res. J. Kerala,* **24**(1):66-72.

14

Intraclonal Variation in Musa (AAB) Palayankodan

C. Rajamanickam, K. Rajmohan and S. Balasubramanyan

Banana is cultivated in tropical and subtropical regions (Singh, 2000). In Kerala, it is cultivated in 99,412 ha with an annual production of 7,31,650 tonnes (FIB, 2003). The centre of origin of banana is South-East Asia, especially Assam and Myanmar (Rao, 1998). India has been considered as one of the important centres of biodiversity for banana, having about eight species and 120 distinct clones. Banana is well adapted to region varying from humid tropics to humid subtropics and semi-arid subtropics up to an elevation of 2,000m above mean sea-level. Thc wide range of adoption of banana is due to its appreciable genetic diversity and suitability to varying agroclimatic conditions. The diverse agroclimatic conditions have encouraged the development and sustenance of a large number of varieties catering to the local needs. Biodiversity in banana cultivars, which consist of triploids is complex especially Palayankodan with combination of different degrees of expression of parental species. In Kerala, several land races of Palayankodan were available and cultivated in different parts, known in different local names. Palayankodan (AAB group) comprises several popular dessert types. Of which, Palayankodan syn. Mysore Poovan is most widely cultivated single clone, because of its drought tolerance and suitability for ratooning. The vast difference in agroclimatic conditions in which clone is grown and India is original home of genomic group are likely to generate numerous mutants in this clone. However, only one mutant, Motta Poovan, has been reported so far. Therefore, assessing intraclonal variation, if any, in banana clone Palayankodan with respect to growth, yield and quality traits was the need.

INTRACLONAL VARIATION IN MUSA PALAYANKODAN

This study was carried out at the Department of Pomology and Floriculture, College of Agriculture, Vellayani, Thiruvananthapuram, Kerala. A total of six Palayankodan intraclones were assembled from different sources and planted in a completely randomized block design with five replications as per Panse and Sukhatme (1967). The cultural practices were followed (KAU, 1996). The intraclones of Palayankodan banana are presented in Table.14.1 Data on plant height, pseudostem girth, leaves/plant, leaf length, leaf width, suckers/plant, days taken from planting to flowering, days taken from shooting to harvesting and crop duration were recorded. Bunches were harvested when they fully mature and traits like bunch weight, hands/bunch, fingers/hand, fingers/bunch and hand weight were recorded. From bunch, middle finger in top row of second hand (Gottriech *et al., 5*) was sampled to record length, girth, weight and volume of finger. Sugar: acid ratio, pulp: peel ratio and shelf-life of fruits were also recorded. The chemical analysis of fruits was done with ripe fruits. Total soluble solids (TSS) content (^{0} Brix) of fruits was recorded with hand refractometer (AOAC, 1984). Total sugars, reducing sugars and titrable acidity content were determined by using standard estimation technique (Ranganna, 16).

Table 14.1. Intraclones name, types, ploidy and genomic composition of banana Palayankodan

Ecotype	*Type*	*Ploidy*	*Genomic composition*
Palode Palayankodan	Dessert	3x	AAB
PKNNR	Dessert	3x	AAB
Chandra Bale	Dessert	3x	AAB
Pisang Ceylon	Dessert	3x	AAB
Motta Poovan	Dessert	3x	AAB
Vellapalayankodan	Dessert	3x	AAB

THE RESULTS

In six intraclones of Palayankodan, certain salient differences were observed with respect to vegetative characters, yield and quality traits. Plant height and pseudostem girth differed significantly. Vellapalayankodan recorded highest plant height (417.2cm) and lowest height (264.2 cm) in Motta Poovan (Table 14.2). Higher pseudostem girth (96.10cm) was observed in Vellapalayankodan, while Chandra Bale showed minimum girth (56.50 cm). The growth habit, strong overlapping of functional leaves, height of plant and long crop duration were involved in growth of pseudostem (Jacob, 1952). There was a significant variance among Palayankodan intraclones. The maximum number (11.20) of leaves/plant was recorded in Pisang Ceylon, which was significantly superior to other Palayankodan intraclones. The lowest number of leaves/plant was observed in PKNNR (6.80), followed by Chandra Bale (7.20). The maximum leaf length was found in Pisang Ceylon (231.34 cm) and lowest (199.04 cm) in Chandra Bale. Rajeevan (1958) reported that, leaf production was continuous till shooting in banana. Retention of leaves at functional level is a clonal trait, which is influenced to some extent by seasonal difference. He also stated that, data relating to functional leaves indicated that this cannot be taken as a criterion to unravel intraclonal variation. According to Stover and Simmonds (1987), growth in height and circumference of pseudostem are closely related to foliar growth, since pseudostem consists of overlapping of leaf sheaths. Number of suckers/plant recorded highest (15.8) in Vellapayankodan and the lowest (6.2) in Chandra Bale. Sreerangaswamy *et al.* (1980) reported that banana though clonally propagated plants showed significant variation in plant height and girth within population.

Table 14.2. Mean values of vegetative traits of banana Palayankodan intraclones

Ecotype	*Plant height (cm)*	*Pseudostem girth (cm)*	*Number of leaves/plant*	*Leaf length (cm)*	*Leaf width (cm)*	*Number of suckers/plant*
Palode Palayankodan	313.2	67.8	10.0	209.0	76.0	11.6
PKNNR	278.4	57.1	6.8	216.3	82.1	8.4
Chandra Bale	292.6	56.5	7.2	199.0	78.7	6.8
Pisang Ceylon	300.6	60.8	11.2	231.3	87.6	6.2
Motta Poovan	264.2	56.6	7.6	211.4	76.5	7.6
Vellapalayankodan	417.2	96.1	8.4	214.0	72.9	15.8
SED	23.7	6.3	0.7	4.4	2.1	1.5
CD (P=0.05%)	60.9	16.2	1.7	11.2	5.4	3.8

Mean values of duration of six Palayankodan banana intraclones are presented in Table 14.3. Among Palayankodan intraclones evaluated, maximum days taken from planting to shooting (300.0 days) was recorded in Vellapalayankodan which was significantly superior to other Palayankodan intraclones. The lowest duration (205.2 days) was recorded in PKNNR. The maximum days taken from shooting to harvesting were observed in Vellapalayankodan (107.0 days) and lowest duration were in Chandra Bale (90.6 days). The highest (407.0 days) crop duration was recorded in Vellapalayankodan and minimum duration (301.2 days) was observed in PKNNR. Simmonds (1987) reported that varietal characters and growing conditions influenced the crop duration in banana. Crop duration of banana is very much influenced by agroclimatic conditions and seasonal variation (Valsalakumari, 1985). Rajeevan (1985) reported significant variation in crop duration of Palayankodan accessions.

Table 14.3. Mean values of crop duration of banana Palayankodan intraclones

Ecotype	*Days taken from planting to shooting (days)*	*Days taken from shooting to harvesting (days)*	*Crop duration (days)*
Palode Palayankodan	227.4	93.0	320.4
PKNNR	205.2	95.0	301.2
Chandra Bale	226.0	90.6	315.2
Pisang Ceylon	219.8	92.8	312.6
Motta Poovan	209.6	104.8	314.4
Vellapalayankodan	300.0	107.0	407.0
SED	14.2	2.8	15.9
CD (P= 0.05%)	36.5	7.3	40.9

The bunch weight, number of hands/bunch, number of fingers/hand, hand weight and fingers/bunch varied significantly (Table 14.4). Vellapalayankodan produced heaviest bunches (23.0 kg), followed by Palode Palayankodan which recorded a bunch weight of 20.10 kg. The lowest bunch weight was recorded in Motta Poovan (10.6 kg). The Vellapalayankodan recorded highest number of hands/bunch (14.60) and lowest (10.20) value was observed in Pisang Ceylon. Vellapalayankodan recorded maximum number of fingers/bunch (254.20) and lowest in Pisang Ceylon (145.80). The maximum number of fingers/hand was observed in Vellapalayankodan (18.9) and the lowest value in Pisang Ceylon of 14.30. Hand weight was recorded highest (2.80 kg) in Palode Palayankodan. The lowest hand weight (1.50 kg) was recorded in Motta

Poovan. This might be due to clonal variation. Teaotia *et al.* (1970) observed that, bunch yield was strongly correlated with pseudostem circumference and its contribution to variation in yield. Babu (2001) reported that variation in bunch weight due to change in location or inherent genetic variations. Clones have been reported to show difference in their performance under different agroclimatic conditions (KAU, 1996). Simmonds (1962) reported that, hands/bunch in banana were affected by environmental factors.

Table 14.4. Mean values of bunch traits of Palayankodan intraclones of banana

Ecotype	*Bunch weight (kg)*	*Number of hands/bunch*	*Number of fingers/ bunch*	*Number of fingers/hand*	*Hand weight (kg)*
Palode Palayankodan	20.1	11.4	207.8	18.2	2.8
PKNNR	12.9	10.6	179.4	16.9	2.0
Chandra Bale	14.1	10.4	179.0	17.2	1.8
Pisang Ceylon	16.4	10.2	145.8	14.3	2.2
Motta Poovan	10.6	10.6	172.2	15.8	1.5
Vellapalayankodan	23.0	14.6	254.2	18.9	1.9
SED	2.1	0.8	17.4	0.7	0.1
CD (P=0.05%)	5.3	2.1	44.7	1.7	0.3

The accessions differed significantly in length, girth, weight and volume of fruit (Table 14.5). Among Palayankodan intraclones, Vellapalayankodan recorded highest finger length (13.140cm) while lowest (8.40cm) in Motta Poovan, followed by Chandra Bale (9.10 cm). Pisang Ceylon recorded highest (10.90cm) girth of finger, followed by Vellapalayankodan (10.50 cm) Motta Poovan (8.10 cm). Significant variance in weight of finger was also observed among Palayankodan intraclones. The highest finger weight (109.20 g) was observed in Chandra Bale, followed by Palode Palayankodan (94.10 g) lowest being in Vellapalayankodan (90.10 g). The highest fruit volume (101.70 cc) was recorded in PKNNR and lowest in Vellapalayankodan (82.50 cc). Among Palayankodan intraclones, highest pulp: peel ratio (3.60) was recorded in Motta Poovan and lowest in Vellapalayankodan (2.53). Variation in finger length has already been described as a varietal character (Lodh *et al.*, 1971; Lenka *et al.*, 2002a). Rajeevan and Mohanakumaran (1993) observed that yield traits varied significantly among Palayankodan accessions in Kerala.

Table 14.5. Mean values of the finger traits of banana Palayankodan intraclones

Ecotype	*Length of finger (cm)*	*Girth of finger (cm)*	*Weight of finger (g)*	*Volume of finger (cc)*
Palode Palayankodan	13.0	10.4	94.1	88.4
PKNNR	11.3	10.3	107.3	101.7
Chandra Bale	9.1	8.3	109.2	97.6
Pisang Ceylon	11.8	10.9	97.6	92.6
Motta Poovan	8.4	8.1	96.7	92.0
Vellapalayankodan	13.1	10.5	90.1	82.5
SED	1.2	0.8	13.8	13.5
CD (P=0.05%)	3.0	2.0	35.5	34.7

The TSS, total sugars, reducing sugars, sugar: acid ratio, shelf-life of fruits and pulp: peel ratio varied significantly among Palayankodan intraclones. The PKNNR recorded highest (27.90%) TSS while lowest (21.34 %) being in Palode Palayankodan. Rajamony *et al.* (1994) found that TSS varied from 22.0% (Motta Poovan) to 30.0% (Kodapanilla Kunnan) under Thrissur condition. There was no significant difference among Palayankodan intraclones with respect to tritable acidity. The highest acidity (0.44 %) was in Motta Poovan and lowest (0.24 %) in Vellapalayankodan. Acidity in fruits has been described as a varietal trait in banana (KAU, 1984; Rajeevan, 1985; Ram *et al.*, 1994 and Devi, 1996).There was no significant difference in total sugars among Palayankodan intraclones. The maximum total sugars (17.29%) was in Chandra Bale and lowest (16.57%) in Motta Poovan. The maximum reducing sugars was in PKNNR (17.09 %), lowest (16.39%) being in Vellapalayankodan. Significant difference in sugar: acid ratio was also observed among the Palayankodan intraclones. The maximum (68.53) sugar: acid ratio was observed in Vellapalayankodan which was significantly superior to other Palayankodan intraclones. The lowest (37.35) was recorded in Motta Poovan. The maximum shelf-life (8.65 days) was in Vellapalayankodan, which was significantly superior to other Palayankodan intraclones. The lowest shelf-life (5.24 days) was observed in Motta Poovan.

Table 14.6. Mean values of the quality traits of banana Palayankodan intraclones

Ecotype	*TSS (%)*	*Acidity (%)*	*Total sugars (%)*	*Reducing sugars (%)*	*Sugar: acid ratio*	*Shelf-life of fruit (days)*	*Pulp:peel ratio*
Palode Palayankodan	21.3	0.41	17.21	17.05	46.60	6.72	3.39
PKNNR	27.9	0.40	17.33	17.09	44.70	5.81	3.42
Chandra Bale	27.4	0.42	17.29	17.06	43.70	5.88	3.45
Pisang Ceylon	23.4	0.38	17.25	17.08	47.10	5.45	3.30
Motta Poovan	27.2	0.44	16.57	16.40	37.40	5.24	3.58
Vellapalayankodan	25.4	0.24	16.60	16.39	68.53	8.65	2.53
SED	0.8	0.0	0.3	0.2	4.3	0.6	0.2
CD (P=0.05%)	2.1	0.1	0.8	0.5	11.2	1.6	0.4

SUMMARY

In Kerala, several landraces of Palayankodan are cultivated in different parts. They are known in different local names. Palayankodan (AAB group) comprises several popular dessert types. Of which, Palayankodan syn. Mysore Poovan is most widely cultivated single clone, because of its drought tolerance and suitability to ratooning. The present investigation was aimed at assessing the intraclonal variation, if any, in banana clone Palayankodan with respect to growth, yield and quality traits. The results revealed that Vellapalayankodan recorded highest plant height (417.2 cm), pseudostem girth (96.10 cm), suckers/plant (15.8), days taken from planting to shooting (300 days), shooting to flowering (107 days), total crop duration (407 days), bunch weight (23 kg), hands/bunch (14.6), finger/bunch (254.2), fingers/hand (18.9) and finger length (13.1 cm). Motta Pooovan and PKNNR were also important ecotypes which need further evaluation.

References

AOAC, Washington. 1984. *Official Method of Analysis.* Association of Official Analytical Chemist, 14th edn. Washington D.C., 8p.

Babu, N. 2001. Performance of some genotypes of banana under foot hills of Nagaland. *New Agriculturist,* **12**(1):13-16.

Devi, B.V. 1996. Evaluation of Nendran (*Musa* AAB group) intraclones. M.Sc. (Hort.), thesis, Kerala Agricultural University, Vellanikkara, Thrissur.

FIB. Thiruvananthapuram, 2003. *Farm Guide 2003.* Farm Information Bureau, Government of Kerala, Thiruvananthapuram, 76p.

Gottriech, M., Bradu, D. and Halery. 1964. A simple method for determining average fruit weight. *Ktvani,* **14:**161-62.

Jacob, K.C. 1952. *Madras banana - A monograph.* Government Press, Madras.

KAU. 1984. *Research Report 1982- 83.* Directorate of Research, Kerala Agricultural University, Thrissur, 38p.

KAU, 1996. *Package of Practices Recommendations: Crops 1996* 11th edn. Directorate of Extension, Kerala Agricultural University, Thrissur, 115p.

Lenka, P.C., Dash, D.K., Ray, D.P. and Jamal, S. 2002a. Evaluation of banana genotypes under agroclimatic condition of Orissa. *In: Global Conference on Banana and Plantain,* 28 - 31 October, Singh, H.P. and Dadlani, N.K. (Ed.), Association of the Improvement in Production and Utilization of Banana and University of Agricultural Sciences, Bangalore, pp.37.

Lodh, S.B., Ravel, P., Selvaraj, Y. and Kohli, R.R. 1971. Biochemical changes associated with growth and development of Dwarf Cavendish banana. *Indian J. Hort.*, **28**:38-45.

Panse, V.G. and Sukhatme, P.V. 1967. *Statistical Methods for Agricultural Workers,* 2nd edn. Indian Council of Agricultural Research, New Delhi, 108p.

Rajamony, L., George, K.C., Anitha, N. and Radhakrishnan, T.C. 1994. Assessment of banana (*Musa* spp.) clones of AAB groups based on stability and adaptation. *Indian J. Agric. Sci.*, **64**:521-26.

Rajeevan, P.K. 1985. Intraclonal variations and nutritional studies in banana cv. Palayankodan. Ph.D. thesis, Kerala Agricultural University, Vellanikkara, Thrissur.

Rajeevan, P.K. and Mohanakumaran, N. 1993. Intraclonal variation in *Musa* (AAB)Mysore. *South Indian Hort.*, **41**:307-14.

Ram, R.A., Prasad, J. and Pathak, R.K. 1994. Grow banana for table purpose. *Indian Hort.*, **39**:8-9.

Ranganna, S. 1986. *Hand book of Analysis and Quality Control for Fruit and Vegetable Products,* 2nd edn, Tata Mc Graw Hill Publishers Company !iimited, New Delhi.

Rao, V.N.M. 1998. *Banana.* 3rd edn, ICAR, New Delhi, 48p.

Simmonds, N.W. 1962. *The Evaluation of the Bananas.* Longmans, Green and Company Limited, London.

Singh, H.P. 2002. Scenario of banana production, utilization and trade. *In: Souvenir - Global Conference on Banana and Plantain,* 28-31, October, 2002. Singh, H.P. and Dadlani, N.K. (Eds), Association of the Improvement in Production and Utilization of Banana and University of Agricultural Sciences, Bangalore, pp. 1-17.

Sreerangasamy, S.R., Sambandamurthi, S. and Murugesan, M. 1980. Genetic analysis in banana. *In: Proceedings of National Seminar on Banana Production Technology,* Muthukrishnan, C.R. and Abdulkhader, J.B.M. (Eds), Tamil Nadu Agricultural University, Coimbatore, 50-56.

Stover, R.H. and Simmonds, N.W. 1987. *Bananas.* 3rd edn. Longman, Scientific and Technical, Harlow, Esse, England.

Teaotia, S.S., Bhati, D.R. and Phogat, P.S. 1970. Simple partial and multiple correlation of quantitative characters of banana. *Prog. Hort.*, **1**:17-24.

Valsalakumari, P.K. 1985. Cytotaxonomical studies on banana cultivars. Ph.D. thesis, Kerala Agricultural University, Vellanikkara, Thrissur.

15

Induced Variation in Diploid South Indian Banana Cultivars

G. Praveen, S. Mukunthakumar, P.N. Krishan and S. Seeni

Genetic system of *Musa* is extremely complicated to study because of its serious problems of sterility, interspecific hybridity, heterozygosity and polypoidy that are common in most of the clones. The asexual behaviour is often an indispensable barrier in using cross-breeding as a tool for genetic improvement. A broad array of applied cell and molecular techniques are increasingly used worldwide to enhance the improvement of banana and plantain germplasm. Therefore, feasible methodology of improving selected diploid *Musa* germplasm through exposure to gamma rays and application of Fluorescent Activated Cell Sorting (FACS) in tracing out of plant mutants was known.

INDUCED VARIATION IN DIPLOID SOUTH INDIAN BANANA

Two diploid South Indian cultivars of *Musa*, Matti and Ambalakadali originally procured from Banana farm, Peringamala, Thiruvananthapuram, were maintained at mist house of Tropical Botanic Garden and Research Institute (TBGRI), Palode, Thiruvananthapuram. Healthy and 3 months old sword suckers of both cultivars were taken for *in-vitro* propagation under the standardized media formulations up to second subculture stage (Mukunthakumar and Seeni, 2003). The well-proliferated independent shootlet clones were transferred to MS basal nutrient media and were taken for Gamma irradiation experiment.

The young shootlets of 7-10 cm height incubated in MS basal media were taken for Gamma ray induction studies. Each culture bottle consisting of five healthy shootlets were considered as an experimental unit and six sets of such units of both the cultivars were exposed for ionizing irradiation from ^{60}Co source with varied intensities of 0.2, 0.25, 0.3, 0.35, 0.4 and 0.45 kilo rads (KR) (Novak *et al.*, 1990). The plants were incubated in the same media for one month (Figs15.1a and b) and later on hardened. The hardened plants were maintained in shade house maintained at 70-80 % RH for 3-4 weeks and subsequently taken to nursery for 4 weeks.

The sample processing and isolation of nuclei was performed according to the procedures of Otto (1990); Dolezel and Göhde (1995). About 20 mg of leaf tissue was chopped with a sharp razor blade in a petri-dish with 0.5 ml ice-cold Otto I buffer (0.1 M citric acid monohydrate; 0.5% Tween 20) to release cell nuclei. Another 0.5 ml Otto I buffer was added to suspension, which was filtered through 42-µm nylon mesh and kept at room temperature. The suspension of released cell nuclei were pelleted by spinning at 150g for 5 minutes. Supernatant was removed leaving about 100 µl of liquid above the pellet. The pelleted nuclei were then stained by addition of 2 ml Otto II buffer (0.4 M anhydrous Na_2HPO_4) containing 50 µg/ml Propidium Iodide and 50 µg/ml RNase. Fluorescence detection was carried out with a FACSARIA flow cytometer with sorter (BD-Biosciences, USA) whereby relative fluorescence intensities were translated into histograms corresponding to relative DNA content, thus ploidy status were tested among samples (Dolezel, 1997). The analytical instrument was calibrated using standard Chicken erythrocyte cell lines (CRBC) so that the G1 peak of nuclei isolated from the diploid plants will be within the channel of 50 and 100. This setting was kept constant throughout the experiment for the comparative analysis of all the six samples used.

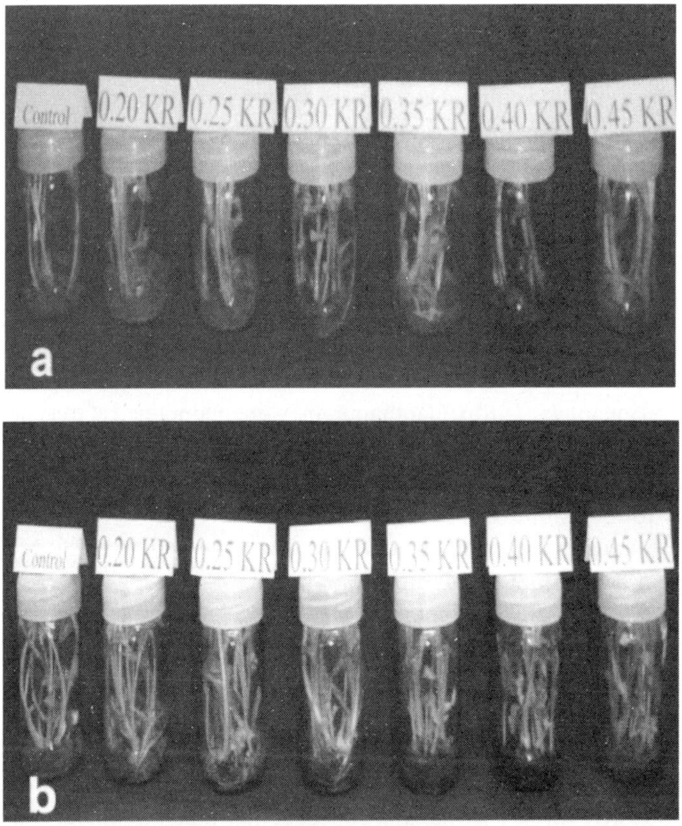

Fig. 15.1. a and b. Explants of matti and ambalakadali subjected to gamma exposure

THE RESULTS

In post-irradiated incubation period, independent assemblies of propagules of both cultivars showed distinct morphological alterations compared to the control explants. The explant of Matti that are subjected for 0.30 KR gamma irradiation shows immense rooting (Fig. 15.1c) in basal agar medium itself. In Ambalakadali, all the explants of the group irradiated by 0.20 KR intensity exhibited irregularity in directional shoot growth (Fig 15.1d). Because of these peculiarities the explants of these two groups along with the control non-irradiated explants were subjected for flow cytometric analysis inorder to estimate the possible cytogenetic variations in irradiated explants. The results from flow cytometric estimation of explants of Matti revealed a slight shift of fluorescent emission peak of G1 nuclei towards the haploid zone (Figs 15.1e and f). This directly indicates the effect of low intensity gamma rays in inducing variations in plant specimens. The earlier works done by Novak *et al.* (1990) on similar streams has also reported the effect of trace intensity of gamma rays on induction of physical mutations in *Musa.*

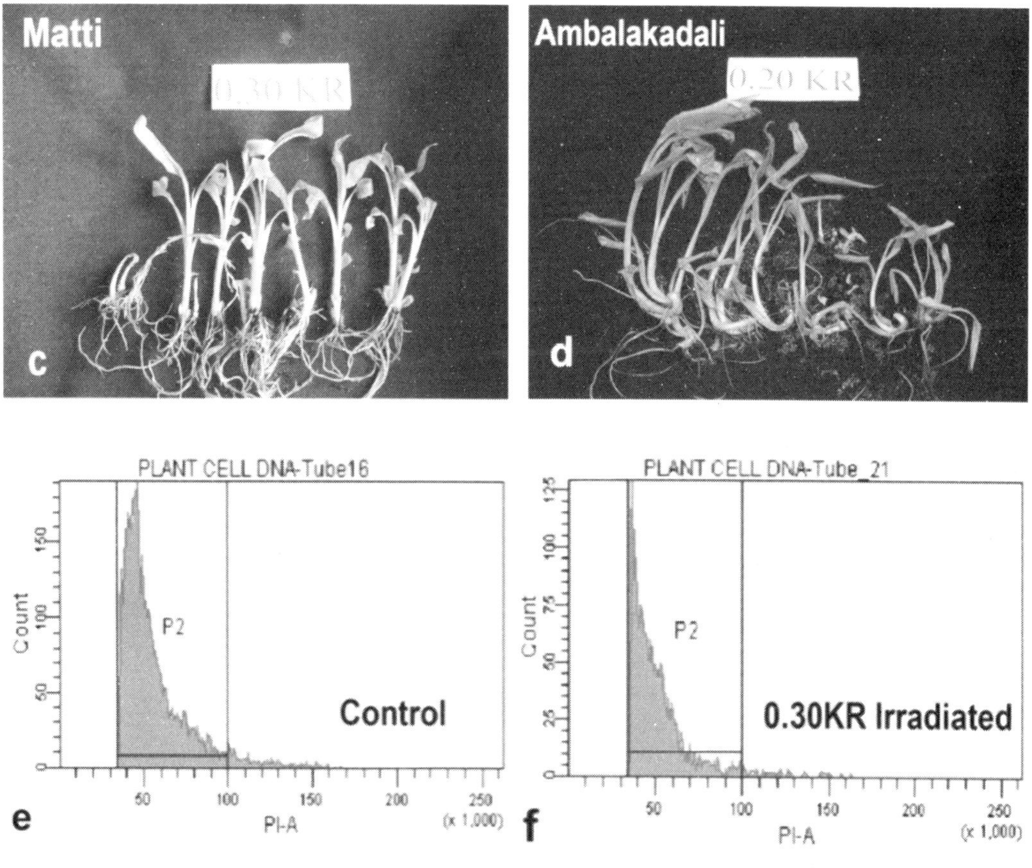

Fig. 15.1c, d, e and f. c and d plantlets showing distinct morphological alterations comparative FACS histogram of Matti explants e and f).

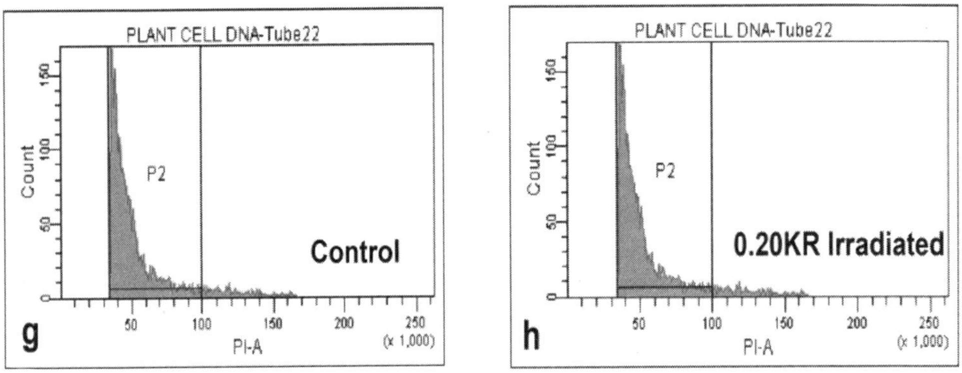

Fig. 15.1 g and h. Comparative FACS histogram of Ambalakadali explants (g and h)

The FACS profile of Ambalakadali does not show any alteration to the G1 peak (Fig. g and h) even though the irradiated explants have distinct morphological irregularities. It is also estimated that the variation induced here was only extra nuclear and hence not stable. However, a general conclusion of this study is that *in vitro* derived plantlets may not be desirable for genetic variation induction. It is also suggested that indirect regenerative tissues, calli or embryonic masses can be used ideally for selection of positive and heritable somaclonal variants in *Musa*.

SUMMARY

The present study discusses the feasibility of improving selected diploid *Musa* germplasm through exposure to gamma rays at the range 0.20 to 0.45 KR. The irradiated *in vitro* propagules of cultivars Matti and Ambalakadali showed certain morphological variation and were subjected to detailed analysis by fluorescent activated cell sorting (FACS) of isolated nuclei. The data generated through FACS analysis revealed that variation induced was only extra nuclear and not stable. The study concludes that *in vitro* derived plantlets may not be desirable for genetic variation induction. It is also suggested that indirect regenerative tissues, calli or embryonic masses can be used ideally for selection of positive and heritable somaclonal variants in *Musa*.

References

Dolezel, J. and G^hde, W. 1995. Sex determination in dioecious plants *Melandrium album* and *M. rubrum* using high-resolution flow cytometry. *Cytometry,* **19**:103-106.

Dolezel, J .1997. Application of flow cytometry for the study of plant genomes. *J. Appl. Genet.,* **38**:285-302.

Mukunthakumar,S. and Seeni, S. 2005. *In vitro* cloning and homestead cultivation of primitive *Musa* cultivars (Matti & Chematti). *Indian J. Exp. Biol.,* **43**:90-95.

Novak, F.J., Afza,R., Van Duren, M. and Omar, M.S .199 and plantain (*Musa* cvs). *Tropical Agriculture (Trinidad),* **67**:21-28.

Otto,F. 1990. DAPI staining of fixed cells for high-resolution flow cytometry of nuclear DNA. *In: Methods in Cell Biology.* Academicz Press, New York, pp.105-10.

16

Screening Banana Germplasm for Drought Tolerance

I. Ravi and S. Uma

WATER deficit is one of the principal environmental limiting factors for crop production throughout the world. It is a consequence of either intermittent or terminal period of drought, that causes significant yield reduction on presently cultivated lands. Depending on evapotranspiration bananas require 25-50 mm water weekly or 1.2-1.4 time class A panevaporation for maximum production. The banana plants need a continuous source of vast amount of soil moisture for optimum growth. Since banana is a tropical giant herbaceous monocot, it has no natural dormant phase and has a high leaf area index and a very shallow root system (Robinson, 1995). These factors may influence the banana plant predisposing to soil moisture deficit stress. Banana and plantain with AAB genomes are more drought tolerant than AAA clones and ABB clones are most tolerant of all (Stover and Simmonds, 1987). Water deficits reduce the productivity of banana by 30-50% (Turner, 1995). Bananuka *et al.* (1999) reported that cultivars, which indicated small reductions in gaseous exchange and leaf area loss and high leaf, high water retention capacity and assimilation rates showed more resistance to drought stress. Difficulties in the past have included the identification of physiological characteristics that are correlated with drought stress that could be used as indicators of drought tolerance. Therefore, it is vital for identification of source for such trait to incorporate in desired plant material or a variety. Hence, study was conducted to screen and identify drought tolerant and susceptible genotypes of banana based on leaf water retention capacity.

SCREENING BANANA GERMPLASM FOR DROUGHT TOLERANCE

About 112 genotypes maintained at NRC for Banana farm were selected for screening drought tolerance. The germplasm was grown following standard cultivation practice. These genotypes, at third month, were not irrigated for three weeks so as to reach the available soil moisture of 65-69% (average of 6 reading in different parts of plot) in field. The three-month-old suckers of uniform growth of plants were selected and screened for drought tolerance by Leaf Water Retention Capacity (LWRC) following Clarke and McCaig (1982). The third (from the top) fully matured leaves were taken for observation. The leaves were excised carefully with a razor and keep it in a polythene bag to control water loss and immediately brought to the field laboratory and weighed fresh weight then hung on a wire allowed to dry in the room (temperature 30 °C, humidity 80%). The leaf weight was measured after 24 and 48 hours finally the dry weight was measured after oven drying at 65 °C for 48 hours (Table 16.1). Then LWRC was calculated following to those of Clarke and McCaig (1982).

Table 16.1. Leaf weight measured 24 and 48 hours after drying

Banana genotype	*Acc.No.*	*LWRC after 24 hours*	*LWRC after 48 hours*
Nattu poovan (AB)	186	91.00	22.75
Thella chakarakeli (AAA)	166	90.50	22.63
Teraben (Mysore AAB)	362	88.00	22.00
Mannan (AAB)	499	86.00	21.50
Muas Balbisiana (BB)	508	85.50	21.38
Ladan (AAB)	721	82.00	20.50
Pisang Berlin(AA)	PB	81.50	20.25
Erode kai (ABB)	415	80.00	20.00
Susceptible variety			
Sirumalai (AAB)	202	37.00	9.25
Kaliban (AAB)	133	36.00	9.00
Nendran (AAB)	296	28.00	7.00
False Horn Plantain(AAB)	733	22.00	5.50
Pisang Seribu (AAB)	579	22.00	5.50

THE RESULTS

About 112 germplasm were screened for drought tolerance during vegetative stage (3 month-old suckers) in field using Leaf Water Retention Capacity. Two genomic representative groups have been identified. The drought tolerant genotypes are: Nattu Poovan (AB), Thella Chakarakeli (AAA), Teraben (Mysore AAB), Mannan (AAB), Musa balbisiana (BB), Ladan (AAB), Erode Kai (ABB), Pisang Berlin (AA); and drought Susceptible genotypes: Sirumali, Kaliban, Nendran, False Horn Plantain and Pisang Seribu.

There was no clearcut demarcation among genomic contribution towards drought tolerance of AA,AB,AAA,AAB,ABB,BB genotype make-up. Therefore, preliminarily selected genotypes must be filed evaluated for drought and then the empirical drought tolerant traits to be validated. This is a continuous evolutionary process till we conclude the traits to screened universally like in other cereal crops.

SUMMARY

The growth and productivity of banana plants are adversely affected by water stress. In traditional banana-growing areas, long term drought is not common even though it is a potential abiotic stress particularly under short dry season existing in different parts of the world. Moreover banana being a long duration crop (10-12 months), drought is a potential threat to banana. In Banana, owing to its high LAI and a very shallow root system make the plant extremely susceptible to water shortage. For any improvement on varieties for drought environment we must identify tolerant genotypes. To identify drought tolerant genotypes, we have screened

112 germplasm collections using Leaf Water Retention Capacity. The drought tolerant genotypes are Nattu Poovan (AB), Thella Chakarakeli (AAA), Teraben, (Mysore, AAB) and Mannan, while Nendran, Pisang Seribu, and False Horn Plantain are susceptible. These genotypes will be highly useful in understanding the drought tolerance mechanism at cellular and molecular levels. Besides, these varieties can be directly used for breeding through conventional and non- conventional breeding.

References

Bananuka J.A., Rbaihayo, R. and Tenywa, M.M. 1999. Reactions of *Musa* genotypes to drought stress. *African Crop Science journal,* **7**:(4)333-39.

Clarke, J.M. and McCaig, T.N. 1982 Evaluation techniques for screening for drought resistance in wheat. *Crop Science,* **22**:503-06

Hallman, D. 1995. The world banana economy, *In: Bananas and Plantains;* Gowen, S. Chapman and Hall, London, U.K., pp. 509-33.

Tover, R.H. 1972. *Banana, Plantain and Abaca Diseases.* Commonwealth mycol. Inst. London, 316 pp.

Stover, R.H. and Simmonds, N.W. 1987. *Bananas,* 3rd edn. Longman, London, 468pp.

Turner, D.W. 1995. The response of banana plant to the environment. *In: bananas and Plantains.* Gowen, S. (Ed.), pp. 206-29. Chapman and Hall, London.

17

Screening Banana Hybrids for Resistance to Fusarium Wilt

P.S. Kavitha, T.N. Balamohan, R. Samiyappan and L. Rajendran

Banana is known to adapt very quickly and produce higher yield under favourable conditions. However, it is prone to different pathogens like fungi, viruses, bacteria and nematodes. The wilt caused by *Fusarium oxysporum* f. sp. *cubense* is most widespread and destructive disease (Simmonds, 1966) and still remains as a major constraint to banana production worldwide. The *Fusarium* wilt often occurs along with plant parasitic nematodes such as *P. coffeae* and *M. incognita,* causing economic losses in India. Thus, the foremost aim in banana and plantain improvement is to enhance the quantitative and qualitative traits, and also to develop hybrids with increased resistance/tolerance to major diseases (Simmonds, 1966). The nematodes often act as a predisposing factor. Therefore, screening for *F. oxysporum* f.sp. *cubense* was necessary.

SCREENING BANANA HYBRIDS FOR RESISTANCE TO FUSARIUM WILT

The hybrids of both phase I and II were tested in pots for resistance to *Fusarium* wilt. The pot mixture was sterilized with 4% formaldehyde (fumigant) and filled in pots. The suckers of uniform size were selected and pared with a knife before planting to remove the outer layers to eliminate nematode infested tissue and treated in hot water at 55°C for 10 minutes. These suckers were planted in pots containing sterilized soil, replicated three times with 10 suckers/ replication. The plants were provided nutrition with Hoagland's No.2 nutrient solution (Hoagland and Arnon, 1950) at fortnightly intervals.

The hybrids were screened for resistance to *Fusarium* wilt by multiplying the race 1 fungus in sand maize medium (Ploetz *et al.*, 1999), which served as *F.oxysporum*. f. sp. *cubense* inoculum. Thirty days after nematode inoculation (primary infection in roots), *F.oxysporum*. f. sp. *cubense* inoculum from the sand maize medium was incorporated in the root zone of the plants at the rate of 10g/pot. Wilt incidence and severity were recorded and a scale of corm discolouration index was observed four months after inoculation.

BANANA HYBRID

The hybrids of phase I and II evaluation were subjected to pathogenecity of *Fusarium oxysporum* f. sp. *cubense* by soil inoculation methc '. Symptoms of fungal infection were noticed one month after inoculation. Scoring was done 4 months after inoculation. Among 24 hybrids of phase I evaluated, H-04-06, H-04-21, H-04-22 and H-04-23, had minimum score of 1 and exhibited resistance to *Fusarium* wilt, while all other hybrids were susceptible (Table 17.1). Of the 19 phase II hybrids, H-03-05, H-03-10, H-03-13, H-03-15, H-03-17 and H-03-19 showed resistance to *Fusarium* wilt, while rest of the hybrids were susceptible (Table 17.2). A critical analysis of inheritance pattern of both phase I and II hybrids revealed that most of the resistant hybrids had either Pisang Lilin or Anaikomban (resistant sources) as one of the parents. This indicated that dominant genes govern resistance.

Table 17.1. Scoring of phase I hybrids for resistance to *Fusarium* wilt 90 days after inoculation under pot culture

Hybrid	*Parentage*	*Genome*	*Score*	*Status*
H-04-01	Peykunnan (op)	AAABB	3.2	S
H-04-02	H-02-32 x Pisang Lilin	AAABB	3.0	S
H-04-03	H-02-32 x Pisang Lilin	AAABB	4.3	HS
H-04-04	H-02-32 x Pisang Lilin	AAABB	4.0	S
H-04-05	H-02-32 x Pisang Lilin	AABB	3.2	S
H-04-06	H-02-32 x Pisang Lilin	AABB	1.0	R
H-04-07	H-02-32 x Pisang Lilin	AAABB	4.0	S
H-04-08	H-02-31 x Pisang Lilin	AAABB	3.3	S
H-04-09	H-02-31 x Pisang Lilin	AAABB	4.0	S
H-04-10	Peykunnan (op)	AAB	3.0	S
H-04-11	H-02-32 x Pisang Lilin	AAABB	4.0	S
H-04-12	Pisang Saba x Pisang Lilin	AABB	5.2	HS
H-04-13	H-02-32 x Pisang Lilin	AAABB	5.0	S
H-04-14	H-02-32 x Pisang Lilin	AAABB	5.0	S
H-04-15	H-02-32 x Pisang Lilin	AAABB	5.3	HS
H-04-16	H-02-32 x Pisang Lilin	AAABB	4.0	S
H-04-17	H-02-32 x Pisang Lilin	AAABB	5.0	S
H-04-18	H-02-32 x Pisang Lilin	AAABB	4.3	S
H-04-19	Peykunnan (op)	AAABB	6.0	HS
H-04-20	H-02-32 x Pisang Lilin	AAABB	6.0	HS
H-04-21	H-02-32 x Pisang Lilin	AAB	1.2	R
H-04-22	Anaikomban x Pisang Lilin	AA	1.0	R
H-04-23	Senna Chenkadali (op)	AA	1.0	R
H-04-24	Peykunnan (op)	AABB	3.3	S

R, Resistant; S, susceptible; HS, highly susceptible

Further, NPH-02-01, a resistant hybrid to nematodes, had both the parents H201 and Anaikomban as resistant sources. Some hybrids were resistant for both nematodes and *Fusarium* wilt, while some hybrids being resistant to nematodes were susceptible to *Fusarium* wilt. This finding indicated that both nematode and *Fusarium* may be individually pathogenic. Mateille and Folkerstsma (1989) also reported similar findings, which were again proved by Gunavathi (2000). In some hybrids, susceptibility to both nematode and wilt pathogen was found. Here nematode infection might have enhanced the fungal activity (Damodaran, 2003).

Table 17.2. Scoring of phase II hybrids for resistance to *Fusarium* wilt 90 days after inoculation under pot culture

Hybrid	*Parentage*	*Genome*	*Score*	*Status*
H-03-04	NPH-02-03 x Pisang Lilin	AABB	3.0	S
H-03-05	Peykunnan OP	AABB	1.0	R
H-03-06	H-02-32 x Pisang Lilin	AB	3.2	S
H-03-07	Peykunnan OP	AABB	3.0	S
H-03-08	H-02-32 x Pisang Lilin	AB	5.0	HS
H-03-09	Peykunnan x Erachivazhai	AABB	6.0	HS
H-03-10	Peykunnan x Pisang Lilin	AABB	3.2	S
H-03-11	H-02-32 x Pisang Lilin	AABB	5.3	HS
H-03-12	Poovan x Pisang Lilin	AABB	5.8	HS
H-03-13	Peykunnan x Erachivazhai	AABB	1.2	R
H-03-14	H-02-31 x Pisang Lilin	AABB	4.0	HS
H-03-15	H-02-32 x Pisang Lilin	AAB	1.0	R
H-03-16	Poovan x Pisang Lilin	AABB	3.5	S
H-03-17	Peykunnan x Pisang Lilin	AABB	1.0	R
H-03-18	H-02-31 x Pisang Lilin	AABB	4.3	HS
H-03-19	Peykunnan x Erachivazhai	AABB	1.0	R
H-03-20	H-02-31 x Pisang Lilin	AABB	5.6	HS
H-03-21	Peykunnan OP	AABB	5.3	HS
NPH-02-01	H201 x Anaikomban	AAB	1.2	R

R, Resistant; S, susceptible; HS, highly susceptible

SUMMARY

Among 24 hybrids of phase I evaluated, H-04-06, H-04-21, H-04-22 and H-04-23 had minimum score of 1 and showing resistance to *Fusarium* wilt, while other hybrids were susceptible. In evaluation of 19 phase II hybrids, H-03-05, H-03-10, H-03-13, H-03-15, H-03-17 and H-03-19 showed resistance to *Fusarium* wilt while the rest of the hybrids were susceptible.

References

Damodaran, T. 2003. Breeding for resistance to *Fusarium* wilt and nematodes in banana (*Musa* spp.). Ph.D. (Hort.) thesis, Tamil Nadu Agricultural University, Coimbatore.

Gunavathi, J. 2000. Studies on resistance to Panama wilt *(Fusarium oxysporum* f.sp. *cubense*) in certain banana hybrids. M.Sc. (Hort.) thesis, Tamil Nadu Agricultural University, Coimbatore.

Mateille, T and Folkerstsma. 1989. Study of nematode fungi associations on banana in the Ivory coast. *FRU T- Revista de Fruticulatura*, **4**(3-4):153-58.

Pleotz, R. C., Haynes, J. L. and Vazgnez, A. 1999. Responses of new banana accessions in South Florida to Panama disease. *Crop Protection*, **18**(7):445-49.

Simmonds, N. W. 1966. *Bananas.* Longman, London.

18

Ploidy Assessment in Banana Hybrids

P.S. Kavitha, T.N. Balamohan and L. Nalina

Breeding of banana is complicated and time consuming (Shepherd, 1954 and 1960), as the degree of sterility is particularly high in edible cultivars. A minimum of two years is required to complete a seed-to-seed crop cycle. The ultimate aim of breeding programme is to develop secondary triploids *via* tetraploid × diploid crosses as suggested by Rowe (1984).The evolved hybrids have to be assessed for their ploidy levels for their utility in further breeding programme. Simmonds (1948) and Borges (1971) observed that stomata size was proportional to ploidy in banana, while stomatal density had the expected complementary relationship. In Jamaica, breeding programme, stomatal densities of two-month-old seedlings were employed to screen the progenies ploidy levels and genomic constitutions. The ploidy status can be assessed by three methods, viz. estimation of stomatal density and size at cellular level as postulated by (Sathiamoorthy 1973), morphological scoring described by Simmonds (1952) and by flow cytometry analysis of nuclear DNA (Dolezel, 1998). Further, use of flow cytometry enables rapid and precise measurements on whole cells, isolated nuclei or chromosomes in a monodisperse suspension. Van Duren *et al.* (1996) used this technique to identify *in vitro* induced tetraploids of SH-3362 banana clone.

ASSESMENT OF PLOIDY IN BANANA HYBRIDS

The cigar leaves were cut using a sharp sterile blade up to 15-20 cm length from top, cleaned gently with sterile distilled water and wrapped with partially wetted sterilized Whattman No.3 filter paper. The samples were then packed in zipped polyethylene cover and analysis was done at the Laboratory of Molecular Cytogenetics and Cytometry, Czech Republic for ploidy. The flow cytometry analysis is considered as the recent and reliable because it is precise, rapid and does not require dividing cells. Precision is more here because of the analysis of nuclear DNA itself, which is least disturbed by environmental factors. Flow cytometry ploidy assay involves preparation of suspensions of intact nuclei from small amounts of leaf tissue and analysis of fluorescence intensity after staining. Diploid banana (2x) nuclei were included in every sample as an internal reference standard.

THE RESULTS

Morphological scoring for banana hybrids was done along with stomatal density and confirmed by flow cytometry analysis. In present study, analysis of ploidy level by stomatal density and morphological scoring revealed of the 24 banana hybrids, two were found to be diploids (AA), three triploids (AAB), five tetraploids (AABB) and 15 unpredictable. Hence these hybrids were subjected to flow cytometry analysis for the assessment of their ploidy level. Young cigar leaves of selected hybrids were analysed for their ploidy level by measuring the size of the nuclear genome.

Histograms representing different ploidies of selected hybrids compared with nuclei isolated from diploid hybrid (2x) used as internal reference standard reveals that H-04-04 and H-04-01 are clear pentaploids (the ratio of peak positions sample: standard was about 2.5 (2x multiplied by 2.5 = 5x). Determination of nuclear DNA increases the reliability of ploidy and easy detection of mixiploids (Dolezel *et al.,* 1997). The results from flow cytometry confirmed the ploidy level (Fig 18.1). The 15 unpredictable hybrids were pentaploids. Determination of nuclear DNA increases the reliability of ploidy and easy detection of mixiploids (Dolezel *et al.*, 1997). Damodaran (2003) observed that two hybrids, NPH-02-01 and H-02-08, which were doubtful in their ploidy status with reference to their general stature, were subjected to flow-cytometry analysis. Both hybrids were found to be triploids (AAB), instead of diploids as evident through stomatal density analysis and morphological scoring.

Table 18.1. Ploidy assessment of banana hybrids based on morphological scoring, stomatal size and stomatal density

Hybrid	*Parentage*	*Genome*	*Mark scored*	*Ploidy*	*Stomatal size (μm²)*	*Stomatal density (no./mm²)*
H-04-01	Peykunnan (op)	AAABB	54	5x	2428.80	6.57
H-04-02	H-02-32 × Pisang Lilin	AAABB	56	5x	2608.85	5.92
H-04-03	H-02-32 × Pisang Lilin	AAABB	56	5x	2892.83	6.57
H-04-04	H-02-32 × Pisang Lilin	AAABB	51	5x	2057.89	7.23
H-04-05	H-02-32 × Pisang Lilin	AABB	44	4x	1836.77	13.15
H-04-06	H-02-32 × Pisang Lilin	AABB	35	4x	1455.30	16.44
H-04-07	H-02-32 × Pisang Lilin	AAABB	57	5x	2812.84	6.57
H-04-08	H-02-31 × Pisang Lilin	AAABB	59	5x	2343.29	5.26
H-04-09	H-02-31 × Pisang Lilin	AAABB	57	5x	2146.55	7.89
H-04-10	Peykunnan (op)	AAB	30	3x	1229.80	32.89
H-04-11	H-02-32 × Pisang Lilin	AAABB	58	5x	2312.87	5.92
H-04-12	Pisang Saba x Pisang Lilin	AABB	62	4x	1473.27	13.15
H-04-13	H-02-32 × Pisang Lilin	AAABB	57	5x	2149.94	5.60
H-04-14	H-02-32 × Pisang Lilin	AAABB	57	5x	2700.26	7.23
H-04-15	H-02-32 × Pisang Lilin	AAABB	57	5x	2295.42	6.78
H-04-16	H-02-32 × Pisang Lilin	AAABB	58	5x	2402.87	5.26
H-04-17	H-02-32 × Pisang Lilin	AAABB	54	5x	2142.18	8.50
H-04-18	H-02-32 × Pisang Lilin	AAABB	59	5x	2487.40	6.24
H-04-19	Peykunnan (op)	AABB	64	4x	1350.56	15.60
H-04-20	H-02-32 × Pisang Lilin	AAABB	58	5x	2370.00	6.57
H-04-21	H-02-10 × Pisang Lilin	AAB	44	3x	1260.01	47.14
H-04-22	Anaikomban x Pisang Lilin	AA	17	2x	701.32	63.74
H-04-23	Senna Chenkadali (op)	AA	20	2x	371.58	59.20
H-04-24	Poovan (op)	AABB	61	4x	1063.90	43.18

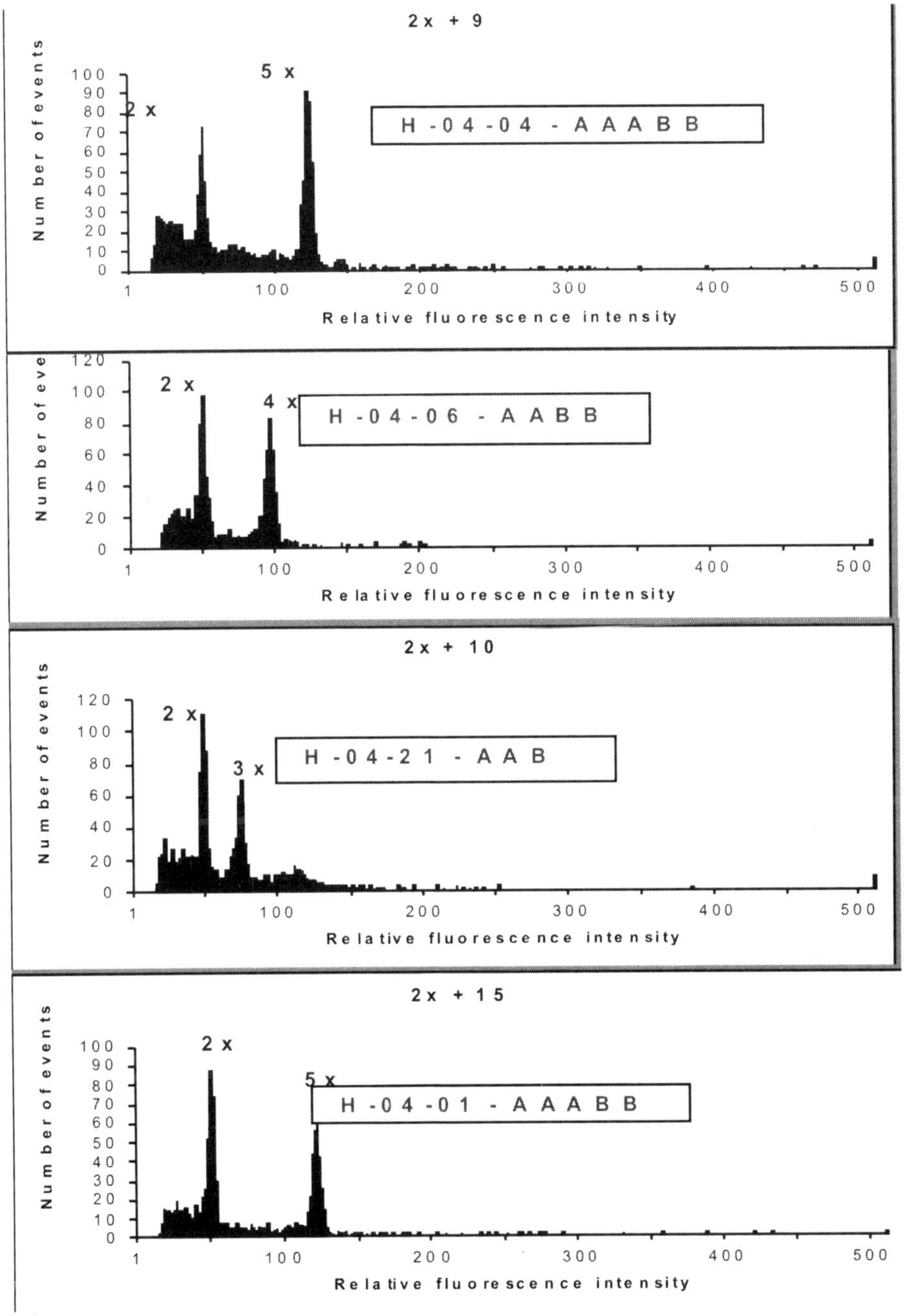

Fig 18.1. Flow cytometry analysis of phase I hybrids for ploidy confirmation

The origin of pentaploids can be explained easily by a fusion of unreduced gametes from tetraploid parent with reduced gametes from diploid parent. The frequency of occurrence of unreduced gametes is genotype-dependent. Furthermore, it may be influenced by external factors.

The pentaploid hybrids might have resulted as follows from the cross with H-02-32 (AABB) as female and Pisang Lilin (AA) as male parent.

H-02-32	×	Pisang lilin
(AABB)		(AA)
↓		↓
AABB		A

AAABB
(pentaploids)

Occurrence of pentaploids in 4n × 2n cross was also earlier reported by Krishnamoorthy (2002). Pentaploids are normally devoid of shooting Sathiamoorthy (1987). All these pentaploid hybrids had bunch emergence but were not economical. Shooting and seed seting on artificial crossing was observed in a pentaploid hybrid by the former author.

The analysis of ploidy level through flow cytometry is an important recent tool which has to be exploited for confirmation of complicated ploidy levels. The assessment was made from the single plant in its first generation (seed to sucker) and hence no replicated data were used. Hence for further confirmation the assessment have to be made from the hybrids in their second generation (sucker to sucker). By assessing the ploidy status the unwanted hybrids can be eliminated for further improvement programme.

SUMMARY

Confirmation of ploidy status for 24 banana hybrids evolved by hybridization programme was done by flow cytometry analysis which enables rapid and precise measurements on whole cells, isolated nuclei or chromosomes in a monodisperse suspension. The results confirmed the ploidy levels and classified 24 hybrids into two diploids (AA), two triploids (AAB), five tetraploids (AABB) and the rest 15 pentaploids (AAABB).

References

Borges, F. 1971. Study of female fertility in clones of plantains and cultivated bananas. *Agronomica Trop.*, **21:** 135-37.

Dolezel, J. 1998. "Flow cytometry, its application and potential for plant breeding", *Current Topics in Plant Cytogenetics Related to Plant Improvement.* Lelley, T. (Ed.), Universitätsverlag, Vienna.

Dolezel, J., Dolezelova, M., Roux, N. and Houwe, I.V. 1997. Use of flow cytometry for rapid ploidy determination in *Musa* species. *InfoMusa,* **6**(1):6-9.

Krishnamoorthy, V. 2002. Breeding for resistance to Sigatoka leaf spot and nematodes in banana (*Musa* spp.). Ph.D. (Hort.) thesis, Tamil Nadu Agricultural University, Coimbatore.

Rowe, P.R. 1984. Breeding bananas and plantains. *In: Plant Breeding Review,* **2:**135-55.

Sathiamoorthy, S. 1973. Preliminary investigations on breeding potential of some banana clones. M.Sc. (Ag.) thesis, Tamil Nadu Agriculture University, Coimbatore.

Sathiamoorthy, S. 1987. Studies on male breeding potential and certain aspects of breeding bananas. Ph.D. (Hort.) thesis, Tamil Nadu Agriculture University, Coimbatore.

Shepherd, K. 1954. Seed fertility of the Gros Michel banana in Jamaica. *J. Hort. Sci.*, **29**:1-11.

Shepherd, K. 1960. Seed fertility of edible bananas. *J. Hort. Sci.*, **35**:6-20.

Simmonds, N.W. 1948. The effects of ploidy upon the leaf of *Musa. Ann. Bot.*, **12**:441-53.

Simmonds, N.W. 1952. Experiments on the pollination of seeded diploid bananas. *J. Genet.*, **51**:32-40.

Van Duren, M., R. Morpurgo, J. Dolezel and R. Afza. 1996. Induction and verification of autotetraploids in diploid banana (*Musa acuminata*) by *in vitro* techniques. *Euphytica,* **88**:25-34.

19

Overcoming Barriers in Seed Setting and Germination in Banana

S. Rekha , D.S. Ambika and M.R. Dinesh

Several improvement programs are being taken up for the improvement in banana. However, banana is difficult for breeding as it is one of the highly evolved plant species and a number of cultivars have been selected by human beings for cultivation. This process has resulted in selection for parthenocarpy, seedless and polyploid group of cultivars, which are high-yielding and vegetatively-propagated. In spite of the difficulties, attempts are being made by breeders for improvement with special reference to disease resistance. Conventional breeding has advantage to introduce desirable genes in a population and selection of heterozygous hybrid, which combines number of desirable traits (Vuylsteke, 2000). In *Musa,* parthenocarpy is due to several dominant genes (Simmonds, 1952a). Seed setting was found to be low due to high levels of sterility in a triploid × diploid cross. Seed setting within clones is influenced by both production of 2n gametes and climatic factors (Oritz and Vuylsteke, 1995). Simmonds, (1952b) observed that crossing among closely related clones gave higher seed yield than self-pollination. In present study, crossing among different diploid genomic groups and its effect on seed setting and germination is reported.

OVERCOMING BARRIERS IN SEED SETTING IN BANANA

The banana accessions used for study were wild species like *M.acuminata* ssp. *burmannicoides* (Calcutta 4), Annara kunnan a wild collection from Kerala, *M.acuminata* wild collection from TNAU, Coimbatore, *M.acuminata* ssp. *malaccensis* and *M.balbisiana* collection Bee hee kela from North-East India. Apart from wild *Musa* species, there were also cultivars like Kadali, Pisang lilin, Anaikomban, Matti (AA group) and Elakki Bale, Putta Bale, Vadakkan Kadali, Chakkara Kadali, Mittili Bale and Hoo Bale of AB group. These species and cultivars were selected for hybridization to generate F_1 populations required for development of molecular markers and *Musa* genome mapping. The selected accessions were cultivated in a compact collection block of IIHR, Bangalore. It was observed that AB group cultivars produced fertile pollen in these conditions hence these accessions were included in hybridization. As and when shooting was observed entire inflorescence was covered to protect it from insects. The pollen from selected male parent was smeared on to stigma of female parent. This process was repeated until all female flowers opened in a banana bunch. Mature bunches were harvested and seeds extracted from individual fruits. The seeds were sown in normal potting mixture.

THE RESULTS

The number of seeds and seedlings obtained in different cross combinations are given in Table 19.1. There were nine cross combinations involving AA × AA genomic group accessions. This combination included wild × wild, cultivar × wild and wild × cultivar of *M.acuminata* ssp. *burmannicoides* (Calcutta 4), Annara kunnan, *M.acuminata* (wild), *M.acuminata* ssp. *malaccensis* and Kadali, Pisang lilin, Anai komban and Matti. There was a large number of seed setting and seedling germination was 11.42%.

In AA × AB cross combination, the AB group cultivars were used as pollen parent and wild *M.acuminata* species as female parent. The AB group cultivars were female sterile and male fertility was 10-30%. Though a number of crosses were attempted in 8 cross combinations seed setting was very low and germination was lowest 4.97%.

The AA × BB cross-involved both wild types *M.acuminata* ssp. *burmannicoides* (Calcutta 4) and *M.balbisiana* - Bee hee kela. A large number of seeds were obtained from a single cross and rate of seedling germination was 8.93%.

The most successful cross combination was BB × BB where two contrasting types of wild highly fertile *M.balbisiana* was used. The accessions included were Bee hee kela and Bhimaithia. Very high seed setting was observed by crossing 30-35 fruits (female flowers), more than 200 seeds were found in each fruit and germination was 90%.

These studies show variation in pattern of seed setting and germination among different genomic groups. Hybridization and seed germination are complicated proceses in breeding banana to develop mapping populations and cultivars resistant to diseases. Apart from physiological and climatic factors affecting the seed setting as reported by earlier workers, present study also indicated effected of genomic groups.

Table 19.1. Seed setting and germination in various cross combinations

Cross combination	*Number of seed set*	*Seedlings germinated (%)*
AA × AA	2469	11.42
AA × AB	443	4.97
AA × BB	2184	8.93
BB × BB	>200/fruit	90.00

SUMMARY

Conventional breeding in banana is difficult because of the problem involved in hybridization. Banana being highly evolved crop is also adapted to vegetative propagation with parthenocarpic fruits. Seed setting is further complicated by its polyploid nature, which also causes seedlessness in cultivated banana fruits. However, natural seed setting is observed among diploid wild *Musa* species, which are male and female fertile. The fertility among cultivated bananas varies. In present studies, crosses made among wild *Musa acuminata*-

Calcutta 4 with AA genomic group cultivars like Kadali, Pisang Lilin, Anaikomban and Matti; the AB group cultivars Elakki Bale, Putta Bale, Vadakkan Kadali, Chakkara Kadali, Mittili Bale and Hoo Bale and its wild relative *Musa balbisiana* of BB genomic group. Seed setting varied among different cross combinations. Seed germination was highest in AA × AA cross, followed by AA × BB cross. In AA × AB cross, seed set was very less and also showed poor germination. In another cross involving BB × BB cross, of wild *Musa balbisiana* types resulted in heavy seed seting and germination. These observations reveal the fact that there are barriers among the genomic groups though banana is said to have evolved from two main *Musa* species that is, *Musa acuminata* and *Musa balbisiana.*

References

Oritz, R. and Vuylsteke, D. 1995. Factors influencing seed set in triploid *Musa* spp. L and production of euploid hybrids. *Annals of Botany,* **75**:151-55.

Simmonds, N.W. 1952 (a). Segregation in some diploid bananas. *J. Genetics,* **51**:458-69.

Simmonds, N.W. 1952 (b). Experiments on the pollination of seeded diploid bananas. *J. Genetics,* **51**:32-40.

Vuylsteke, D. 2000. Breeding bananas and plantains: from intractability to feasibility. *Acta Horticultural,* **540**:149-56.

20

Nutritional Dynamics in Banana for Enhancing Productivity

S.C. KOTUR

BANANA feeds heavily on soil. To sustain high production with quality banana, it is of paramount importance to ensure proper nutrition of the crop. A large variety of soils, climates and production systems grow different varieties of bananas. Therefore, site-specific nutritional management needs to be generated to meet actual needs of growers. An excellent review on this subject has been made by Lahav and Turner (1983), Mitra and Dhua (1985), Fox (1989) Lahav (1995) and Kulasekaran (2003). Therefore, to know nutritional dynamics of banana for enhancing productivity is of vital importance.

NUTRITIONAL DYNAMICS

In one stool of banana, there are always plants and suckers of different stages of growth and they are inter-dependent. Nutrients within banana plantation are located in a number of pools each characterized by a certain total amount of nutrient and release rate and available reserve in soil. Generally, proportional distribution of nutrient between pools depends largely on total amount present. In any dynamic system, losses and gains balance it. For major nutrients 49-56%, for secondary nutrients 32-45 % and for micronutrients 4-55% account for nutrient removal from fruits (Table 20.1). There is an imminent need to replenish these quantities removed. Other losses from system include run-off, leaching and gaseous losses, especially of N.

Gains accrue from rain and irrigation water. The gain can contribute N significantly from first monsoon showers which the latter contribution Ca and Mg depending upon the quantity of irrigation water. A large amount of leaf litter left behind, after decomposition, returns the nutrients contained back to the system. Since yield depends directly on nutrient pool in plant involving large pools in rhizome and pseudostem, it has implication on later fertilization of the ratoon crop. To achieve maximum fertilizer-use efficiency of fertilizers, it should maximize productivity in any given site. In other words, nutrients should be added until the content in crop is sufficient for the desired level of production and further supplements should be need based.

Table 20.1. Nutrient removal by banana of 50tonnes/ha production (kg/ha)

Nutrient	Amount removed in fruits	Amount remaining in plants	Total	Proportion removed in fruits (%)
N	189	199	388	49
P	29	23	52	56
K	778	660	1438	54
Ca	101	126	227	45
Mg	49	76	125	39
S	23	50	73	32
Fe	0.9	0.5	5.9	15
Mn	0.5	12	12.5	4
Zn	0.5	4.2	4.7	12
B	0.7	0.57	1.27	55
Cu	0.2	0.17	0.37	54
Mo	-	0.003	0.003	-

Nutrient Dynamics within System and Plant

According to Turner and Barkus (1973), removal of pseudostem of mother plant after harvesting accounts for 40 % of the need of suckers for N, P, K Ca, Mn and Cu over 10 weeks period. It is beneficial to retain pseudostems with leaves intact 45-60 days after harvesting. This led to a faster growth of suckers compared to either when leaves are cut-off or entire pseudostem is cut-off at the ground level. This can lead to early flowering suckers (ratoon) and saving of fertilizer applied to succeeding crop.

Distribution of Nutrients within Plant

Generally leaves are abundant in B, Mn, Ca and Mg, while N,P and K are evenly distributed between leaves and pseudostems. Rhizome accumulates Na and Zn, while roots store Cu and Fe. The nutrient dynamics between different parts, however, is a function of temperature and stage of plant growth. In stool, rhizome operates as a pump and its phloem vessels as conduits within banana stool.

Kinetics of Nutrient Absorption

The ability to absorb nutrients is genetically controlled and varies widely among crop varieties. Studies on nutrient kinetics are useful to identify varieties which: (i) respond to high level of nutrients under fertile, irrigated and high input intensive production system and (ii) extract native nutrients efficiently from poor/ marginal soils under arid/dry land and low input systems. In respect of nitrates, maximum rate of absorption (Imax) that confers the roots the capacity to respond to higher concentration of nitrate in rhizosphere solution very high in all varieties of banana compared to other fruit crops (Keshava Murthy and Kotur, 2001). Among cultivars Cavendish can be considered most responsive and Ney Poovan the least (Table 20.2). The affinity of nitrate ions to absorbing sites on root is lowest in Ney Poovan, indicating that it can draw these ions even when nitrate concentration is low in rhizosphere solutions prevalent in marginal or unfertilized soils. The highest average net inflow of nitrate into roots is found in Cavendish crop compared to the rest showing its efficiency to absorb N from highly fertile soils or when lavishly supplied with fertilizer. Dwarf Cavendish is least responsive to P in solution, while Nendran was the most responsive. The former possessed low I max but higher Km and *vice versa* was true in the case of the latter. Nendran also appeared to respond better to P fertilization compared to rest of the cultivars. Verification of these parameters needs to be carried out in field to fine-tune genotypic adaptability as a substitute to N or P amendment on marginal soils.

Table 20.2. Kinetic parameters of nitrate and phosphorus absorption of banana varieties

Cultivar	*Imax(pmoles/cm/s)*	*Km(µmoles/litre)*	*(pmoles/cm/second)*
Nitrate			
Cavendish	1312	2017	62
Robusta	680	2464	27
Monthan	734	2613	27
Ney Poovan	217	982	20
Phosphorus			
Cavendish	1.39	86.80	0.14
Robusta	6.95	115.00	3.02
Nendran	13.78	35.62	0.69
Monthan	3.88	46.61	0.80
Ney Poovan	5.61	53.69	0.88

Time and Placement of N and P fertilizers

The time of application of fertilizers should synchronize with the needs of crop. The uptake of N, P and K is 7-10 fold during vegetative stage than after flowering. Therefore, bulk of fertilizers should be applied during vegetative stge. Studies conducted using isotope-labelled N and P fertilizers in Robusta banana showed that absorption of fertilizer N and P was faster when applied at early and late vegetative stages than at later stages of plant growth. Fertilizer N applied at late vegetative stage gives highest yield (60%); Keshava Murthy and Kotur, 2001; Table 20.3) However, N applied at later stages, viz. at bud differentiation and shooting could maintain higher proportion of fertilizer N (18.5 – 22.5 %) in plant till bunch harvesting compared to early and later vegetative stage (6.0 – 16.3 %). Uptake and utilization of fertilizer N also increased at later stages of application and the highest N utilization was obtained from application at shooting (33.91%) and more of it is also partitioned into fruits (22.49%).

Thus, application of N in 4 equal splits, viz. 50, 100, 160- 180 days after planting and at shooting is recommended. Since N is not stored in plants even application of N at shooting is found beneficial. Further, increase in number of split applications may push up the cost of fertilizer application. In case of P, application at early and later vegetative stages results in higher phosphorus derived from fertilizer (8.38 and 12.9%) as against at shooting (2.79%, Table 20.4). It is best to incorporate the fertilizer despite practical difficulty of dense mat of roots. At early and late vegetative stages, placement in 25 cm wide circular bands around the plant between 25 and 50 cm radial distances results in higher proportion of phosphorus derived from fertilizers (8.36 – 12.9 %) compared to placement at 50 and 75 cm (8.16%). Thus, it is recommended to apply P in 2 equal doses once on 50th day (8th leaf stage) and again 160-180 days after planting (bud differentiation stage).

Table 20.3. N use parameters of Robusta banana as influenced by stage of plant growth

Stage ofplant growth	*N dff (%, 60 days after application)*	*At harvesting*			
		N dff (%)	*Total N uptake (g/plant)*	*Fertilizer N uptake (g/plant)*	*Fertilizer N utilization (%)*
Early vegetative	37.2	6.1	96.2	5.9	11.7
Late vegetative	60.0	14.7	91.7	13.5	29.9
Flower initiation	36.8	18.5	78.1	14.4	28.8
Shooting	6.8	22.5	87.3	19.4	38.9

Table 20.4. Phosphorus derived from fertilizer (Pdff, %) in Robusta banana as influenced by time and placement of superphosphate

Stage of growth (days after planting)	*Distance of placement (cm)*	*Days after fertilizer application*			
		30	45	60	75
Early vegetative (50)	25-50	2.04	3.49	6.78	8.38
Late vegetative (100)	25-50	8.19	11.09	11.90	12.90
	50-75	2.27	4.65	6.44	1.22
Shooting	25-50	1.31	2.09	-	2.76
	50-75	1.32	1.77	-	2.87

Other Modes of Nutrient Supplement

Fertigation has gained popularity due to saving of irrigation water and fertilizer nutrients. The application of 100g N/plant through drip is on a par with the yield obtained with 200-300g N/plant applied to soil (Srinivas, personal communication, Table 20.5). Heavy needs if nutrients of banana require exogenous nutrient supplements that can also be made through foliar spraying and feeding the bunch through its distal end after de-navelling. Foliar spaying is useful to supply nutrients quickly and effectively. However, it is best to use this mode to correct nutrient needs that are difficult to attain through soil application. Multi-nutrient spray is best to apply specific nutrients based on visual symptoms or as advised from soil/leaf analysis considering the economic and ecological considerations.

Table 20.5. Response of Ney Poovan to N applied through drip and to soil

Treatment (g/plant/crop)	*Fruit yield (tonne/ha)*	*Bunch weight (kg)*	*Number of fingers/bunch*	*TSS (%)*
100, drip	29.9	12.9	147	32
200, drip	30.0	12.7	137	31
300, drip	29.7	12.2	142	31
100, soil	27.2	11.7	141	33
200, soil	29.3	12.9	138	31
300, soil	28.4	11.8	148	31

De-navelling and feeding N, K and S blended in cowdung increases the size of fingers and hence the weight of bunch significantly in Robusta (Kotur and Keshava Murthy, 2008, Table

20.6) and Ney Poovan. The movement of nutrients in blend into the bunch is demonstrated using ^{15}N isotopic label. In addition, composition of fruits of Robusta in respect of K, Ca, Mg, S, Fe and Mn is also influenced. This technology is cost-effective and is becoming popular in Karnataka and elsewhere.

Table 20.6. Effect of de-navelling and feeding nutrients to the bunch on fruit yield of Robusta banana

Treatment	*Robusta*	*Ney Poovan*
Control (male bud retained)	16.724	5.250
De-nevelled	19.041	5.623
Denavelled + cow dung applied	19.904	5.980
Denavelled + enriched cow dung applied	27.993	9.362

Nutrient Diagnostics

Soil testing is largely used: (i) prior to taking up new banana orchards; (ii) to use soil test crop response (STCR) based fertility equations to achieve targeted yields and to optimize fertilizer use; and (iii) to monitor problems of soil salinity/alkalinity/acidity and of acute nutrient deficiency or toxicity that may arise from time to time.

Leaf analysis is another important diagnostic tool to monitor nutrient status of plant and take up mid-term correction to achieve optimum nutrient balance leading to high yields. Leaf nutrient norms have been standardized for 2 varieties under Bangalore conditions by Anjaneyulu and Raghupathi (2007) and Raghupathi and Anjaneyulu (2007). However, leaf nutrient guides vary according to genome, planting material (sucker/tissue culture raised), water supply (rainfed/ irrigated/ fertigated) and the incidence of fungal, nematode and viral diseases (Lahav 1995). Therefore, situation-specific nutrient standards need to be evolved to make diagnostic and advisory efforts effective.

Banana is a heavy feeder of nutrients and therefore banana growing may lead to mining of nutrients from soil in the long run. This aspect is especially crucial in case of potassium and needs to be addressed urgently. To maximize the benefits of nutrient dynamics in enhancing banana production nutrient applications should synchronize with crop needs. It is best to integrate all factors like: nutrient efficient variety, copious application of manures, appropriate time and placement of fertilizers, judicious supplementing with fertigation, foliar and bunch-feeding of nutrients. Adequate nutrient diagnostics need to be adopted to avoid adverse deficiency/excess of nutrients. Banana is produced under a variety of soil, climatic and production systems and therefore situation-specific strategy need to be evolved.

References

Anjaneyulu, K. and Raghupathi, H.B. 2007. Diagnosis recommendation integrated system norm for Ney Poovan banana. *National Conference on Banana,* Trichy.

Fox, R.L. 1989. Detecting mineral deficiencies in tropical and temperate crops. *In: Westview Tropical series 7.* Pluncknett, D.L. and Sprague, H.B. (Eds). Westview Press, Boulder, CO.

Keshava Murthy, S.V. and Kotur, S.C. 1998. Effect of ammonium sulphate and superphosphate on the comparative efficiency of N and P absorption and N utilization in Robusta' banana (*Musa x paradisiaca* using labeled fertilizers. *Indian Journal of Agricultural Sciences,* **68**:765-68.

Kotur, S.C. 2008. Enhancing the fruit yield of Robusta banana (*Musa* × *paradisiaca*) by de-navelling and feeding nitrogen, potassium and sulphur through the distal stalk-end of the bunch. *Indian Journal of Agricultural Sciences,* **78**:109-15.

Keshava Murthy, S.V. and Kotur, S.C. 2001. Phosphorus absorption kinetics in citrus (*Citrus* spp), grape (*Vitis vinifera*), mango (*Mangifera indica*) and banana (*Musa paradisiaca*). *Indian Journal of Agricultural Sciences,* **70**:186-88.

Keshava Murthy, S.V. and Kotur, S.C. 2001. Kinetics of nitrate absorption in grape (Vitis vinifera), mango (Mangifera indica) and banana (*Musa* × *paradisiaca*). *Indian Journal of Agricultural Sciences,* **71**:650-51.

Kulasekaran, M. 1993. Manana nutrition. *In: Advances in Horticulture,* vol. 2, Fruit Crops: Part 2, Chadha, K.L. and Pareek, O.P. (Eds.) Malhotra Publishing House, New Delhi.

Lahav, E. 1995. Banan nutrition. *In: Banana and Plantains*, pp. 258-316. Chapman and Hall, London, UK.

Lahav, E. and Turner, D.W. 1983. *Banana Nutrition. Bull.* 7. International Potash Institute, Worblaufen-Berne, Switzerland.

Mitra, S.L. and Dhua, R.S. 1988. Banan. *In: Mineral Nutrition of Fruit Crops.* Bose T.K., Mitra, S.K. and Sadhu, M.K. (Eds), pp. 185-220. Naya Prokash, Calcutta, India.

Raghupathi, H.B. and Anjaneyulu, K. 2007. Development of leaf nutrient norms for banana cv. Rbusta using diagnosis recommendation integrated system. *National Conference on Banana, Trichy.*

Turner, D.W. and Barkus, B. 1973. Loss of mineral nutrients from banana psuedostems after harvest. *Tropical Agriculture, Trinidad,* **50**:229-33.

21

High-Tech System for Production of Quality Banana and Plantain

B.M.C. Reddy and V. Kumar

India has retained its top position for more than a decade in both area and production of banana in the world. In the Indian fruit industry, banana ranks first in production with a total annual production of 19.2 million tonnes from 5.65 lakh ha and 32% of the total fruit production. Though India is the largest producer in the world, Costa Rica has the highest productivity of 52.5 tonnes/ha, followed by India (34 tonnes/ha). Maharashtra has maximum productivity (63 tonnes/ha), followed by Gujarat (42.7 tonnes/ha) and Tamil Nadu (42.5 tonnes/ha.). The highest productivity in Maharashtra is attributed to planting of high-yielding varieties, using of tissue culture planting material, high-density planting, drip irrigation and fertigation, proper weed management and timely plant-protection measures. By bridging the vast gap in banana productivity existing among banana-growing states, the national productivity could definitely be increased close to that of Costa Rica.

BANANA PRODUCTION SYSTEM

Banana is the crop where high-tech production is successfully practised. The high-tech production technologies include:

- Introduction of improved variety
- Micropropagation
- Crop geometry
- Drip and fertigation
- Green manuring
- Mulching
- Recycling of banana wastes (vermicompost)
- Proper hygiene through integrated disease and pest management
- Proper harvesting and handling of bunches with minimum or nil damage.

In India, banana is cultivated under four different systems. They are:

Garden Land Cultivation

In many banana-growing states almost all the commercial varieties are grown under this system.

Wetland Cultivation

This system is in practice especially in delta regions of Cauvery (Tamil Nadu), Godavary (Andhra Pradesh), River Tapti (Maharastra and Gujarat) and Ganges (East Uttar Pradesh and Bihar). Banana Poovan (Mysore-AAB), Nendran (Plantain- AAB), Rasthali (Silk-AAB), Karpuravalli (Pisang Awak-ABB) and Robusta (AAA) are commercially grown under this system of cultivation.

Perennial Banana Cultivation

- Hill Banana cultivation in lower Pulney hills and Kolli hills of Tamil Nadu.
- Cultivation of wild and edible bananas in North-Eastern States of India.

The choice of spacing depends upon the cultivar and further varies from region-to-region, depending upon the cultural practices. The planting density/spacing has significant effect on growth, yield, fruit quality, pest and disease incidence. The factors that are influenced by plant density or spacing are:

- plant canopy and light interception
- plant height and girth
- leaf number and leaf area index
- sucker production and root growth
- flowering and fruit maturity
- crop cycle interval
- bunch weight and quality
- fruit yield
- pests and disease incidence.

There has been a considerable emphasis on high-density planting, wherein yield of an individual plant cannot be increased beyond certain limit. The total yield and net returns could be increased per unit area by adopting closer spacing.

Many experiments have been conducted to standardize optimum spacing and to enhance productivity and profitability of different cultivars under different agroclimatic conditions (Table 21.1, 21.2 and 21.3).

Table 21.1. Yield performance of certain banana cultivars under different spacing/plant density

Cultivar	Spacing (m)	Population(plants/ha)	Yield (tonnes/ha)
Robusta	1.2 x 1.2	6,944	174.39
Robusta	1.5 x 1.5	4,444	145.44
Basrai	1.2 x 1.2	6,944	130.13
Basrai	1.3 x 1.3	5,917	105.90
Basrai	1.5 x 1.5	4,444	86.60
Lacatan	1.2 x 1.2	6,944	113.30
Poovan	1.0 x 1.0	10,000	99.10
Giant Governor	1.0 x 1.0	10,000	100.45
Robusta	1.8 x 3.6 (3 plants/pit)	4,630	110.35(86.41*)
Nendran	2.0x 2.0 (3 plants/pit)	7,500	58.65(26.38*)

* Conventional planting

Table 21.2. Plant population under different planting systems

Method of planting	*Spacing (m)*	*Plants/ha*
Conventional Planting: Dwarf Cavendish	1.5 x 1.5	4,444
Robusta and Nendran	1.8 x 1.8	3,086
Rasthali, Poovan, Karpuravalli, Monthan.	2.1 x 2.1	2,267
High-Density Planting: Paired row planting system		
Dwarf Cavendish	1.2 x 1.2 x 2.0	5,208
Robusta, Grand Naine, Poovan, Rasthali and Ney Poovan	1.5 x 1.5 x 2.0	3,806
3 suckers/hill (1foot apart in t pit)	1.8 x 3.6	4,630
(Robusta, Nendran)	1.8 x 3.0	5,550

Table 21.3. Effect of high-density planting on banana Robusta

Spacing(m)	*Population (plants/ha)*	*Fertilizer dose (g NPK/hill)*	*Bunch weight (kg)*	*Yield (tonnes/ha)*	*Benefit: cost ratio*	*Increase over control (%)*
1.8 x 1.8 (1plant/hill)	3,086	200:60:300	28.00	86.41	1:2.75	-
1.8 x 1.8 (3plants/hill)	3,086	300:90:450	24.38	110.35	1:3.85	27.70

However, major limiting factor is sunlight, which affects flowering, crop duration and subsequent performance of ratoon crops.

Irrigation Management

Maintaining optimum moisture at all stages of growth is very critical and the total water requirement of banana plants is about 900- 1200 mm for its entire life-cycle. In banana plantations different systems of irrigation is in practice and each system has its own merits and demerits.

Flood/furrow irrigation is most common method of irrigation under garden land cultivation. In this system, water consumption is very high and is costly, with uneven distribution of water and fertilizers; there is a chance for fast spread of nematodes.

Trench irrigation a very popular method of irrigation under wetland system of cultivation wherein water is allowed to stand in trenches opened between every two rows of plants and the same is used for draining out the excess water during rainy days.

In drip irrigation, water is allowed to reach the root zone of the crop in small quantities. It is one of the best methods of irrigation in areas of water scarcity. Drip irrigation has the advantage of large area coverage with less water, requiring very less labour but more uniform, fertigation and salinity friendly. In India, drip irrigation is superior to conventional basin irrigation in terms of more vigorous growth, higher yields and minimized weed growth and high water-use efficiency. In addition, it activates the uptake of nutrients. For banana, replenishment of evaporation losses up to 80% is optimum to get higher yield. The efficiency of conventional irrigation, sprinkler irrigation, drip irrigation are 30-40, 40-60 and 90-95% respectively.

Every rupee spent on drip system for banana results in an additional benefit of Rs 6.9 (Tables 21.4: 21.5).

Table 21.4. Drip irrigation schedule for Robusta banana

Stage	*Duration (weeks)*	*Quantity of water (litres/plant)*
After planting/ratoon	1-4	4
Juvenile phase	5-9	8-10
Critical growth	10-19	12
Flower bud differentiation	20-32	16-20
Shooting	33-37	20 and above*
Bunch development	38-50	20 and above*

*Depending on the season

Table 21.5. Comparative performance of Grand Naine banana under drip/ flow irrigation

Particulars	*Irrigation system*	
	Drip irrigation	*Flow irrigation*
Spacing (m)	1.8 m x 1.5 m	1.8 m x 1.5 m
NPK (kg/ha)	550, 111, 550	740, 148, 740
Days to harvesting	385	426
Bunch weight (kg)	23.6	15.54
Yield (tonnes/ha)	87.5	57.5
Number of hands	10.6	8.9
Number of fingers	159	138
Finger length (cm)	19.3	16.5
Finger girth (cm)	12.2	11.7
Water applied (cm)	97	176
Water saving (%)	45	-
Water-use efficiency (kg/ha/year)	902	327

Drip irrigation has major advantage of using soluble fertilizer through irrigation water. This technique is called 'Fertigation'. Fertigation gives flexibility in application of fertilizers, which enables to meet the specific crop requirements at various stages of growth. Easily and completely soluble nitrogen and potasic fertilizers feeding at the rate of 150g/plant/crop is sufficient to meet the requirement (Table 21.6).

Table 21.3. Fertigation schedule for banana (g/plant/week)

Weeks after planting	*Urea*	*Total(g/plant)*	*MOP*	*Total(g/plant)*
9–18	15	150	8.0	80
19–30	10	120	10	120
31–40	7.0	70	12	120
41–46	Nil	Nil	10	50
Total	-	340	-	375

The response to applied inputs of nutrients and water differs not only between different genotypes but also among cultivars belonging to the same genotypes. Among different genotypes, triploids respond much better to inputs as compared to tetraploids or diploids. Within triploids, Cavendish bananas are more efficient as compared to other triploid cultivars. The response of cultivars for different sources of nutrients vary significantly among cultivars depending on their genomic classification. Hence, nutrient requirement of each variety should be met properly depending on their genomic constitution and their nutrient uptake pattern (Table 21.7; 21.8).

Table 21.7. Optimum leaf nutrient concentration for banana

Element	*Low*	*Sufficient*	*High*
Nitrogen (%)	2.00 – 2.49	2.50 – 3.00	> 3.0
Phosphorus (%)	0.14 – 0.17	0.18 – 0.40	> 0.5
Potassium (%)	2.00 – 2.29	2.30 – 4.00	> 4.00
Calcium (%)	0.40 – 0.69	0.70 – 1.4	> 1.4
Magnesium (%)	0.20 – 0.24	0.25 – 0.4	> 0.4
Sulphur (%)	0.20 – 0.25	0.26 – 0.5	> 0.5
Iron (ppm)	80 – 99	100 – 300	> 300
Zinc (ppm)	10 – 12	13 – 50	> 50
Manganese (ppm)	150 – 199	200 – 2000	> 2000
Copper (ppm)	4 – 5	6 – 30	> 30

Table 21.8. Nutrient uptake/removal by Robusta banana

Element	*kg/ha*
Nitrogen	388.0
Phosphorus	52.0
Potassium	1500.0
Calcium	227.0
Magnesium	125.0
Sulphur	73.0
Manganese	12.0
Iron	6.0
Zinc	4.7
Boron	1.27
Copper	0.37

Application of N in the form of urea and K in the form of muriate of potash (MOP) through the drip system could be advantageous. The MOP may be dissolved in water and kept overnight to facilitate complete dissolving while urea can be dissolved instantly. These fertilizers could be allowed into the system and the fertilizers may be supplied into the system either on daily basis or weekly basis and it may be stopped 10-15 days prior to harvesting of bunches. Nowadays, there are several formulations of water-soluble fertilizers available in the market. A specific formulation needed for banana crop based on the crop growth stage can also be chosen for fertigation. The commercially available fertigation grade water-soluble fertilizers are highly effective in banana and even in places where the quality of irrigation water is fully not suitable for drip system.

Denavelling and Feeding Nutrients to Bunch

- In 'Robusta' banana, feeding 15g of ammonium sulphate and 7.5g of sulphate of potash blended in 500g of fresh cowdung through the distal end of rachis soon after fruit setting increase fruit yield by 68% over retention of male bud.

- Similarly, feeding 5g of ammonium sulphate and 2.5g of SOP increases bunch weight by 78% in Ney Poovan.
- The movement of nutrient into the bunch has been demonstrated using 15N isotopic label.

The bunch grade as well as fruit quality could be increased by denavelling of bunch a week after the emergence of the last hand by leaving about 10-15 cm long pedicel below the last hand. In addition to the applied nutrients, application of enriched cow dung paste along with effective surfactant to aid effective the absorption enhances the bunch grade and bunch weight besides improving fruit quality (Table 21.9).

Table 21.9. Comparison of different treatments for enhancing fruit grade and quality

Treatment	*Robusta*	*Ney Poovan*
Denavelled	19.041	5.623
Denavelled + cowdung	19.904	5.980
Denavelled + enriched cowdung	27.993	9.362
Control (male bud retained)	16.724	5.250

Besides, treated fruits also show enhanced nutrient composition of N, P, Ca, Mg, S, Fe and Mn which may have nutraceutical implications.

Removal of Male Bud and Bunch Spraying

About a week after the emergence of the last hand, remove the male buds leaving about 15cm long stalk and spray the bunches with 2% potassium sulphate (20g/litre of water) solution added with surfactant by thoroughly drenching the bunch. Repeat the spray one month later and this would help in improving the bunch grade by increasing the fruit size and quality.

Use of Bunch Covers

From bunch emergence to harvesting, fruits are subjected to attack by sucking pests and loose their consumer appeal. The fruits often fetch low price in the market causing lose of income to farmers. It has been proved that bagging the bunch with 100-gauge thick white or blue polythene sleeves with 6% ventilation is effective to avoid any insect attack. It is done after the emergence of the last hand in the bunch. The wrapping material is applied as a sleeve and tied to the peduncle leaving the bottom of the cover open so that there is no accumulation of floral remnants. Polythene covering enhances bunch weight as well as advanced fruit maturity by 7-10 days. The covered bunches produce blemishless fruits with attractive colour development and as the fruits have good eye appeal, thus fetches premium price in market.

Green Manuring

Growing of intercrops during the early stages of growth of banana plants would not only fetches additional income to farmers but also drastically reduces weed growth and by ploughing/mulching back the crop wastes after the harvesting of intercrop adds fertility to soil. One month after planting and immediately after first monsoon in June-July, *dhaincha*/cowpea is sown as a green manure crop and is ploughed before flowering. Growing and mulching back of leguminous crop adds about 15-25 tonnes of green matter/ha and 60-90 kg N/ha.

Mulching

Mulching is useful in conserving soil moisture, increasing number of feeder roots and thereby improving the nutrient-use efficiency as well as water-use efficiency. In addition, mulching also helps in suppression of weed growth and thereby enhances banana yield by 30-40%. Use of black polythene mulch significantly increases the bunch weight of Robusta banana compared to sugarcane trash and dried banana leaves and sheaths.

Integrated Insect Pest and Disease Management

Removing crop debris and diseased plants, collection and destruction of egg masses, pruning infected parts, frequent shallow ploughing, flooding, crop rotation, trap cropping, intercropping, and adjusting time of sowing / planting, avoiding monoculture and soil solarization are important cultural methods to reduce the incidence and severity of insect-pests, diseases and weeds in field. Controlling the insects to such a level that the economic loss is averted rather than aiming at destroying them completely and it could be achieved through various practices as discussed below.

Growing of Intercrops

Sunnhemp, in addition to being a green manure crop, when incorporated into the soil produces some organic acids which are antagonistic to several nematodes. Growing of sunnhemp and incorporating it on 45th day checks the nematode population. There is a two-fold increase in the population of predatory nematodes when green manure is used in soil. The plant parasitic nematode level decreases because of increased predators level as well as plant acids of certain green manures having nematode repellent effect.

Comparison between conventional and high-tech methods

Operation	*Conventional method*	*High-tech method*
Planting material	Suckers	Tissue cultured plants
Spacing	1.5m x1.5m	1.5m x1.8m
Planting density	4,444 plants/ha	3,703 plants/ha
Irrigation	Flood irrigation	Drip irrigation
Fertilizers	Hand broadcasted (2 or 3 dosages)	Applied through drip irrigation frequently
Flowers	Appear in 9-10 months; not uniform.	Start appearing uniformly in 7 months
Harvesting	13-15 months	First crop is harvested in 11-12 months. Second crop (first ratoon) is 20-21 months. Third crop (second ratoon) is 28-30 months
Bunches	Small to medium (12-20 kg), 10% plants do not produce bunches	Large bunches (25-60 kg). All the plants produce bunches
Ratooning	Not practiced	One main crop and two rations
Quality	Fingers are small to medium	Fingers are big and high quality
Period and yield	One crop in 18 months50-60 tonnes/ha	Three crops in 30 months;Yield: 80 tonnes/ha/crop 240 tonnes/ha in 30 months
Price realisation	Rs 3-5/kg	10-15% higher compared to conventional method
Total revenue	Rs 0.8-1.0 lakh/ha	Rs 5 lakh and above/ha

Maturity and Harvesting

The maturity of fruits can be assessed by:

- Change of peel colour from dark green to pale-green
- Disappearance of angularity and fullness of fingers
- Ringing sound upon tapping of the fruits
- Complete drying and dropping of floral remnants from the tip of fruits.

Tips for Hygienic Harvesting and Handling

- The matured fruits should be harvested with a sharp knife leaving sufficient length of peduncle for easy handling and transportation.
- Avoid keeping the harvested fruits directly on soil as it cause mechanical damage and result in spoilage of fruits. In stead, fruits should be kept on banana leaves and spread on the soil.
- Proper care is required while handling and transporting of bunches and this would avoid any damage of fruits and ensures better price in the markets.

Thrust Area

Banana is produced under a variety of soils, climatic conditions, production systems and therefore situation-specific strategy need to be evolved for each variety of commercial importance having export potential.

CONCLUSION

In banana, adoption of advanced production technologies has been proved to be a success story. Utmost care is required right from the selection of high-yielding varieties and healthy planting material preferably the ideal tissue cultured plants for planting. As the plant spacing and maintenance of ideal plant population play a vital role in successful banana cultivation, it is important to choose ideal planting density depending on cultivar and management level to bridge the gap between actual yield and the potential yield from a unit area. It is also essential to adopt adequate nutrient diagnostics and apply nutrients at correct proportion and in time to avoid deficiency/excess of nutrients. The productivity of high-quality bunches could be enhanced to 100 tonnes/ha, with high-tech inputs like tissue culture plants, high-density planting, fertigation with liquid fertilizers and organic manures as basal dose and integrated nutrient and pest management practices as well as careful post-harvest handling of bunches.

22

Quality Management System for Banana Production in India

M.M. Mustaffa and H.P. Singh

Quality is a term defined by consumers, buyers, graders or any other client based on a number of subjective and objective measurements of food products. The Codex Alimentarius Commission has adopted the codex committee on "Fresh Fruits and Vegetables". The codex standards are a combination of grading for quality and inspection for wholesomeness, safety and freedom from economic fraud. Individual countries establish the systems for creating standards for imported foods

Globally, banana and plantain are cultivated in 8.834 million ha with a total production of 97 million tonnes (2005) in 130 countries in tropical and subtropical regions of the world. In most of the banana and plantain-growing countries, either single or two cultivars dominate the whole economy. In banana republic countries, Cavendish group of banana is grown commercially for export to USA and UK markets. In Brazil, Pome group of fruits is the most important type grown commercially. In Western African countries, plantain is most preferred and grown for local trade and export. In East African countries, high land cooking type AAA cultivars are commonly grown along with Cavendish bananas. In the South East Asian countries including India, polyclonal system of cultivation is adopted for both local as well as export markets. In this region, next to India, Philippines, Indonesia and Malaysia are major banana-growing countries while in Australia, mainly Cavendish bananas are grown (Tables 22.1; 22.2).

Table 22.1. India's share in world area and production

Country	Area (million ha)	Production (million tonnes)
World	8.84	97.15
India	0. 622	21.76

Table 22.2. Area and production in banana-growing countries

Country	Area ('000 ha)		Production (tonnes)		Productivity (tonnes/ha)	
	2001-02	2006-07	2001-02	2006-07	2001-02	2006-07
India	466	622	14.20	21.76	30.50	34.98
Brazil	513	491	5.74	6.58	11.18	13.41
Philippines	400	415	5.06	5.63	12.65	13.56
Indonesia	285	315	3.60	4.87	8.14	15.49
China	259	269	5.40	6.25	20.84	23.20

India, inspite of being the largest producer of banana, it does not export much due to its huge domestic demand. India exports about 12.8 thousand tonnes, ranking 44th in the world export trade with a value of Rs 134 million. Ecuador is the major export of banana (28%), followed by Costa Rica (15%), Philippines (11.7%) in Asia. From India, banana is exported mainly to Nepal, UAE and Saudi Arabia.

NATIONAL SCENARIO

Banana is an important fruit crop, accounting for 31.7% of total fruit production from 12.6% arable area with a total annual production of 21.76 million tonnes from 0.622 million ha. There has been a significant increase in production of banana during the last 25 years while the last decade had witnessed a steep growth due to banana research and development. The area has increased from 0.2 million ha in early 70s to 0.622 million ha in 2007s, while banana production has taken a giant leap of 650% from 3 to 21.76 million tonnes (Fig 22.1).

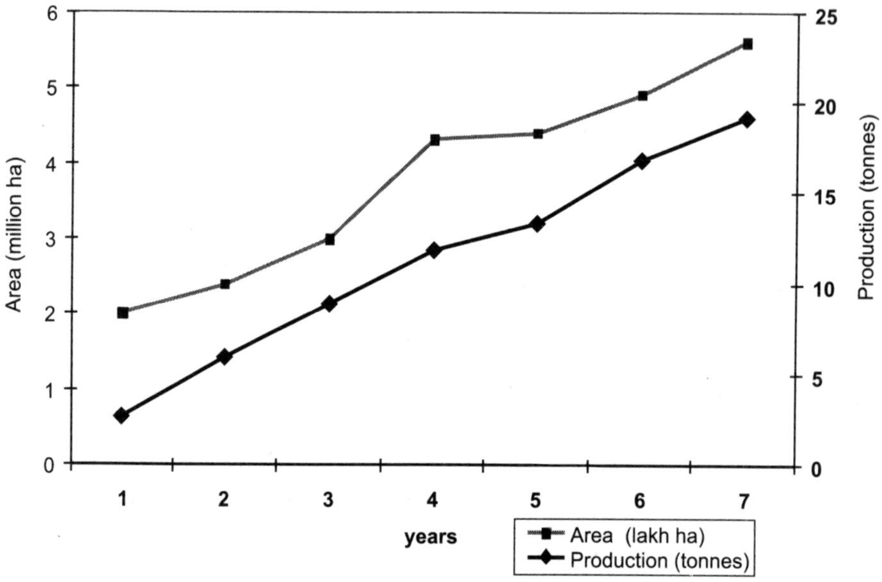

Fig 22.1. Area and production

In India, banana and plantain are widely grown in tropical, subtropical and coastal regions, comprising Tamil Nadu, Kerala, Karnataka, Andhra Pradesh, Maharashtra, Gujarat, Orissa, Bihar, eastern Uttar Pradesh, West Bengal, Assam and North-Eastern states with considerable socio-economic and cultural significance (Table 22.3).

Table 22.3. Area and production of banana in India

State	Area ('000ha)	Production ('000 tonnes)	Productivity (tonnes/ha)
Andhra Pradesh	72.4	2173.3	23.0
Assam	43.3	598.9	13.8
Bihar	29.0	1125.1	32.9
Gujarat	53.4	2912.6	42.7
Karnataka	58.13	1441.1	24.6
Kerala	59.1	463.77	7.6
Maharashtra	73.4	4621.9	62.9
Madhya Pradesh	14.9	773.0	40.0
Tamil Nadu	102.2	5019.4	49.28
West Bengal	31.7	802.1	19.2
Others	76.60	1538.5	11.5
Total	622.3	21767.8	34.98

India is home of a wide range of Musa cultivars with varying genomic status and diverse agroclimatic conditions, which has encouraged the development and sustenance of a large number of varieties catering to local needs. Though, more than 20 varieties have been grown commercially, Cavendish groups form the mainstay of the India banana Industry, owing to its high yield, wide market acceptability, short crop duration and high economic returns per unit area. Poovan is another cultivar grown commercially in different regions for its wider adaptability and tolerance to biotic and abiotic stresses. Rasthali is grown in some specific regions due to its premium price in the market. Ney Poovan is also grown commercially in many parts of the country. Virupakshi varieties are also grown to a large extent in some states. Nendran is especially grown in Kerala and Tamil Nadu due to the local preference. The varietal situation prevailing in different states/regions indicated the regional adaptation sustainability (Figs 22.2 and 22.3).

Commercial Banana Varieties

- Dwarf Cavendish /Robusta/ Grand Naine (AAA)
- Red Banana (AAA)
- Rasthali (Silk, AAB)
- Poovan (Mysore, AAB)
- Nendran (French Plantain, AAB)
- Karpuravalli (Pisang Awak, ABB)
- Ney Poovan (AB)

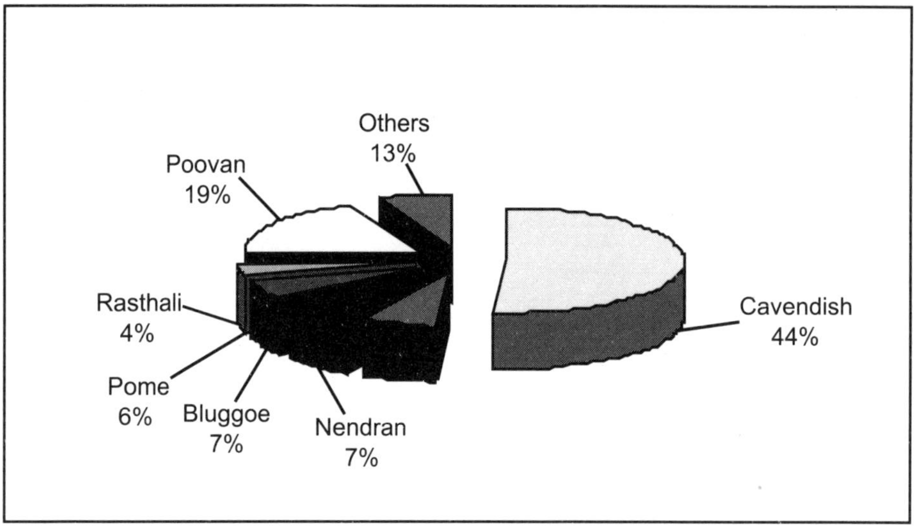

Fig. 22.2. Area under different banana cultivars (%)

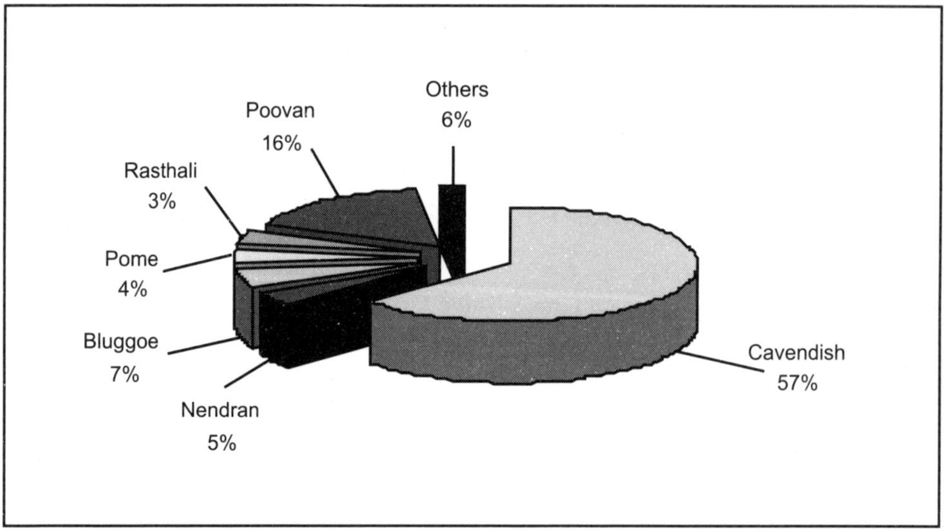

Fig. 22.3. Production under different banana cultivars (%)

Production System

Banana as a group of different genomes consisting of dessert banana, plantain, silk, mysore, pome and cooking bananas with different size, colour, flavour, taste and quality of fruits. These are grown in different production systems under different agroclimatic regions, which influence the quality parameters. Traditionally, banana and plantain are propagated by suckers. Due to continuous use of poor quality and diseased suckers, its production has severely affected bunch production. The size and quality of Nendran fruits are reduced by the use of traditional suckers as compared to tissue cultured plants. The tissue cultured plants of Cavendish produce quality fruits, meeting the export standards. The finger size and bunch weight are higher in tissue culture plants compared to sucker plants in Poovan and Nendran bananas. The plant population also plays an important role in determining the bunch weight and finger quality in most of the varieties. The optimum population of 4,500 plants/ha in Robusta and Grand Naine produce bunches which are suitable for export and also for the elite super markets where quality is very important. Similarly for Rasthali, Ney Poovan, Red Banana, Nendran and Poovan, plant population has a definite role in determining the size of fingers and bunch quality. A population of 4,500-4,800 plants/ha is optimum for production of quality bunches with desirable traits.

Banana needs 900-1200 mm water annually. Drip and fertigation in Robusta and Grand Naine are significantly influenced by producing good quality fruits suitable for export. Similarly, bunch parameters of Poovan, Rasthali, Ney Poovan and Red Banana are greatly influenced by drip and fertigation technologies which enhance the production of exportable grade banana fruits as compared to flood and other methods of fertigation.

Banana is a heavy feeder of fertilizer especially of potassium. Optimum fertilization is very essential for the production of bunches with maximum bunch weight with maximum fruit size, which is of exportable grade. Especially, Cavendish group of bananas require heavy dose of fertilizers (250:50:350 NPK) in 4-5 split applications for maximum bunch weight with exportable grade of fruits. Banana and plantain grown under low-input especially K fertilizers reduces the bunch weight, fruit size, quality enhancing the pre-mature ripening of fruits during transportation. The micronutrients especially magnesium, boron and zinc have a definite role in reducing fruit cracking and increasing finger size.

Banana plants grown under organic farming also improves the shelf-life of fruits. For tissue culture Robusta plants, higher levels of NPK (350:50:450 g/plant) in 6 split doses of N and K need to be applied to get maximum bunch weight with good quality fruits of export grade. The optimum fertilizer application is a must for production of high quality banana fingers with maximum bunch weight. Bunch weight can be increased by spraying of 2% potassium sulphate during last hand opening and 30 day after first spray there by improving finger size by 30% and bunch weight by 42% in Nendran, Rasthali, Ney Poovan and Poovan. The use of 2,4-D at 25 ppm bunch spray increases finger size and bunch weight by 20% in Rasthali and Karpuravalli.

The application of vermicomposts, composts, neem cake improves substrate dynamics of soil, thereby the beneficial soil microorganism population increases and banana plants are protected from serious soil-borne diseases like Fusarium wilt and nematodes, thereby the soil become self-sustainable. The organically produced banana fruits have better shelf-life than inorganically grown fruits. Studies to produce blemish-free banana fruits have indicated that covering of banana bunches with 100-gauge polyethylene sleeves with 4-6% ventilation produce blemish-free fingers and enhance early maturity by 15 days. The colour and size of fingers are very attractive green colour with exportable size and grade fingers. This technology is suitable for Cavendish and other indigenous type bananas to produce export grade banana fingers.

Harvesting of bananas for market depends on maturity of fingers. The fruits intended for distant destinations with Voyage time greater than 15 days, are harvested at 3/4 grade, while fruits to be distributed much closer to production areas are often harvested at 'full three-fourths grade' stage of maturity. Since maturity indices based on stage of fruits may be misleading, the time taken to reach desired maturity, estimate using degree-days have to be developed for every variety and location in which they are grown. Bunches harvested based on the degree days are best suited for long distance shipment since premature ripening is not witnessed in such treatments. The removal of female flower remnants enhances the maturity and also prevents the incidence of Cigar end rot especially in Nendran and Ney Poovan bananas, thereby improving size and quality of fingers. Removing male buds and also the last one to two hands also improve finger size, bunch weight and quality of fingers as compared to central fruits.

Effective control of leaf spot diseases plays a crucial role in bunch development and quality of fingers for export. For proper bunch development and finger size, 12-14 leaves are required at the time of flowering and 8-10 leaves at the time of harvesting. Reduction in leaf number due to severe incidence of leaf spot diseases, affect the proper finger filling due to reduction in photosynthetic activity and also enhances the pre-mature ripening of fingers in the bunch itself, deteriorating the quality of fruits and unsuitable for export purposes. The finger size, bunch weight and quality of fruits are severely affected by the infestation of soil-borne nematodes and corm weevil. The Cavendish group of bananas are highly prone to nematodes, especially root-knot and root-lesion nematodes. Effective chemical and biocontrol agents like soil application of *Trichoderma viride,* VAM, neem cake and growing of either sunnhemp or marigold reduce the population of nematodes. The corm weevil incidence can be effectively controlled by pre-treatment of suckers with Monocrotophos and trapping the weevils using pseudostem traps applied with *Beauveria bassiana.* Scarring beetle and thrips affect the quality of fruits and have poor market value. These insects can be controlled effectively by use of bunch covers impregnated with insecticides, without affecting the quality of fingers.

Fusarium wilt and bacterial head rot are serious diseases in Rasthali, Ney Poovan, Red Banana etc. Bacterial head rot is becoming a serious problem in tissue culture Cavendish group of bananas, which severely affect the bunch size and quality. Cavendish group of bananas, Nendran and Poovan bananas are tolerant to Fusarium wilt disease.

The virus diseases, viz. banana bunch top, bract mosaic, Cucumber mosaic and streak virus diseases causes considerable loss. Due to planting of virus infected plants and late infection, production of bunch is severely affected. The bunch size and quality of fingers are reduced, causing pre-mature ripening of fruits in the transit. To control the virus diseases, planting of clean, diseases-free plants from certified tissue culture companies are recommended. In India, 30 tissue culture companies are producing about 40 million tissue cultured plants annually. Mainly Robusta and Grand Naine are multiplied and to a smaller extent Ney Poovan, Nendran and Red Banana are multiplied depending on the demands by farmers.

Post-Harvest Management

Banana being a climacteric fruit is highly perishable. The fruits after harvesting have a green life of 4-5 days and a yellow life of 3-4 days depending on the stage of harvesting and variety, at the ambient conditions. Though, India's production is 21.76 million tonnes of banana and plantain per year, the extent of post-harvest losses is about 20-24%, amounting to a loss of around 3-4 million tonnes of fruits, valued at Rs 3,000 corer per annum. The post-harvest losses which include the losses, which occurs at farmers' fields (5-10%), transportation (20-25%), wholesale market (20-25%) and retail market (40-50%). The major causes of losses are improper harvesting, poor handling, and packaging during transportation and improper storage under un-controlled temperature. Post-harvest diseases like Anthracnose, crown rot and peduncle rot also affect the fruits during storage and ripening. These diseases are to be controlled using systematic fungicides, botanicals to obtain quality fruits.

In banana-processing sector, only 3-4% of total production is processed. The banana-processed products mainly consist of banana chips, which are produced to a tune of over one lakh tonne a year. Majority of chips are produced from Nendran banana, however, Robusta and Monthan bananas are also being processed into chips to some extent in Maharashtra and Tamil Nadu. Other popular banana products are puree powder, fig pickles, juice, jam, sauce health drink, banana wine etc.

Trade and Exports

After cereals, sugar, coffee and cocoa, banana is the fifth important agricultural produce globally. Out of total 97 million tonnes of banana and plantain produced, only 20% is exported, while the rest is consumed domestically. Ecuador, Costa Rica, the Philippines and Columbia alone contribute 64% of the total banana exports in the world (2004). Total value of banana exports is valued $ 4.7 billion. The Philippines is a leading Asian exporter with 74% of its production. The production is primarily exported to Japanese market. (Fig. 22.4).

Banana and plantain consumption is highest in African countries with 40-45 kg/year/individual as against 10.5 kg/year in India. Banana and plantain are the main staple food in East and West African countries. Most of the plantain grown in West Africa is exported to

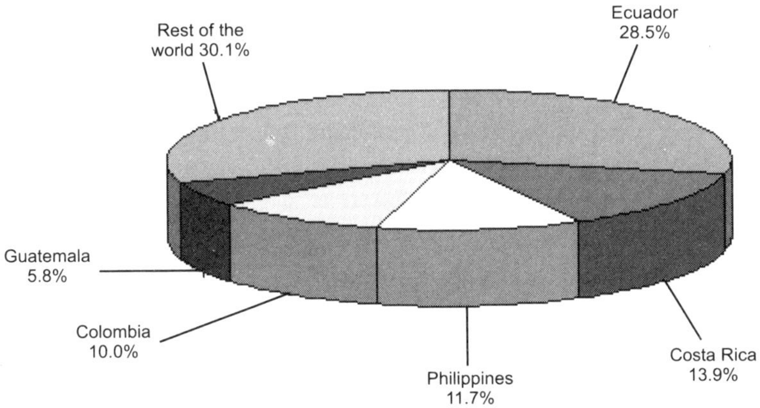

Fig. 22.4. Major banana and plantain exporting countries

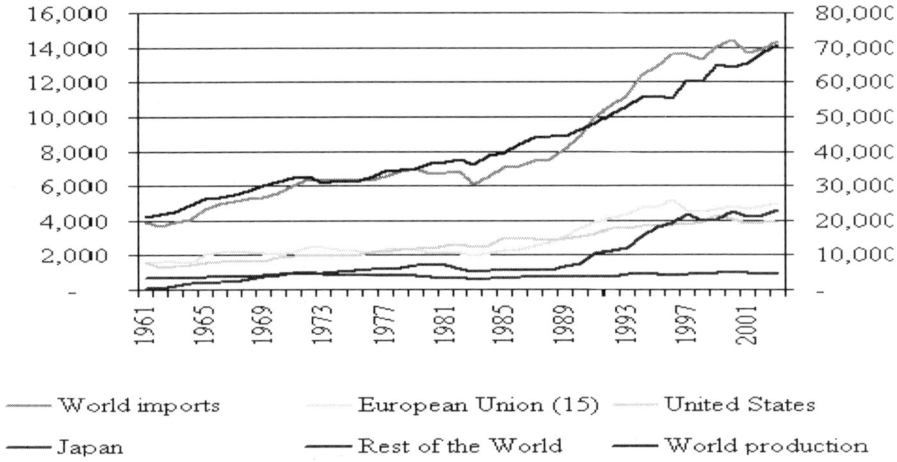

Fig. 22.5. Major banana and plantain-importing countries

France and European markets. Thus, banana and plantain cultivation forms the food, nutrition and social security for millions of people in African and Asian countries.

Earlier, banana *mandies*, farmers' cooperative societies and commission agents undertook marketing of banana. Now, big corporate houses like Reliance, ITC, Adhitya Birla Group, Godrej and Walmart have entered into the retailing of fresh fruits including banana. Due to the entrance of these companies, forward-backward linkages for the supply chain have to be developed leading to availability of good quality bananas in retail niche market.

India, being a land of small and marginal farmers, it is important to mobilize them to produce good quality banana for meeting the standards of niche domestic and export markets. Linking small farmers with high value urban niche and export markets would lead to development and growth of banana industry in rural sector.

There should be diversification in banana industry into agro-processing and value-added products of exportable quality. There is a strong need to strengthen research in this area to develop demand-driven technology by improved variety, production management etc. in both public and private sectors. This sector needs to be developed as an organized industry. There is a need to set up institutional agencies at the micro level to advance credit so as to get quality production of banana. Developments in competitive international transportation linked to domestic air transport, road or rail transport has to be improved to enhance the domestic and export trade of quality banana produced and also to reduce the post-harvest losses.

There is a need to develop farmers' cooperatives with cluster approach/contract farming for the production of quality prodbuce and also awareness should be created to understand the quality standards and requirements of importing countries. Quality control and longer shelf-life are crucial for export. There is a scope to export organic bananas to Europe and Japan markets.

23

Blemish-Free Quality Banana and Plantain Production

P. SUNDARARAJU

BANANA is an important fruit crop in India, accounting for 32% of total fruit production with a total annual production of 19.19 million tonnes from 0.565 million ha (Singh, 2007). Though India is the largest producer of banana in the world with highest productivity, the share in the international trade is not visible due to inadequate quality banana. Pests and diseases may be important constraints to fruit production and quality in all systems of banana cultivation. The quality requirements by the international fruit trade are strict and any damage inflicted to fruit even if only on skin is not acceptable. Among insect pests, fruit flies, red rust thrips, flower thrips and fruit scarring beetle are serious pests, particularly for those growers producing high quality fruits for competitive, discriminating markets. However at present, skin blemishing pests are not a major consideration in less regulated markets closer to areas of production, although it is probable that the trend towards quality improvement will be maintained where consumers become more discriminating. Besides insect pests, plant parasitic nematodes are most damaging pests of all banana varieties causing severe crop losses in commercial Cavendish bananas and seriously limiting productivity wherever non-export dessert and cooking bananas are grown (Sarah, 1989; Gowen and Queneherve, 1990). Hence, there is an urgent need to exploit the implementation of advanced tools for improving productivity and reducing losses.

PESTS OF BANANA

Banana Fruit Scarring Beetle

Banana faces great setback due to attack of scarring beetle and pseudostem weevil. Of these, scarring beetle *(Besilepta subcostatum)* has been recognized as a serious insect pest in eastern India often causing considerable damage to leaves as well as fruits during summer and rainy season. In most of the banana-growing tracts of Bihar and West Bengal scarring beetle poses serious threat (Kumar, 1989; Singh,2001). The beetle lives within the roll of the central leaf, flower bracts and feeds the epidermis of leaves, skin of newly emerged young and tender fruits, upper and lower surfaces of flower bracts, causing enumerable scars on them (Fig.23.2).

In severe attack, leaf shows burnt up appearance and fruits because full of scars, looks ugly and fetch low price in market. The clean cultivation, particularly removal of grass hosts of scarring beetle, bunch cover with polyethylene (Chaudhury *et al.*, 1996) or cloth and spray of Endosulfan (0.07%) are management practices (Sarma and Devroy, 1995).

Thrips

Thrips are very small insects that can occur in large numbers on fruits as flowering axis emerges. Rust blemishes that develop on fruit are a result of nymphs and adults feeding on skin, usually on concealed surfaces between adjacent fingers. The damage caused by most species of rust thrips can be prevented by enshrouding the fruit in polyethylene sleeves soon after flower emergence.

Red Rust Thrips

The *Chaetanaphothrips orchidii* is an American species which feeds on fruits causing a reddish stain between the fingers. The stain is oval-shaped where the fingers touch but can extend the entire length of the finger. In severe cases the peel cracks. Control is preventive, Continuous bagging with perforated polyethylene bags when the fruit is 2-3 weeks old controls the red rust thrips (Lachenaud, 1972). The *C. signipennis* thrips differs from other red rust thrips in that it feeds on pseudostems and suckers as well as on fruit. This causes a rust-coloured ark on suckers which is an indicator of thrips populations. The *C.signipennis* is controlled by bagging 3-week-old stems with Diazinon-dusted bags or by bags made from plastic impregnated with Dursban at the time of extrusion. Occasionally, it is necessary to spray the pseudostems, suckers and unbagged fruits with Diazinon or Sevin. Plants such as elephant-foot yam, plantains, Abaca and *Heliconia* sp.are to be eliminated around the edges of the plantation because they host thrips in the axils of leaves.

Corky Scab Thrips

One of the *Thrips florum* complexes causes a corky scabbing of fruit in Australia (Swaine & Corcoran 1970), India (Sivakumar and Mohanasundaram 1971) and Philippines (Ostmark 1974). It causes a grey-brown to reddish roughening of the skin sometimes with cracking. The roughening is heaviest on the outer curve of the apical hands and extends to the base of the fingers and onto the cushion. These thrips begin to feed on flower tips, fruit and adjacent bracts before the bud emerges from the plant and up to 14 days after the emerging fruit is first seen. This combined heavy feeding damage and puncturing, later turns brown, and as the bunch approaches maturity assumes the characteristics of corky scab (Swaine and Corcoran, 1970). Damage is greatest when the weather is hot and dry.

Flower Thrips

There are five *Frankliniella* species whose oviposition marks cause the raised pimples found on almost every fruit cluster from Central America and West Indies (Ostmark 1974). Flower thrips are found mainly on recently shot to two-week-old fruit. The adult lays its eggs individually in the peel of fruit less than two weeks old causing a small, raised pimple to form at each oviposition site.The only area where flower thrips caused sufficient damage to warrant attempts to control was in the Dominican Republic (Harrison 1963b). Elsewhere the blemishes are too inconspicuous, especially after ripening, to warrant attempts to control. Since the thrips are present on the fruit from the time the bud emerges, and oviposits before the fruit is bagged, there is at present no practical control.

Fruit Flies

Fruit flies are serious pests of fruit crops, but are relatively less important in banana. Tephritids and drosophilids attack ripening and ripe fruits of banana.

Management of Insect pests

Proper bunch management technologies, particularly bagging of bunch with polyethylene covers have been found to improve the yield and quality of banana (Robinson, 1996). While bagging is a common practice in West Indies, Latin America and Africa, it is rarely followed in India. Bagging bunches not only improves size and weight of fingers but increases blemish-free bunch and shortens the shooting to harvesting interval. Providing ventilation in bunch covers is essential to have proper circulation of air, maintenance of temperature and prevent premature ripening of fingers. The level of ventilation requirement varies from region-to-region depending upon temperature and humidity. Immediately after emergence of last hand, the bunches are covered with 100- gauze polyethylene sleeves. Fruit bagging and tagging are cultural practices originally adopted by export-oriented operations. Bagging is carried out to protect fruits from pest invasion and diseases. Bags may be dusted with insecticides/fungicides. Early bagging is recommended only when peel scarring insect attack of fruits is serious. Apart from protecting fruits, bagging creates a microclimate, which has been observed to accelerate growth and increased productivity without affecting the physiological operation of fruits. Tagging facilities organized harvesting and marketing of the producer. High quality bananas must be green and free of unsightly insect blemishes on arrival. None of the insects enter the pulp and most of the blemishes are confined to the outer skin. Thus, eating quality is not affected.

Advantages by Using Bunch Covers

The bunches covered with polyethylene sleeves are matured in 100 days while the control fruits take 120 days for full maturity. The bunch covers raised the average temperature around the bunches which might cause a reduction in shooting to harvesting interval. This may be due to better and advanced filling up of fingers without any blemish on the fruits by assimilates with the creation of optimum favourable microclimate condition inside the cover.

Insecticide Treatment of Bags

Insecticide-impregnated or dusted bags are required for the control of some fruit pests (red rust thrips, Colaspis, mealy-bugs). These can be purchased from the manufacturer of the polyethylene bags where Dursban insecticide is injected into the plastic when it is is extruded. These are called Dow poly-D bags. Sometimes it is cheaper for the grower to apply an insecticide dust - usually Diazinon.

The inner surface of each bag is dusted with about 2 g of 10% Diazinon dust. Diazinon 40% wettable powder is reduced, 0 -10% by mixing with four parts of talc, diatomaceous earth, kaolinite or any other inert diluent.

Hand rotary dusters may be used to apply the powder to the inner surface of the bags. Bags should be placed over the discharge tube of the duster and the handle given one rapid turn. A

Rust thrips infestation on banana fruits

Banana scarring beetle : symptoms on banana

Bunch protected with polyethylene cover

single turn of this type deposits the powder evenly throughout the bag. An electric-powered blower with two discharge tubes greatly facilitates this operation. Bag dusting should be done in a well-ventilated building or shed.

Dusted bags should be handled with care. Only the quantity that can be used in a normal day should be taken by the worker; bags should not be left in the field overnight. Dusted bags may be held for 3 weeks before being used if stored in a dry, dark location. Dow's Poly-D bags may be stored for 1 year under the same conditions.

Diazinon dust may irritate the eyes, nose and throat, therefore, workers dusting bags should be furnished with goggles and masks, and wear protective clothing. Workers should always wash thoroughly before eating. In order to prevent excessive Diazinon exposure, workers dusting bags should work one week followed by three weeks on other duties before returning to bag dusting.

Nematode Pests of Banana

Root infecting plant parasitic nematodes can devastate banana plantations causing plants to fall over because of disrupted root anchorase. Yield losses may be quite exaggerated as marketable fruit is rarely salvaged from uprooted plants. The relationship between destruction of roots by nematodes and crop losses can be influenced by the management of the plants. When deprived of the support of lateral shoots, fruit bearing stems may be more vulnerable to uprooting. As a corollary to this, the formation of heavier fruits causes extra stresses as the pseudostem and roots which may break if there are strong winds or heavy rains.

Heavy nematode burdens affect plant vigour which results is smaller bunches with shorter fingers and longer periods elapsing between the ratoon crops (Gowen, 1979; Sarah, 1989). Foliar symptom (above ground symptom) of nematodes damage may not be dissimilar from those attributed to waterlogging, weed competition, nutrient deficiencies or other pests and diseases.

Integrated Nematode Management

Single treatment is not effective to control the nematodes infesting banana. Hence, it is recommended to integrate various treatments such as (i) paring of suckers and dipping in Nimbecidine @ 1.5% (15ml/litre water) or Juerken @ 1.5% or Neewin @ 1.5% for 30 minutes or hot water treatment @ 50°C for 20 minutes prior to planting (Sundararaju and Thenmozhi,2007) followed by (ii) application of Carbofuran @ 50 g/plant or Rugby 10G @ 10g / plant or Caldan 4G @ 10g/plant third months after planting (Sundararaju, 2004) or use of oil cakes (neem cake @ 500 g/plant) and any one of the biocontrol agents (*Trichoderma viride* or *Paecilomyces lilacinus* or *Pseudomonas fluorescens* @ 20 g/plant) or botanicals (Sundararaju and Kiruthika, 2007; Sundararaju, and Sudha,2007) with (iii) marigold or sunnhemp as an intercrop (Sundararaju, 2005) have been proved to be the most effective treatments, resulting in enhanced plant growth with significant reduction in nematode populations or absolute control of nematodes by way of producing blemish-free quality banana and plantain.

References

Choudhury, H., Chandra, K. and Baruah, K. 1996. Influence of bunch over treatments on infestation of fruit scarring beetle and crop duration in Dwarf Cavendish banana. *Crop Research, Hisar* **12**:50-55.

Gowen, S.R. 1979. Some considerations of problems associated with the nematode pests of bananas. *Nematropica* **9**:79-91.

Gowen, S.R. and Queneherve , P. 1990. Nematode parasites of bananas , plantain and abaca. *In: Plant Parasitic Nematodes* in *Subtropical* and *Tropical Agriculture*, Luc,M., Sikora ,R.A.and Bridge J. (Eds), pp.431-460. CAB International, Wallingford.

Kumar, K.K. 1989. Distribution and extent of damage of scarring beetle *(Colaspis* sp.*)* to banana in Bihar . *In:Proceedings of Sixth Biennial Workshop of the All India Co-ordinated Research Project on Tropical Fruits*, held from13 to16 September, 1989 at Tirupati (India) , pp. 402-405.

Ostmark, H.E. 1974. Economic insect pests of bananas. *Ann. Rev. Entomol.* **19**:161-176.

Robinson, J.C. and Nel, D.J. 1984. Influence of polyethylene bunch covers on yield and fruit quality of winter developing banana bunches. *Hortic .Science* (South Africa) **1:** 26-28.

Sarah, J.L. 1989. Banana nematodes and their control in Africa. *Nematropica* **19:**199-216.

Sarma, M. and Devory, T.C. 1995. Efficacy of certain insecticides against banana leaf and fruit scarring beetle, *Basilepta subcostatum* (Jac). *Indian Journal of Entomology* **57**:59-62.

Singh, P.P. 2001. Pha ke pramukh khet aur Rakshan. Baghwani Margadarshika, Department of Horticulture, RAU, Bihar, Pusa (Samastipur), pp. 173-75.

Singh, H.P. 2007. National and International Scenario of Banana and Plantain. *In: Banana : Technological Advancements.* Singh, H. P. and Uma, S. (Eds), National Research Centre for Banana,Tiruchirapallipp, pp. 1-19.

Sundararaju, P. 2004. Evaluation of newer nematicides for the control of root-lesion and root-knot nematodes infesting banana. *International Congress on Musa;Harensing* Research to Improve Livelihoods, held at Penang, Malaysia, on 6-9 July 2004, pp.159.

Sundararaju, P. 2005. Effect of marigold, *Tagetus erecta* intercropped with banana cv. Nendran against root-lesion nematode, *Pratylenchus coffeae. Indian J.Nematol.* **35**:123-26.

Sundararaju, P and Kiruthika, P. 2007. Effect of compatible biocontrol agents along with neemcake and botanicals on *Meloidogyne incognita* infecting banana. National *Symposium on Nematology in 21 st Century : Emering Paradigms* on 22 and 23 Nov. 2007 at AAU, Jorhat, Assam, pp. 44-45.

Sundararaju, P. and Sudha,R. 2007. Efficacy of organic amendments on root-lesion nematode, *Pratylenchus coffeae* infecting banana in cv. Nendran. *Third National Symposium on Plant Protection in Horticulture: Emerging trend and Challenges*, Organized by Association for Advancement of Pest Management in Horticultural Ecosystems (AAPMHE) at IIHR, Bangalore on 7-9, March, 2007, p.06.

Sundararaju,P. and Thenmozhi,R. 2007. Efficacy of neem based pesticides used as sucker treatment against root-knot nematode infesting banana cv.Robusta. Ibid.p.51.

24

Biological Techniques for Nutrient Management in Banana

GEORGE V. THOMAS

ORGANIC horticulture is a holistic production management system which promotes and enhances agro-ecosystem health, including biodiversity, biological cycles and soil biological activities. The system not only safeguards the health of human beings but also protects soil, water and environment. The organic production system does not require external application of inputs but emphasizes upon management practices where agronomic, biological and mechanical methods are used to achieve sustainable production avoiding the use of synthetic materials.

The adverse environmental and soil effects of modern agriculture characterized by intensive use of fertilizers, pesticides and other off-farm inputs have been documented worldwide. There is more and more evidence that chemical-based fertilizers, herbicides and pesticides are extremely harmful to environment. Conventional agriculture with the use of large quantities of synthetic chemicals in farming resulted in rapid deterioration of soil, lowering of organic matter and humus content, poor aeration and water-holding capacity, problems of residual toxicity, large-scale outbreak of pests and diseases and pollution of soil and water. The production systems are relying heavily upon technology rather than biology and ecology. A gradual transition from modern to organic agriculture has been accepted as a viable option for overcoming the present crisis in agriculture.

CONCEPT OF LIVING SOIL

In organic farming, soil is recognized not as an inert material, but a medium supporting the growth of a large diversity of macro-and micro-flora and fauna. Soil harbors a variety of insects, earthworms, enchytraeids, nematodes, fungi, actinomycetes, bacteria and hosts other known and unknown organisms, which together support plant life through a variety of physical and biochemical activities. The activity and numbers of soil flora and fauna are increasingly considered as true indicators of soil quality or soil health. Biologically mediated processes are central to the ecological functioning of soil and soil biological activity is the driving force in the decomposition of organic matter, formation of humus, nutrient transformations, evolution and maintenance of stable soil structure, biological fixation and solubilisation of nutrients and biological control of soils-borne diseases. Only those soils with high diversity of flora and fauna can continuously support the growth of healthy crops and are termed living soils and are considered as the basis of organic farming.

NUTRITION THROUGH ORGANIC AND BIORESOURCES

In order to build up the health of soil, sound planning and appropriate interventions are essential. A system- approach, involving different crops in combinations or in rotations, livestock integration, crops residue recycling, augmenting organic matter availability through legumes, green manures and off-farm organic wastes, adoption of zero tillage, supplementing the biological fauna and fostering their growth and avoiding the use of chemical fertilizers and plant-protection chemicals are the key factors needing concerted attention.

Proper coordination between natural resources like soil, organic matter, biotic life and water are vital for plant nutrient supply management. Organic farming systems rely on the management of soil organic matter to enhance the chemical, biological and physical properties of soil. One of the basic principles of soil fertility management in organic systems is that plant nutrition depends on 'biologically-derived nutrients' instead of using readily soluble forms of nutrients; less available forms of nutrients such as those in bulky organic materials are used. This requires release of nutrients to the plant via the activity of soil microbes and soil animals. In organic horticulture, nutrients are supplied through organic sources, i.e. FYM, compost, vermicompost, poultry manure, green manures, oil cakes, etc.

Coconut-based Banana Production System

Crop diversification is the key to sustainability in organic farming. Banana has been found to be a highly profitable intercrop in coconut and a recanut gardens with good irrigation facilities. The experiments conducted at CPCRI, Kasaragod, indicated that Gros Michel and Peda Pacha are highest yielders followed by Dwarf Cavendish (Gopalasundaram and Nelliat, 1979). Studies conducted by Uma *et al.* (1989) indicated that, Palayankodan, Robusta, BB Batheesa, Karpooravally and Poovan were suitable varieties for intercropping in coconut garden. In coconut-based high-density multispecies cropping system model at different level of recommended fertilizer doses, banana (Kadali) performs well and yields on an average of 8-10 kg fruits per bunch (Reddy *et al.*, 2002). In root (wilt) affected coconut garden, among banana varieties studied as mixed crop in high-density multi-species cropping system, Palayankodan and Karpooravalli varieties produced highest per plant yield followed by Poovan, Njalipoovan, Nendran and Robusta at Kayamkulam, Kerala (Maheswarappa *et al.*, 2003). Under this system, banana crop alone could contribute on an average gross return of Rs 10,000-15,000/year in one ha of coconut garden. In farmers garden, Nendran banana could be profitably cultivated as an intercrop in coconut garden in early phase of coconut garden, and yield obtained was 10-12 kg/plant which provided net profit of Rs 6,400/- from 0.1 ha coconut garden in Pallikkara, Kasaragod (Thamban and Arulraj, 2007).

Crop Residue Availability for Recycling

Perennial horticultural crops like banana produce a lot of biomass that can be recycled to replenish the soil organic carbon. Banana produce annually 13.94 and 8.17 tonnes dry matter/ha in Robusta and Ney Poovan, respectively. Banana is an important component crop in cropping systems of perennial plantation crops like coconut and arecanut. These perennial crops produce large quantities of biomass, viz. husk, dried leaves and coir dust, which can be effectively utilized as a source of organic manure to improve the productivity of the crops.

The annual biomass produced from banana in coconut-based cropping system at Kasaragod model ranged from 716 to 1767 kg/ha. The part of banana available for recycling is pseudostem and leaves. The annual nutrient export from banana ranged from 3.6 to 16.1, 0.6 to 1.9 and 19.0 to 85.7 kg/ha of N, P and K, respectively. Out of this 2.9-13.2 kg N, 0.4-1.5 kg P and 14.9-73.7 kg K/ha could possibly be ploughed back into the system(Subramanian *et al.*, 2005).

Crop residues recycling by composting methods can form an important source of nutrients in organic farming. The quantity and quality of compost applied clearly influence the build-up of organic matter and availability of nutrients to crops. Vermicomposting is an effective technology for recycling biomass (Table 24.1).

Table 24.1. Nutrient content of recyclable biomass from banana

	Nitrogen	Phosphorous	Potassium
Nutrient content (%)	1.54	0.32	1.88
Kg/ha/year	199.00	23.00	660.00

Source: Tony Pattison, 1999

Banana pseudostem biomass is conserved and recycled by solid state fermentation (SSF) into plant growth stimulating soil conditioner (Phirke *et al.*, 2001). Recycled biomass alone or combined with biofertilizers reduced mortality (10 and 12%) of planted suckers, enhancing chlorophyll content (593 and 661 micro g/g), and gave biomass (35.0 and 36.6 kg) at par and improved yield (54.2 and 55.0 tonnes/ha), with that of the control (23%, 312 micro g/g, 36.3 kg and 49.4 tonnes/ha, respectively).

On-farm Waste Recycling Through Vermicomposting

Vermicomposting enables production of high-quality organic manure for use in organic culture from crop residues using compost worms. Earthworms like *Eisenia foetida, Perionyx excavatus* and *Eudrilus euginae* are efficient in making vermicompost. A trial conducted at CPCRI, Kasaragod, in vermicomposting of recyclable biomass of coconut + banana biomass waste mixture in a 3: 1 ratio using *Eudrilus* sp. of earthworm results in production of good quality compost with a C: N ratio of 12.27: 1. The compost is a good source of major nutrients, *viz.* N, P and K at levels of 1.51, 0.24 and 0.29 %, respectively. The application of banana biomass also facilitates multiplication of the local species of *Eudrilus* species earthworm, to ten-fold level at the end of the vermicomposting cycle. In another trial, different crop residues including banana, coconut, pineapple, sugarcane and glyricidia are composted using *Eudrilus* sp. of earthworm. The characteristics of vermicompost differs significantly (Table 24.2).

Table 24.2. Characteristics of vermicompost produced from different crop residues

Crop residue	N (%)	P (%)	K (%)	OC (%)
Glyricidia	2.14	0.79	2.69	20.53
Banana	1.23	0.72	2.91	16.64
Pineapple	1.24	0.51	2.09	16.79
Sugarcane	1.44	0.48	0.45	23.97
Coconut leaf	1.38	0.27	0.29	20.92

Vermicomposting is low-cost technology in cement tanks (Prabhu *et al.*,1998; Thomas *et al.*,2001). The weathered wastes obtained during rainy seasons are preferred. These organic wastes are to be treated with cowdung at the rate of 10% by weight in the form of slurry and must be allowed to undergo a preliminary decomposition for about 2–3 weeks. The earthworms at the rate of 1,000 worms per tonne of coconut leaves are to be introduced. The compost bed should be mulched properly using any locally available plant material or gunny bag and has to be protected from direct sunlight. Watering is to be done to maintain enough moisture. In 60-75 days, compost is ready, leaving behind only a portion of under composed material. Composting is faster in heaps and earthworms are seen feeding and casting from all sides, may be due to better aeration and exposure. On an average, 70% recovery of vermicompost is obtained. The same technology for vermicomposting is also tested in large pits taken in inter-spaces of four coconut palms in sandy loam and coastal sandy soils and is found to work well. When composting is done in field itself, a lot of labour required for transportation of biomass and compost can be saved. This technology may be tried even in plantations with very limited irrigation facilities, as only a limited number of pits or trenches need to be watered.

Total microbial counts and beneficial microbial population are also more in the compost compared to the base material. The C:N ratio of organic matter ingested by earthworm decreases and bound nutrients are converted into easily available forms. A small portion of digested organic matter is absorbed by earthworm and the leftover, which is relatively stable humus-like organic matter is egested in the form of pellets or granules, frequently referred to as vermicastings. The gut passage of organic matter results in enhanced availability of nutrients, increased counts of microbes and enrichment with a number of bioactive compounds. Two types of active nitrogen-fixing bacteria not commonly isolated from soils have also been found regularly associated with vermicasts.

Coir pith which is an organic material resistant to decomposition, can be converted into compost either through the use of local species of *Eudrilus* or through microbial degradation, and the final product can be used to improve soil physical properties and moisture-holding capacity to a great extent (Thomas *et al.*, 2001). The composted biomass is superior in quality to the content of major and micronutrients and plant beneficial microorganisms, particularly bacteria involved in nutrient transformation and growth-promotion properties.

Properties of Vermicomposts

Vermicastings have different physico- chemical and biological properties from the base materials. The vermicompost has following advantages over conventional composts.

- Vermicompost is a loosely packed, granular aggregate of enzymatically-digested organic matter containing nutrients in easily available or mineralizable form. It is also rich in earthworm mucus containing energy sources. So, when vermicompost is added to soil, it acts as top soil and enhances soil microbial activity, thus making soil alive.

- Because of its granular nature, vermicompost improves soil aeration, water-holding capacity and thus root growth.
- Earthworms and associated microbes mediate degradation of chemicals such as hydrogen sulphide, and ammonia, which cause foul odours in composting yards and for composts. The final vermicompost is quite odourless and has earthy smell.
- Vermicomposts contain more counts of plant beneficial microbes such as nitrogen-fixing bacteria, phosphate-solubilizing bacteria, fungi, actinomycetes, plant growth-promoting bacteria, and microbes capable of degrading a number of biopolymers such as cellulose and lignin.
- Vermicomposts contain a number of bioactive compounds such as auxins, gibberellins, cytokinins, vitamins and amino acids, which have the capacity to influence plant growth, development, reproduction and yield.
- Vermicomposts contain a number of humic acids which have plant growth-promoting activity. They also improve cation exchange capacity and soil physical conditions.
- Vermicompost is easy to handle, store, and incorporate into soil compared to the conventional composts. Since it is less bulky, it contains higher amount of nutrients.

Vermiwash for Growth Promotion and Quality Produce

Another important product of vermicomposting technology is vermiwash, which is rich in nutrients in readily available form, plant growth-promoting hormones, beneficial microorganisms and can be used for improving the productivity of crops. Vermiwash is effective as a foliar spray for growth promotion and bio-suppression of pathogens in crop plants. All the physiologically active water-soluble components of vermicompost such as humic acids, plant growth regulators, amino acids, vitamins, micronutrients and microbial cells are extracted in water and is known as vermiwash. The water-soluble components from vermicompost may be collected by passing water slowly through the worm-beds or by simple suspension of vermicompost in water. This is used for foliar applications as such or after sufficiently diluting, based on the need. Vegetables and ornamental plants respond very well to this treatment.

***In situ* Biomass Generation Using Leguminous Green Manures**

Green manure legumes having symbiotic association with *Rhizobium*/*Bradyrhizobium* can be an important source of nitrogen and other nutrients for crops in organic farming. They have potential to contribute 60-200 kg N in about 45-60 days. Growing and incorporation of green manures such as sunnhemp, *dhaincha*, cowpea, greengram, blackgram etc. is one of the principal supplements for the supply of plant nutrients in organic farming. In addition to supply of nitrogen, addition of green manures also supply large quantum of biomass that benefit the soil by improving the physico- chemical properties of soil.

Growing and incorporation of a leguminous crop adds 15-25 tonnes of green matter/ha and 60-90 kg nitrogen/ha depending on crop species. *Sesbania aculeata* tops the list of green manures in accumulating maximum quantum of plant biomass (22.5 tonnes/ha) and nitrogen (125 kg/ha).Cowpea is capable of producing 7.17 tonnes/ha of biomass and 63.30 kg /ha of nitrogen, followed by greengram and blackgram with 6.60 tonnes/ha, 5.12 tonnes/ha and 60.2 kg/ha, 51.0 kg/ha of biomass and nitrogen, respectively (Mustaffa and Kumar, 2008).

The technique for utilization of leguminous cover crops as green manures to supply biologically fixed nitrogen and easily decomposable biomass to coconut involves cultivation of leguminous creepers such as *Pueraria phaseoloides, Mimosa invisa* and *Calopogonium mucunoides* in its basins during monsoon period (June-October) and incorporation of legume biomass in respective basins. During 140-150 days, promising legumes generated 15-28 kg of biomass and 102-197 g of nitrogen in basin of a coconut palm (Thomas and Shantaram, 1984; 1993). There is a feasibility of substituting up to 50% of fertilizer nitrogen with the nitrogen contributed by leguminous green manures. Generation of large quantities of nitrogen rich biomass is also possible through the cultivation of fast growing leguminous tree crop, *Glyricidia*. It can also be grown in littoral sandy soils where no other green manure can establish (Subramaian *et al.,* 2000).

Biofertilizers as Cost-Effective Eco-friendly Inputs

A number of beneficial microorganisms are being increasingly used in crop production for improving the availability of nutrients, increasing nutrient-use efficiency and biological control of soil-borne diseases. The coconut growing soils harbour unusually very high diversity of beneficial microorganisms. These include *vesicular*-arbuscular mycorrhizal fungi, phosphate-solubilizing bacteria and fungi, non-symbiotic nitrogen fixers such as *Beijerinckia indica*, associative nitrogen fixers such as *A. lipoferum, A. brasilense, Azospirillum amazonense, Herbaspirillum frisingense, Bacillus* spp., *Burkholderia* spp., *Azoarcus* spp., *Arthrobacter* spp. and others which are multifunctional (Table 24.3, Thomas and Prabhu, 2003).

Table 24.3. Biodiversity of beneficial microbes in roots and rhizosphere soils of tropical fruit crops

Association	Group of organisms	Important genera
Symbiotic	Arbuscular mycorrhizae	*Glomus* *Gigaspora,* *Acaulospora* *Sclerocystis* *Scutellospora*
Associative symbiotic	Diazotrophs	*Azospirillum* *Herbaspirillum* *Azoarcus* *Burkholderia* *Arthrobacter* *Xanthobacter* *Pseudomonas*
Endophytic	Plant Growth Promoting Rhizobacteria (PGPR)	*Bacillus*
Rhizospheric association	Diazotrophs	*Beijerinckia* *Azospirillum*
	Phosphate mobilisers fungi	*Aspergillus* *Penicillium*
	Bacteria	*Bacillus* *Micrococcus* *Pseudomonas*

Bioinoculants in organic culture is important. Symbiotic (*Rhizobium*) and non-symbiotic (*Azotobacter*) bacteria in association with crop plants fix atmospheric nitrogen. The *Rhizobium*-legume association can fix up to 300 kg N/ha in a single crop season. Phosphate-solubilizing bacteria are used in organic culture for utilizing the fixed phosphorus in soil.

In a study on endophytic nitrogen-fixing bacteria in banana using 16S rRNA and rpoB partial gene sequences, plasmid profiles and plant colonization parameters, a large group of enterobacterial isolates, *Klebsiella* spp. and *Rhizobium* sp. were identified. *Klebsiella* spp. were isolated from inside the roots and stems of plants grown in two geographical regions sampled and from tissue culture-derived plantlets (Martinez *et al.*, 2003).

Microbial inoculants containing efficient organisms supply crop nutrients in soil habitat or root zone through microbial transformations or fixation of nutrient elements like nitrogen from atmosphere. The nitrogen-fixing bacteria of different categories such as asymbiotic, associative symbiotic or symbiotic forms, phosphate-mobilizing micro-organisms and those capable of producing plant growth-promoting substances can form important nutrient inputs.

Poultry manure and banana waste, used as inoculant carriers of a bacterial consortium constituted by strains of *Azospirillum*, *Azotobacter* and P-solubiliser bacteria biofertilizer for a soil cultivated with banana (*Musa paradisiaca* AAA Simmonds), had improved plant performance and soil physical and microbiological properties. Six months after planting, plant growth is increased with increase in dose of biofertilizers applied (Maria *et al.,* 2008). The biofertilizer prepared on BW enhance the density of P-solubilizer bacteria, the concentrations of available P and foliar P to a greater extent than with biofertilizer prepared on PM. The increases produced by the biofertilizer prepared on poultry manure for soil aggregate stability, enzymatic activities and the labile carbon fractions are highly correlated to dose applied. Both biofertilizers can be considered potentially useful as inoculant carriers of PGPR but the usefulness of banana waste appears to be restricted to moderate doses of application (3%).

In an inoculation trial on banana with different biofertilizers, an increase in yield of about 10 % under *Azospirillum* inoculated plants has been recorded as compared to non- inoculated control receiving 100% N (Sukhada, 1996).

A commercial mixture of microorganisms (including N-fixing bacteria, humus producing bacteria, moulds and algae) in a liquid suspension, improved pseudostem circumstance and number of fingers/hand, and advanced flowering time (Fernandez Falco *et al.*, 1998). Biofertilizers including *Azotobacter, Azospirillum*, phosphate-solubilizing bacteria (PSB) increased soil organic carbon, microbial biomass and dehydrogenase activity in rhizosphere of banana (Gogoi *et al.*, 2004).

Studies on *in-vitro*-derived banana plants grown in soil inoculated with a vesicular-arbuscular mycorrhiza (*Glomus fasciculatum*) at planting and/or with a phosphate-mobilizing bacterium (*Pseudomonas fluorescens*) revealed that plant height and dry weight are greater in inoculated plants compared to the control (Alonso Reyes *et al.*, 1995).

Animal Manures and Oil Cakes as Organic Nutrient Source

Animal manures are most common amendments in organic farming. Cattle account for about 90% of total animal dung and nutrient. The application of cowdung, rabbit dung and pig manure fortified with 25% DAP and added with biofertilizers to banana is superior both in terms of vigorous plant growth, early flowering and maturity, higher bunch weight as well as total yield and better fruit quality as compared to inorganic nutrition. Oil cakes like neem, marotti, castor etc. are used due to their fungicidal, nematicidal or insecticidal properties. Other edible cakes like groundnut, mustard etc. are also used as organic manures due to high nutrient content.

Mulching for Prolonging Moisture Retention

The beneficial effects of mulching with crop residues to reduce moisture loss and prolong the moisture retention period and to improve the physico-chemical and biological properties of soil has been well recognized. Banana cultivation produces abundant organic matter consisting of remains of leaves, pseudostems and bunch stalks. This residual organic matter (trash) can be used effectively to benefit the crop as a mulch in banana plantations. Dry matter accumulated more rapidly at the base of pseudostems when trash is placed at the base. There is 23 tonnes organic matter/ha around pseudostems. Coconut leaves, husk and coir pith could be utilised as mulches to reduce the loss of soil moisture and create conditions for proper root growth and proliferation of soil flora and fauna. Decomposition of mulches after a period of time result in enrichment of soil organic matter pool. Coconut husks are also important sources of potash which becomes available to palms over a period of time.

Soil Health Enhancement in Organic Farming

Soil health plays a key role in life support system and therefore viewed as the component of soil quality and ecosystem health that reflects the properties of soil as living system. The impact of organic treatment on physical, chemical and biological properties of soil in coconut-based farming system has been studied. Improved soil fertility status in organic treatments particularly with respect to organic carbon content and available nitrogen status is reported. The improvement in physical properties of soil in organic treatment is reflected in bulk density, water-holding capacity and hydraulic conductivity (Table 24.4;Palaniswami *et al.,* 2008)

Table 24.4. Effect of organic cultivation on soil physical properties

Treatment	*Water-holding capacity (%)*	*Bulk density (g cm-3)*	*Hydraulic conductivity (mm second-1)*
50% organic + 50% Inorganic	19	1.23	0.0064
100% organic	21	1.41	0.0044
100% inorganic	18	1.56	0.0105
Control	19	1.56	0.0037

Organic treatments in coconut-based farming system brought about improvement in soil biological properties as evidenced by the increase in microbial biomass and soil enzyme activities particularly that of dehydrogenase and phosphatase. The increased microbial activities in organic input system is indicative of higher contribution of microbes in nutrient transformation and availability of biologically derived nutrients in the system. The positive impact of organic farming practices on soil health and fertility and productivity of crops on sustainable basis has also been brought out.

The effect of organic farming practices in banana on soil quality compared to conventional farming reveals increasing cation exchange capacity, organic carbon, microbial biomass carbon and nitrogen, lowering bulk density and in turn increasing porosity, increasing population densities of soil bacteria, fungi, non-symbiotic nitrogen-fixing bacteria, and calcium phosphate-solubilizing bacteria, enhancing soil aggregate stability, and adjusting soil pH to mild alkalinity (Chang Chun Mei *et al*., 2007).

Vermiwash application increass bacterial, fungal, actinomycetes, free-living nitrogen fixing, phosphate solubilizing and fluorescent pseudomonad, and dehydrogenase, urease and phosphatase enzyme activities compared to fertilizer treated plots. This clearly indicates that vermiwash has potential to enhance soil health status for production quality crops.

Productivity Under Organic Farming

The application of 2.5 kg compost + 1 kg vermicompost + 1 kg neem cake + 2.5 kg poultry manure/plant 3, 5 and 7 months after planting results in highest growth and yield in Rasthali and Karpuravalli bananas. Organic treatments also improved quality of bananas and low incidence of sigatotka leaf spot disease (Mustaffa and Kumar, 2008).

The input application for banana Nendran including nutrient addition and pest and disease management are done through organics. Five nutrient sources, *viz*. farmyard manure, poultry manure, vermicompost, and coir pith compost along with recommended NPK (190:115:300 kg /plant) with three times of application using ash as the source of potassium for all organic treatments (Pushakumari *et al*., 2008). The results revealed that different treatments did not impart any significant effect on number of hands/bunch, number of fingers/hand, length and girth of fingers and bunch weight. The organic application, viz. FYM, poultry manure, vermicompost and coir pith result in 17.38, 17.93,17.03 and 18.65 tonnes/ha banana yield respectively. Different organic sources could be effectively used as a substitute for chemical fertilizers without any reduction in bunch yield of Nendran banana. Thus, it can be inferred that organic production of banana with organic nutrient sources is technically feasible and there is a great scope for harvesting the potential of organic farming for sustainable production of banana.

Research efforts are to be undertaken to evaluate specific advantages of organic farming in terms of qualitative differences of produce, long-term soil fertility and economic viability and higher economic returns. Intensification of studies needed to develop effective biofertilizer formulations, bioagents with higher persistence and competitive ability, development of on-farm organic manure production through approaches like vermicompost, vermiwash technology, intercropping with N_2 fixing legumes etc.

References

Alonso Reyes, R., Gonzalez Parra, M., Exposito Garcia, L., Crubelo Rodriguez, R., Roque Martinez, L. and Pazos Gonzalez, M. 1995. The influence of mycorrhizae and phosphate solubilizing bacteria on the growth and development of banana vitroplants. *Infomusa* **4**(2):9-10.

Chang ChunMei; Sah JyGau; Chiang ShihChao; Chao ChihPing and Ke DingFang. 2007. Effects of organic farming on soil and banana postharvest quality. *Journal of the Taiwan Society for Horticultural Science* **53**(4):381-39.

Fernandez Falcon, M., Enrique Alvarez, C., Borges Perez, A. and Borges Rodriguez, A. 1998. Bacteria enriched inoculant enhances banana development and influences nutrition. *Fruits Paris* **53**(2):79-87.

Gogoi, D; Kotoky, U. and Hazarika, S. 2004. Effect of biofertilizers on productivity and soil characteristics in banana. *Indian Journal of Horticulture* **61**(4):354-56.

Gopalasundaram, P and Nelliat, E. V. 1979. Intercropping in coconut. pp 6-23. *In: Multiple Cropping in Coconut and Arecanut gardens.* (Nelliat, E. V. and Bhat, K. S. (Eds.). Tech. Bull. No. 3, Central Plantation Crops Research Institute, Kasaragod.

Maheswarappa, H. P., Anithakumari, P. and Sairam, C. V. 2003. High density multi-species cropping system for root (wilt) affected coconut gardens : Its impact on productivity and economic viability. *Journal of Plantation Crops* **31**(1):23-27.

Marla del Carmen Rivera-Cruz, Antonio Trujillo Narcla, Georgina CÛrdova Ballona, Josef Kohler, Fuensanta Caravaca and Antonio Rold·n. 2008. Poultry manure and banana waste are effective biofertilizer carriers for promoting plant growth and soil sustainability in banana crops. *Soil Biology and Biochemistry.* doi:10.1016/j.soilbio.2008.09.003

Martinez, L., Caballero Mellado, J., Orozco, J. and Martinez Romero, E. 2003. Diazotrophic bacteria associated with banana (Musa spp.). *Plant and Soil* **257**(1):35-47.

Mustaffa, M. M. and Kumar, V. 2008. Feasibility and potential of organic farming in banana. In Souvenir *National Conference on Organic Farming in Horticultural Crops with Special Reference to Plantation Crops.* George V. Thomas, Anitha Karun, Maheshwarappa, H. P., Subramanian, P. and Naresh Kumar (Eds). Central Plantation Crops Research Institute, Kasaragod, Kerala 15-18 October, pp 49-57.

Palaniswami, C., George V. Thomas, P. Subramanian, R. Dhanapal and V. Selvamani. 2008. Comparative evaluation of soil health and fertility under organic, inorganic and integrated nutrient management in coconut based mixed farming system. In *National Conference on Organic Farming in Horticultural Crops with Special Reference to Plantation Crops.* Central Plantation Crops Research Institute, Kasaragod, 15-18 October, 2008. George V. Thomas, Krishnakumar, V, Palaniswami. C, Reddy, D. V. S. and Gupta, Alka. (Eds) pp 114.

Phirke, N. V., Patil, R. P., Chincholkar, S. B. and Kothari, R M. 2001. Recycling of banana pseudostem waste for economical production of quality banana. *Resources, Conservation and Recycling* **31**(4):347-353.

Prabhu, S. R., Subramanian, P., Biddappa, C.C. and Bopaiah, B.M. 1998. Prospects of improving coconut productivity through vermiculture technologies. *Indian Coconut Journal* **29**(4):79-84.

Pushpakumari, R., Sheela, K. R., Nandakumar, C. and Annamma George. 2008. Standardization of organic farming practices for banana var. Nendran. *In: National Conference on Organic Farming in Horticultural Crops with Special Reference to Plantation Crops.* George V. Thomas, Krishnakumar, V, Palaniswami. C, Reddy, D. V. S. and Gupta, Alka (Eds.). Central Plantation Crops Research Institute, Kasaragod, Kerala

Reddy, D.V.S., Subramanian, P. and Gopalasundaram, P. 2002. Coconut based high density multi-species cropping system under different levels of fertilizers in red sandy loam soils. 106-111. *In: Plantation Crops Research and Development in the New Millennium.* (PLACROSYM XIV). Rethinam, P., Khan, H.H., Reddy, V. M., Mandal, P.K., Suresh, K. (Eds.). Coconut Development Board, Kochi, India.

Subramanian, P., Srinivasa Reddy, D.V., Palaniswami, C., Gopalasundram, P. and Upadhyay, A.K. 2005. Studies on nutrient export and extent of nutrient recycling in coconut based high density multispecies cropping system. *CORD* **21**(1):20-27.

Subramanian, P., Biddappa, C.C., Maheswarappa, H.P., Dhanapal, R. and Palaniswami, C. 2000. Feasibility studies of growing Glyricidis sepium as intercrop in coconut garden under littoral sandy soil. *J. Plantn Crops* **28**(3):212-17.

Suma, A., Pushkaran, K., Babylatha, A.K. and Nybe, E.V. 1989, Comparative evaluation of banana cultivars in coconut garden. *Indian Coconut Journal* **19**(7)6-9.

Thamban, C. and Arulraj, S. 2007. Coconut based cropping system for sustainable income. *Indian Horticulture* **52**(4):22.

Thomas George , V. Prabhu, S.R., Subramanian, P. And Iyer Rphini. 2001. Organic farming Technologies in Coconut. ATIC series publication No.4, Central Plantation Crops Research Institute, Kasaragod 14 p.

Thomas George V. and Prbhu, S. R. 2003. Association of diazotrophic and plant growth promoting rhizobacteria with coconut palm (*Cocos nufifera* L.). *VIth International PGPR Workshop*, Reddy, M. S., Annadaraj, M., Eapen, S. J.,. Sharma, Y. R and Kloepper, J. W. (Eds) IISR, Calicut, pp. 20-26.

Thomas George V. and M.V.Shantaram 1984. *In situ* cultivation and incorporation of green manure legumes in coconut basins; an approach to improve soil fertility and microbial activity. *Plant and soil* **80:**373-80.

Thomas George V. and M.V.Shantaram 1993. Biomass production and nodulation of green manure legumes in coconut basins in laterite soil and their influence on soil fertility. *CORD* (Jakartha) **IX:**38-47.

Tony Pattison, 1999. Effects of residual banana organic matter on burrowing nematode (*Radopholus similis*) in established plantations (http://nematologists.org.au/Jan99/banana_jan1999.htm).

25

Good Agricultural Practices for Domestic and Export Markets of Banana

T.N. BALAMOHAN

Good Agricultural Practices (GAP) are procedures that improve conventional methods of production, beginning with the choice of cultivation areas reaching until post-harvest procedures with emphasis on the health, well-being and safety of workers and consumers. Primary aims of GAP are: 1) healthy food production with minimum chemical use, 2) safe and safety foods towards zero microbial loads and 3), preservation of environment while exploiting the resources.

In the context of rapid changing and globalizing economy, it is the duty of stakeholders of food chain to evolve a concept aiming at providing safe, quality and healthy food to consumers. Besides, alarming rate of food-borne diseases in recent past, has hastened the process in evolving a 'comprehensive system of food production' called Good Agricultural Practices. The food-borne illness outbreaks by specific agent from 1973-2005 are increasing (*see* Box).

Fruit and vegetable outbreaks by specific agent (1973-2005)

Agent	1973-87	1988-98	2004-05
Bacterial	24	57	208
Parasitic	1	8	6
Viral	4	7	170
Unknown	35	41	572
Total outbreaks	64	112	996

Source: CDC Food Borne Outbreak Surveillance System

The FAO committee on agriculture proposed GAP framework which provides an insight into the scope and wide-ranging objectives of GAPs. The framework identifies 10-15 generic components of GAPs, including soil management, water management, crop and fodder production, crop protection, animal production, animal health and welfare, harvesting and on-farm processing and storage, energy and waste management, human welfare, health and safety, and wildlife and landscape conservation.

FRAMEWORK FOR GOOD AGRICULTURAL PRACTICES

Site History and Soil

To ensure quality produce, site history and soil health are important factors to be considered before taking up any GAP practices. Site history will throw light on the past, enlighten the usage and suitability of the land for agriculture and will provide the needed primary information about the land for further exploitation.

Irrigation Water

Water is an important and renewable resource for effective agriculture. Its quality and quantity will directly reflect upon the produce produced from the farm. Efficient irrigation technologies and its management techniques will minimize the waste and it will avoid excessive leaching and salinization.

Crop and Fodder Production

Crop and fodder production involve the selection of annual and perennial crops, their cultivars and varieties, to meet local consumer and market demands according to their suitability to the site and their role within crop rotation for the management of soil fertility, pests and diseases and their response to available inputs.

Crop Protection

Maintenance of crop health is essential for successful farming for both yield and quality of produce. This requires long-term strategies to manage risks by the use of disease- and pest-resistant crops, crop and pasture rotations, disease breaks for susceptible crops, and the judicious use of agro-chemicals to control weeds, pests, and diseases following the principles of integrated pest management.

Animal Production

Good practices related to animal production include appropriate livestock units to avoid negative effects on landscape, environment, and animal welfare, avoid biological, chemical and physical contamination of pasture, feed, water and the atmosphere, frequently monitor the condition of stock and adjust stocking rates, feeding and water supply accordingly; design, construct, choose, use and maintain equipment, structures.

Animal Health and Welfare

Good practices related to animal health and welfare include those that minimize risk of infection and disease by good pasture management, safe feeding, appropriate stocking rates and good housing conditions; keep livestock, buildings and feed facilities clean and provide adequate, clean bedding where livestock is housed.

Harvests and On-Farm Processing and Storage

Product quality also depends upon implementation of acceptable protocols for harvesting, storage, and processing of farm products. Harvesting must conform to regulations relating to pre-harvest intervals for agro-chemicals. Food produce should be stored under appropriate conditions of temperature and humidity in space designed and reserved for that purpose.

Energy and Waste Management

Good practices related to energy and waste management include those that establish input-output plans for farm energy, nutrients and agro-chemicals to ensure efficient use and safe disposal.

Human Welfare, Health and Safety

Human welfare, health and safety are further components of sustainability. The social and economic welfare of farmers, farm workers and their communities depends upon it. Health and safety are also important concerns for those involved in farming operations. Due care and diligence is required at all times. Good practices related to human welfare, health and safety include those that direct all farming practices to achieve an optimum balance between economic, environmental, and social goals.

Wild-Life and Landscape

Agricultural land accommodates a diverse range of animals, birds, insects, and plants. from the countryside because their habitats have been destroyed. The challenge is to manage and enhance wildlife habitats while keeping the farm business economically viable.

Good practices related to wildlife and landscapes include those that identify and conserve wildlife habitats and landscape features, such as isolated trees, on the farm; create, as far as possible, a diverse cropping pattern on the farm; minimize the impact of operations such as tillage and agrochemical use on wildlife.

Benefits

- The GAP assures approved and appropriate methods of farming for production.
- Small, medium and large-scale farmers will achieve added value for their produce and better access to markets;
- Consumers will be assured of better quality and safer food, produced in sustainable ways;
- Business and industry will gain profit from better products and
- Altogether farmers, consumers and industry will enjoy a better environment.

Banana is grown in all tropical regions as the world's fourth most important food crop after rice, wheat and maize in terms of gross value of production. They are both a staple food and an export commodity. Export bananas represent a significant source of income and employment for several developing countries. Banana is the world's most exported fruit in terms of volume. It ranks second (after citrus) in terms of total value and first in terms of export revenues to developing countries.

Banana cultivation, like any agricultural production activity, may have a negative environmental impact if not properly managed. The banana export industry is highly integrated comparatively to other horticultural crops. Production is generally intensive, with high levels of external inputs, and often takes place in plantations organized along agro-industrial lines. Most farms rely on frequent use of agro-chemicals to maintain fertility and limit losses caused by pests. Because large monoculture crops are prone to increased attacks by pests and diseases, growing quantities of pesticides are generally needed. In turn, the extensive use of agrochemicals has given rise to the emergence of pest strains that are resistant to pesticides. In several cases, inappropriate production practices have led to excessive deforestation, pollution of rivers and ground waters, soil pollution and damage to the flora and fauna surrounding the plantations. Under such circumstances, GAP for banana is very much essential for domestic and export market.

CODEX STANDARD FOR BANANAS DEFINITION OF PRODUCE

This standard applies to commercial varieties of bananas grown from *Musa* spp. of the *Musaceae* family, in green state, to be supplied fresh to consumer, after preparation and packaging. Bananas intended for cooking only (plantains) or industrial processing are excluded. Varieties covered by this Standard are included in the Table 25.1.

Table. 25.1. List of main groups, subgroups and cultivars of bananas for dessert

Group	Subgroups	Main cultivars
AA	Sweet fig	Sweet-fig, Pisang Mas, Amas Date, Bocadillo
AB	Ney Poovan	Ney Poovan, Safed Velchi
AAA	Cavendish	Dwarf Cavendish, Giant Cavendish, Lacatan, Poyo (Robusta), Williams, Americani, Valery, Arvis
	Gros Michel	Gròs Michel, High gate
	Pink fig	Pink Fig, Green pink Fig
	Ibota	
AAB	Apple fig	Apple Fig, Silk
	Pome	Pacovan, Prata Ana
	Mysore	Mysore, Pisang Ceylan, Gorolo

Minimum requirements

In all classes, subject to the special provisions for each class and the tolerances allowed, the bananas must be, whole, firm and sound. Produce affected by rotting or deterioration make it unfit for consumption.

CLASSIFICATION

Bananas are classified in three classes as defined below:

"Extra" Class: Bananas in this class must be of superior quality. They must be characteristic of the variety and/or commercial type. The fingers must be free of defects, with the exception of very slight superficial defects, provided these do not affect the general appearance of produce, quality, keeping quality, and presentation in package.

Class I: Bananas in this class must be of good quality. They must be characteristic of the variety. The following slight defects of fingers, however, may be allowed, provided these do not affect the general appearance of the produce, the quality, the keeping quality and presentation in the package:

a. Slight defects in shape and colour
b. Slight skin defects due to rubbing and other superficial defects not exceeding $2cm^2$ of the total surface area and
c. The defects must not, in any case, affect the flesh of the fruit.

Class II: This class includes bananas which do not qualify for inclusion in higher classes, but satisfy the minimum requirements. The following defects may be allowed, provided the bananas retain their essential characteristics as regards the quality, the keeping quality and presentation.

a. Defects in shape and colour, provided the product retains the normal characteristics of bananas.
b. Skin defects due to scraping, scabs, rubbing, blemishes or other causes not exceeding $4cm^2$ of total surface area and
c. Defects must not, in any case, affect flesh of the fruit.

PROVISIONS CONCERNING SIZING

For sizing bananas of the Gros Michel and Cavendish sub-groups, the length of fingers is determined along outside curve from blossom end to base of pedicel where edible pulp ends and diameter is defined as thickness of a transverse section between the lateral faces. The reference fruit for measurement of length and grade is,

a. for hands, median finger on outer row of hand,
b. for clusters, the finger next to the cut section of the hand, on the outer row of the cluster and
c. minimum length should not be less than 14.0 cm and the minimum diameter should not less than 2.7 cm.

PROVISIONS CONCERNING TOLERANCES

Tolerances in respect of quality and size shall be allowed for produce not satisfying the requirements of the class indicated.

Quality Tolerances

"Extra" Class: Five per cent by number or weight of bananas not satisfying the requirements of class, but meeting those of Class I or, exceptionally, coming within the tolerances of that class.

Class I: Ten per cent by number or weight of bananas not satisfying the requirements of the class, but meeting those of Class II or, exceptionally, coming within the tolerances of that class.

Class II: Ten per cent by number or weight of bananas satisfying neither the requirements of the class nor the minimum requirements, with the exception of produce affected by rotting, major imperfections, or any other deterioration rendering it unfit for consumption.

Size Tolerances

For all classes, 10% by number or weight of bananas not satisfying the requirements as regards sizing, but falling within the size immediately above and/or below those indicated in provision concerning sizes.

PROVISIONS CONCERNING PRESENTATION

Uniformity: The contents of each package must be uniform and contain only bananas of the same origin, variety, and quality. The visible part of contents of package must be representative of the entire contents.

Packaging: Bananas must be packed in such a way it will protect the produce properly. The materials used inside the package must be new, clean, and of a quality such as to avoid causing any external or internal damage to the produce. The use of materials, particularly of paper or stamps bearing trade specifications is allowed, provided the printing or labelling has been done with non-toxic ink or glue. Bananas shall be packed in each container in compliance with the Recommended International Code of Practice for Packaging and Transport of Tropical Fresh Fruit and Vegetables (CAC/RCP 44-1995).

Description of Containers: The containers shall meet the quality, hygiene, ventilation and resistance characteristics to ensure suitable handling, shipping and preserving of the bananas. Packages must be free of all foreign matter and smell.

Presentation: The bananas must be presented in hands and clusters of at least four fingers per hand. Bananas may also be presented as single fingers. Clusters with no more than two missing fingers are allowed, provided the stalk is not torn but cleanly cut, without damage to the neighbouring fingers. Not more than one cluster of three fingers with the same characteristics as the other fruit in the package may be present per row.

MARKING OR LABELLING

Consumer Packages: In addition to the requirements of the Codex General Standard for the Labelling of Pre-packaged Foods (CODEX STAN 1-1985, Rev. 2-1999), the following specific provisions apply,

Nature of Produce: If the produce is not visible from the outside, each package shall be labelled as to the name of the produce and may be labelled as to the name of the variety.

Non-Retail Containers: Each package must bear the following particulars, in letters grouped on the same side, legibly and indelibly marked, and visible from the outside, or in the documents accompanying the shipment.

Identification: Name and address of Exporter, Packer and/or Dispatcher and identification code (optional).

Nature of Produce: Name of produce if the contents are not visible from the outside and name of variety and/or commercial type (optional).

Origin of Produce: Country of origin and, optionally, district where grown or national, regional or local place name.

Commercial Identification: Bananas in fingers (when appropriate), Class and net weight (optional).

CONTAMINANTS

Heavy Metals: Bananas shall comply with those maximum levels for heavy metals established by the Codex Alimentarius Commission for this commodity.

Pesticide Residues: Bananas shall comply with those maximum residue limits established by the Codex Alimentarius Commission for this commodity.

Hygiene: It is recommended that the produce covered by the provisions of this standard be prepared and handled in accordance with the appropriate sections of the Recommended International Code of Practice - General Principles of Food Hygiene (CAC/RCP 1-1969, Rev. 3-1997), and other relevant Codex texts such as Codes of Hygienic Practice and Codes of Practice. The produce should comply with any microbiological criteria established in accordance with the Principles for the Establishment and Application of Microbiological Criteria for Foods (CAC/GL 21-1997).

Agmark Grades for Banana

It is a standard developed by Indian Government which is similar to Codex standards.

Agencies to Take up Certification

The GAP programme are market-driven and in the supply chain-driven systems, the key player namely the retailer introduces a set of proprietary GAP guidelines for its suppliers by involving all the stakeholders including the consumers who are at the other end of the supply chain. However, to make this as an operational one, the retailer along with rest of the stakeholders, engaged agencies, to make sure that the standards are being followed by the suppliers/producers. The following are retailer-led enforcing agencies for GAP across the globe.

- Eurep GAP or GLOBAL GAP
- Nature's Choice
- Farm to Fork.

EurepGAP/GlobalGAP is a private certification system driven by 22 large-scale retail chains and large fresh produce suppliers/producers in Europe that form the core members of the Euro-Retailer Produce Association (EUREP). There are also associate members (mainly suppliers of agrochemicals, certification bodies and consultancy firms) who may participate in meetings but are not part of the Eurep Gap decision-making process. Initially the EHI-Euro Handels Institute acted as international secretariat. Under Eurep Gap, there are about fifteen selected agricultural components to be adhered to which are listed below:

- Traceability
- Record keeping
- Varieties and rootstocks
- Site history and management
- Soil and substrate management
- Fertilizer use
- Irrigation and fertigation
- Crop protection
- Harvesting
- Produce handling
- Waste and pollution
- Worker health and safety
- Environmental issues
- Complaint form
- National legislation

In India, there are a few approved certification bodies namely, Eurocert India, Foodcert India (P) Ltd, IMO Control Private Ltd, Control Union Inspections India Pvt. Ltd, Skal International India and SGS for giving GAP certification.

26

Quality Production of Banana for Domestic and Export Markets

Jai Oberoi

Highly portable, easily peeled, deliciously sweet and packed with nutrients, banana is the king of all tropical fruits. Banana is world's fourth most valuable food crop, behind rice, wheat and potato. Some ten million people in 25 tropical countries depend on banana for their living. Always available regardless of season, bananas are a staple of many households around the world and are now emerging as a convenience food for the on-the-go generation who also want health and nutrition. India leads the world in banana production, accounting for almost one-quarter of the worldwide crop. However, most Indian bananas are consumed domestically with hardly any participation from India in the world trade.

Over 80% of the world banana trade is controlled by five companies: Del Monte, Chiquita, Dole Foods, Fyffes Bananas, and Noboa/Bonita. Most bananas imported to USA and Europe originate on plantations in Latin America, Central America and the Caribbean. Like most tropical fruits, banana is highly perishable, necessitating efficient harvesting, packing and transportation systems. Bananas for export are shipped green in refrigerated vessels that prevent them from ripening before arrival. Once they arrive at their destination bananas are ripen in special facilities. The major banana companies have heavily invested in, and exercise control over every link in the supply chain– from production, to logistics, ripening, marketing and branding. Western consumers have become used to branded but cheap, standardized (shape, colour, size) without any blemishes, uniformly ripened and coloured fruits. But these bananas sadly and utterly lack taste!

MAINTENANCE OF QUALITY OF BANANA

Maintaining high yields of cheap and blemish-free fruit requires frequent and intense applications of agro-chemicals. Fertilizers must be applied to meet the high nutrient requirement of crop and to compensate for nutrient loss to weathering. Herbicides are used to keep the ground-free of vegetation. Nematicides must be applied directly to soil around the base of trees to protect the roots from nematode damage. Aerial applications of fungicides are done up to 50 times a year to battle the destructive Black and Yellow Singatoka fungi.

Chemicals

In packing facility, workers apply fungicides and disinfectants such as formaldehyde to fruits to protect it during shipment. In most banana-growing regions of Central and South America it is virtually impossible to grow bananas without heavy and extensive chemical exposure. It is estimated that approximately 30 kg/ha/year of pesticides are applied to Central American banana plantations, 10 times the amount used for agriculture in industrialized countries. Many of the chemicals used on banana plantations have been heavily restricted or banned for use in the host countries from where most pesticides are manufactured and exported from. Excessive chemical use has contaminated the soil on plantation land. In many regions, such as the South Pacific lowlands of Costa Rica, the land has been so contaminated that future agricultural use is impossible. After heavy rains, drainage canals from banana plantations inundate nearby streams and water supplies with contaminated water and sediment, adversely impacting aquatic life. Plastic bags coated with Chlorpyrifos, a neurotoxic pesticide, are used to cover the maturing fruits and protect it from insect damage.

Safety Norms

The most common method of disposal is in open air dumps from where these bags often blow out and litter the surrounding landscape. This is especially troublesome when they end up in streams or the ocean, where they cause problems for aquatic life. Workers on banana plantations come in daily contact with pesticides and face the greatest risk from chemical contamination. A majority of the accidental poisonings that occur on plantations are partially due to lack of safety instruction and equipment. Independent growers are persuaded by chemical companies and transnationals to use pesticides to increase productivity, with little instruction on the dangers of these chemicals. High temperatures and humidity prevent many workers from wearing the necessary safety equipment such as respirators, rubber gloves and boots, and heavy coveralls. Many workers cannot read the handling instructions because they are either illiterate or unable to read the English labels. Overall, what the multinationals have achieved over the past century in their desire for meeting consumer expectations is very high production levels but at an equally high environmental, human and social cost.

This is now becoming unacceptable for a large, and vocal community of NGO activists, governments, environmentalists, scientific bodies and affluent consumers who feel a sense of guilt in consuming products cultivated or manufactured through exploitative systems that harm the environment and all those connected with the production processes. This is resulting in a growing movement towards protection of environment and safety of people who produce and manufacture articles of daily consumption for the rich markets of Europe and North America. Consumers are demanding, and are willing to pay for, environmentally safe products, manufactured in safe working conditions by labour who are paid a fair wage and who live in decent habitations.

Scope of Indian Bananas

This is where the opportunity for Indian banana lies. Not in replicating the systems and processes of the multinationals. Not in setting up huge mono crop plantations with sophisticated logistics and handling systems at great cost. But in developing home grown solutions that are rooted in the Indian situation and can take advantage of its many strengths and turn these into competitive advantage. India's strengths lie in its biodiversity and the ready availability of a number of varieties of banana as against just one cloned variety, in the small family owned plots instead of large extensive plantations, in the emphasis on varying tastes, colours, shapes and sizes of bananas, in the absence of disease epidemics, in the entrepreneurial ability of individual farmers and not large insensitive corporate houses. The Indian response to the multinationals must comprise all above elements.

It would be right also at this stage to take note of the various weaknesses in the existing banana systems in India. Most of the weaknesses are concentrated in the post-harvest handling, transporting and ripening stages of the entire value chain. Thus, even good size and shape green bananas at field are full of the bruises, mechanical damages, cuts etc. when they reach the *mandi*. Thereafter, they are ripen under the most unsatisfactory and un hygienic conditions and marketed in open carts, shelters where they undergo further deterioration before they reach the end consumer. There is in-built a tremendous amount of wastage both at the green stage and at the ripe stage, ultimately paid for by the farmer and the consumer. These are the aspects that need to be examined and viable, affordable, sustained solutions developed in keeping with conditions of infrastructure obtaining in our country.

This massive improvement in handling, transporting and ripening systems are needed whether ultimate destination for bananas is domestic market or export markets. It is better to experiment with various solutions for domestic market and once these are perfected to use them for accessing export markets. Thus, our strategy for improving banana value chain should be to bring about significant improvements and reduction of wastages in fruit for domestic markets and consumers and only then venture to tap the export potential. This two-stage process is efficient in use of scarce resources, be less capital intensive and will also be inclusive of all small holders.

Indian Banana Model

An Indian banana model is based on aggregating in virtual clusters hundreds of small farmers, encouraging them to adopt safe, efficient, minimum chemical usage cultivation practices. The fruits are aggregated at local, community owned small-sized boxing stations and thereafter transported to ripening chambers in especially manufactured packing arrangements. The packaging is environment-friendly and recyclable. Appropriate technology and innovation to solve problems in most cost-effective manner will be the key. These models can be based on mutually beneficial partnerships of small holders with business houses who can be responsible for the investment as well as the branding and marketing of fruits. The farmers are not alienated from their holdings but are offered incentives to practice safe, efficient agriculture.

SUMMARY

We need to develop indigenous solutions for the handling, transporting and ripening of bananas and upgrade domestic market before we venture out into the more sophisticated markets of Europe and North America. There is a tremendous opportunity waiting to be tapped as the existing multinational model is under severe criticism for their labour practices and with the rise of disease and pest issues in their original locations also becoming economically unviable.

27

Mechanization for Quality Banana Production

PITAM CHANDRA

BANANA and plantain are cultivated in 130 countries. Global production of sweet banana is about 72 million tonnes annually and India's contribution being 23%. In India, banana production constitutes about one-third of the total fruit production. Around 98% of the world production comes from developing countries and 85% of the total world's production is grown on small farms and consumed locally. The remaining 15% is produced on commercial plantations, mainly in Latin American and the Caribbean countries. In India, about five lakh small and medium farmers are engaged in banana cultivation over an area of about 4.5 lakh ha.

World exports of bananas also show a high level of concentration, with developing countries, accounting for the bulk of exports. Only Latin American and the Caribbean countries supplied about 70% of total exports in 2004. Bananas are imported mainly by the European Union, the USA and Japan, which together accounted for about 67% of total world imports in 2004. Although India's share in global sweet banana production is about 25%, the export is negligible, a meager 0.04%. Some of the best banana varieties are cultivated in India; however, our production and post-harvest practices have constrained the exports. Production costs in India are almost three times as much as those incurred by countries that export banana. The challenges are to reduce costs and improve quality.

A few banana-growing countries have realized the need for improved production technology for enhancing banana productivity and profitability. Australia has indicated that mechanization of banana production is required on account of rising labour costs, work place health and safety, and environmental considerations.

DEVELOPMENT OF TOOLS AND MACHINES FOR BANANA CULTIVATION

The mechanization of banana cultivation has not been attempted consciously. Efforts have been made in terms of digging of holes, leading to the development of a tractor-mounted post-hole digger. Similarly, equipment for land preparation exists in relation to other crops and the same equipment could be used in banana cultivation also. Drip and fertigation systems have been developed for many horticultural crops including banana. Banana nurseries have been successfully developed including tissue culture facilities and hardening facilities on the lines of other crops. Only a few specific efforts have been reported as far as banana crop is concerned. These efforts include banana clump remover, banana stem shredder, and banana hand cutter. Some of these efforts are being briefly explained below.

Mini Power Tiller for Weeding

Weed control is most expensive operation in India. Majority of Indian farmers uses hand-hoe for weeding which requires 40-60 manual labourers for one hectare. With a mini power tiller, in a day about one hectare of land can be weeded by an operator and two labourers for weeding in between plants. The engine-operated mini power-tiller has been developed for weeding and interculture in between rows of coconut, banana, maize, tapioca, cotton, pulses, sugarcane, sorghum, mango and grape.

A depth adjustment wheel is provided at the back of weeding tools to regulate the depth of operation. Distance between the ground wheels is adjustable depending on row-row spacing of crop. Since the machine is light in weight (around 100 kg), it can be easily turned at fields ends. It costs around Rs 30,000. The cost of weeding with this machine comes to Rs700-900/ha as compared to Rs 2,000-2,500 /ha for manual weeding.

Tractor-Mounted Banana Stem Shredder

Banana is mostly check row planted with the spacing varying from 125cm × 125cm to 150cm × 150cm, depending upon the variety. Plant population is about 4,500/ha. After harvesting of banana bunch, stem is manually cut and left in rows. After harvesting of whole field, these are collected and left near the boundary for drying and subsequent burning. This process is tedious and time consuming. The banana stem shredder helps in disposing of stem immediately after harvesting. Shredded material is suitable for mulching in banana garden and also for vermincompost.

The shredding unit consists of 4 blades placed perpendicular to each other at 225 cm distance. Additionally, 12 spikes with flat cutting edge are fitted with a gap of 120 mm between the rows. The stem is cut into small pieces and water and fibre are separated. It takes 1.2

Tractor-mounted banana stem shredder

minutes to shred the stem having average height of 2,400 mm. About 52 stems are required for shredding in one hour. Its cost of operation is Rs 300/hour.

Tractor-Mounted Banana Clump Remover

A tractor-operated banana clump remover has been developed to mechanize the clump removing operation. Two 100mm × 15mm × 1000mm sub-soiler shanks with shares of size 190mm × 40mm × 5mm are fitted in the nine-tine cultivator frame at 225 mm spacing. These two sub-soilers perform as a fork while removing the banana clump. A deflector has been provided to push the soil sideways. The equipment is attached to 3-point linkage of a 26 kW tractor.

For removal of banana clump, sub-soiler shank is positioned behind it, and pressed into soil with the hydraulic system. The tractor is gently moved forward, simultaneously lifting the sub-soiler. This combined action removes the entire clump along with its root portion as a whole mass. The field capacity of machine is 0.5 ha/hour. The cost of manual digging of clumps comes to Rs 2,000-2,500/ha. The cost of operation is Rs 700/ha. Thus, there is a saving of Rs 1,800/ha.

Tractor-mounted banana clump remover

Banana-Comb Cutter

Its serrated cutting blade is attached to handle with a 150-mm round pipe of diameter 12.5 mm. The handle is about 100-mm long which provides easy grip during cutting. The semi-circular blade is made of stainless steel sheet having 75-mm length and 25-mm width to suit banana-bunch stem geometry. The hand-tool facilitates cutting as well as separation of banana-comb in a single push with minimum damage.

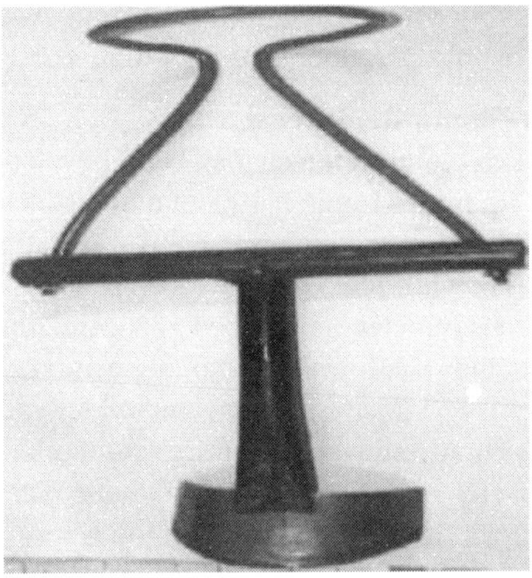

Banana comb cutter

Mist Sprayer

It uses a tractor equipped with either air-blast style mist sprayer or the 4-wheel drive mist sprayer to control foliar diseases. A downward comb shaped herbicide sprayer attached to a tractor may also be used to control weeds.

Bagging Machine

One bagging machine is reported to handle about 20 bunches/worker/hour including removal of bell and flower, thinning of fruit hands, and bagging.

Bunch Transportation Tractor

Sixty bunches arranged vertically on a bunch transportation tractor could be sent from field to the packing station in one shuttle.

Greenhouse

For banana production, the basic greenhouse is a structure 5-6 m high covering from 0.8-1.25 ha. Most houses are 100 m wide and 125 m long permitting a single sheet to stretch from one end to the other. There are many design variations but only two basic types — a wooden pole block-type house or a galvanized steel pipe frame house. The most dominant style is a flat-roofed structure with curved ends. Plastic covers are usually 200-micron thick and last for 2 years. Ventilation in these structures is achieved by separating the strips or by lifting the sides or, in some cases, by dropping the sidewalls.

Earthing-Up

In case of furrow planting, earthing-up should be done during rainy season to avoid water-logging while during winter and summer the plant should be in furrow.

Banana Plant Support

Banana plantations are quite susceptible to wind damage. A wind velocity of 25-30 kg/hour is sufficient to cause significant damage to banana plants; wind speeds of 65km/hour causes considerable loss and higher speeds to a complete crop failure. This problem is mitigated in some subtropical cultivations by growing banana under greenhouses. Normally, mechanical support for plants to avoid toppling due to strong wind or bunch weight is provided. Pseudostems are propped up with bamboo, especially, at the time of bunch emergence.

Leaf Removal

Pruning of surplus leaves helps reduce disease from spreading through old leaves. Leaf pruning can change light and temperature factors of microclimate. Pruning of leaves before bunch initiation delays flowering and harvesting cycle. For maximum yields a minimum of 12 leaves are to be retained.

Covering Bunch with Blue Plastics

Covering the whole banana fruit with a big blue sack when it is about 2 months old is good practice. The plastic bag is 45-micron thick with a perimeter of 1.5-1.6 m and length of 1.15m. Nylon rice sacks work well too. It makes the fruits get bigger and also gives good protection from the birds. Bats and squirrels which like the banana fruit very much. The bag helps to hold available heat (a miniature greenhouse of sorts) as well as trap the ethylene oxide released by the fruit, which accelerates ripening. The plastics bag/sack must have holes at the bottom for drainage of water. The sack is tied with soft twine several inches from the first hand.

Removal of Male Flower Buds

Removal of male buds after completion of female phase is necessary. Once the process of fruit setting is over, inflorescence rachis should be cut beyond the last hand otherwise it grows at the cost of fruit development. This helps in early maturity of the bunch .

Harvesting the Bunch

After 70-80 days of bagging bunch is necessary or if the last short leaf is starting to die (very small last leaf is not considered as this will dry as soon it is opened). Another sign is when the plants has 6 or less of the function/alive leaves. A notch is cut halfway into the tree opposite the side of the bunch. This causes the tree to bend and the bunch is cut-off.

Transportation

Banana bunches are very heavy and need to be transported to storage unit such that the damage to fruits is minimum. There is a global trend in adopting precision agricultural techniques. Precision agriculture includes identifying variables of importance (such as yield, soil type or moisture), mapping data, determining the zones and then implementing a management strategy. These can be linked with other layers such as aerial photos and the variability analysed using GIS and statistical methods. In a nutshell, it is identifying areas of high yield, and limiting inputs into these areas, and boosting areas of lower yield with better management techniques.

Thus, there is a need to consider developing a whole package of tools and machines to facilitate timely farm operations with efficient input resource utilization and drudgery reduction for more profitable banana cultivation. Depending upon the size of land holding and financial availability, banana growers may either own the equipment or may access them on custom hiring. Therefore, mechanization service centres may also be envisaged to facilitate banana growers in this endeavour.

28

Coconut-Based Banana Production System

George V. Thomas

Coconut (*Cocos nucifera* L.) is cultivated in the country predominantly by small holders in tropical regions, occupying 1.93 million ha with a total production of 12,832 million nuts and productivity of 6,632 nuts/ha. Kerala, Tamil Nadu, Karnataka and Andhra Pradesh are major coconut-producing states, accounting for more than 90% of area and production. Monocropping of coconut is unsustainable as it utilizes only 45-50% of solar radiation and 21% of land area, and income derived from such a system is not sufficient to sustain even the small families. Coconut-based cropping/farming systems, involving cultivation of compatible crops in interspaces of coconut and integration with other enterprises like dairying offer considerable scope for increasing production and productivity per unit area, time and inputs by more efficient utilization of resources like sunlight, soil, water and labour.

NATURAL RESOURCE USE IN COCONUT GARDEN

A spacing of 7.5 m × 7.5 m in square system is recommended for coconut (175 palms/ha) but in homestead gardens of Kerala, the density is much higher (200-250 palms/ha), whereas in malnad region of Karnataka much wider spacing is adopted with 120 palms/ha. Experimental evidences have shown that a solo crop of coconut, at the recommended spacing of 7.5m × 7.5 m does not fully utilize the available resources such as land space, aerial space, water and nutrients.

Coconut palm like all monocots has a typical adventitious root system and about 74% of the roots produced by a palm under good management did not go beyond 2 m lateral distance and 82% of the roots are confined to 31-120 cm depth of soil (Kushwah *et al.*, 1973; Maheswarappa *et al.*, 2000).

The venetian structure of coconut crown and orientation of leaves allow part of the incident solar radiation to pass through canopy and fall on the ground. The leaves in a coconut palm crown are not randomly distributed, but clumped around few widely spaced growing points. This non-random distribution also leads to low extention coefficient of around 0.65 for PAR. Age, spacing, soil fertility, varietal characteristics, leaf area and time of day influence light penetration through canopy. The nature and amount of sunlight ransmitted through coconut canopy and falling on ground shows temporal as well as spatial variations. Nair and Balakrishnan (1976) estimated that as much as 56% of sunlight is transmitted through canopy during peak hours (10-16 hours) in palms aged around 25 years. The angle of sunrays (and thus the time of the day) influences the amount of light passing through coconut canopy. With the movement of sun and movement of coconut fronds in wind, the light and shade patterns under the palms are constantly changing. The distribution of light at different positions in canopy zone of coconut varies very much because of non-random distribution of leaves. This causes differences in growth and yield of intercrops at different positions of plantation floor.

Based on the growth habit of palm and amount of light transmitted through its canopy, banana can be grown as a mixed crop palm in following phases:

- Planting till full development of canopy (about 8 years): good transmission initially, but decreasing with age suitable for growing annuals/biennials.
- Mature trees (more than 25 years): increase in trunk height; reduction in crown size - light transmission increasing with age - ideal for raising annual and /or perennial crops.

Therefore, performance of banana as a mixed crop in coconut garden is described.

Banana prefers tropical humid climate. For planting banana, select 3-4 months old disease-free sword suckers from healthy clumps. The rhizomes are to be smeared with cowdung solution and ash, and dried in sun for about 3-4 days and stored in shade up to 15 days before planting. Planting should be done in a pit of 50 cm × 50 cm × 50 cm at a spacing of 2.5 m × 2.5 m. Apply compost/farmyard manure @ 10 kg /plant in pits and plant suckers upright in centre of pits with 5 cm of pseudostem remaining above the ground level. The recommended dose of N:P:K for banana is 100:200:400 g/plant/annum to be applied in equal 2 splits during May-June and September-October.

BANANA AS MIXED CROP IN COCONUT GARDEN

Yield and Returns

Banana is a highly profitable intercrop in areas with good irrigation facilities. Gros Michel and Peda pacha are high-yielding followed by Dwarf Cavendish (Gopalasundaram and Nelliat, 1979). Suma *et al.* (1989) reported that, Palayamkodan, Robusta, BB Batheesa, Karpooravally and Poovan are identified as suitable varieties for intercropping in coconut garden. Under high-density multispecies cropping system model at different levels of recommended fertilizer doses, banana (Kadali) performs well and yielded on an average of 8-10 kg bunch (Reddy *et al.*, 2002).

Table 28.1. Banana yield in different coconut-based cropping/faming systems

Cropping/farming system	*Banana yield (kg/bunch)*
HDMSCS	8.6
Mixed faming	7.5
Coastal sandy soil	24.6
Mixed cropping	7.4

In farmers' gardens, Nendran banana could be profitably cultivated as an intercrop in coconut garden in early phase of coconut garden, yielding 10-12 kg/plant and a net profit of Rs, 6,400 from 0.1 ha coconut garden in Pallikkara, Kasaragod (Thamban and Arulraj, 2007).

Adoption of coconut-based cropping/farming systems could pave way for providing additional employment over coconut solo crop. The same ranges from 125 mandays/year/ha in coconut + vegetables to 600 mandays/year/ha in coconut-based mixed farming. The additional employment over coconut sole crop in coconut + banana system was 230 mandays/year/ha (Table 28.2).

Table 28.2. Additional employment generation in coconut-based farming systems

System	*Additional employment generated (mandays/year/ha)*
Coconut monocrop	
Coconut + elephant-foot yam	131
Coconut + tapioca	130
Coconut + vegetables	125
Coconut + clove	150
Coconut + banana	230
Coconut-based HDMSCS	191
Coconut-based mixed farming	600

Under littoral sandy soil in adult coconut garden, Grand Naine performs well, yielding 24.6 kg bunch under husk burial and coir pith application (CPCRI, 2008). Under root (wilt) affected coconut garden, Palayankodan and Karpooravalli produce highest plant yield followed by Poovan, Njalipoovan, Nendran and Robusta at Kayamkulam, Kerala (Maheswarappa *et al.*, 2003). Under this system, banana crop alone could contribute on an average gross return of Rs 10,000-15,000/year/ha of coconut garden.

Rao *et al.* (2004) studied effect of two cropping system models involving cocoa, banana, elephant-foot yam, pineapple, colocasia and turmeric on productivity. All above crops are productive and remunerative in coastal ecosystem of Andhra Pradesh. The nut yield increases by 18-27% under cropping system models over 4 years (compared to pre-experimental yield), whereas increase in yield over the same period is 3.6% under coconut monocrop. The income and employment generation being Rs.39,496 -47,589/ha and 569-680 mandays compared to coconut monocrop (Rs 9,4770/ha and 166 mandays).

Effect on Soil Nutrient Status

Soil total N, organic carbon content, available P, K, Ca, Mg and micronutrients of surface soil give higher values than subsurface soil of coconut and component crops. The soil nutrient status under various fertilizer treatments is given in Table 28.3 (Palaniswami *et al.,* 2007). The organic matter concent declines with increased fertilizer addition. In coconut, highest organic-matter content (0.712%) is found in no fertilizer treatment, which declins to 0.512 % in treatment, full dose of fertilizers. The increasing fertilizer levels might have led to higher active root biomass production. The exudates so secreted by roots might have led to proliferation of microbes, which would have decomposed the organic matter leading to reduction in organic-matter content at higher fertilizer levels. The total N content in soil increases appreciably with increasing fertilizer levels. At lower fertilizer levels, organic-matter content in soil is high, the total N content is lower suggesting high C:N ratio of organic matter. Similar trends are observed in all component crops except in pineapple where organic matter content is higher at higher fertilizer levels. The available P and K status of soil increased with increasing fertilizer levels in all the main as well as component crops. Thus, P and K when applied to soil have a tendency to get fixed, which becomes slowly available to crop later on.

Table 28.3. Soil nutrient status of coconut and banana rhizosphere under HDMSCS

Treatment	*Organic matter (%)*		*Total N (ppm)*		*Available P (ppm)*		*Available K (ppm)*		*Ca (ppm)*	
	*Coconut**	*Banana***	*Coconut**	*Banana***	*Coconut**	*Banana***	*Coconut**	*Banana***	*Coconut**	*Banana***
No fertilizer	0.712	0.741	265	290	98.71	28.86	76.71	42.01	93.99	81.51
One -fifth	0.700	0.723	365	230	161.73	29.81	96.80	75.49	91.76	65.20
One-fourth	0.702	0.714	755	515	165.67	25.54	117.24	97.59	99.07	88.75
One-third	0.519	0.563	1435	1050	220.44	29.15	112.36	109.79	68.75	96.17
Two-third	0.523	0.598	1410	1215	285.57	29.55	161.21	162.43	76.44	100.54
Full	0.512	0.576	1460	1250	342.65	32.81	229.60	236.92	71.68	101.30

*Average of three depths; ** average of two depths

Treatment	*Mg (ppm)*		*Fe (ppm)*		*Cu (ppm)*		*Zn (ppm)*		*Mn (ppm)*	
	*Coconut**	*Banana***	*Coconut**	*Banana***	*Coconut**	*Banana***	*Coconut**	*Banana***	*Coconut**	*Banana***
No fertilizer	23.72	17.59	86.99	76.80	0.67	0.87	1.46	1.28	33.71	33.74
One ñfifth	17.32	15.17	78.13	93.83	0.35	0.95	1.00	1.21	35.49	56.65
One-fourth	15.86	27.62	113.10	97.90	0.91	1.11	2.04	1.56	53.69	40.01
One-third	16.61	31.39	87.21	81.89	0.55	0.74	1.41	1.35	57.98	68.79
Two-third	13.35	40.34	78.29	103.45	0.65	0.81	1.13	1.73	49.61	80.53
Full	12.96	33.36	97.71	73.97	0.78	1.03	1.16	2.31	51.29	47.62

Effect on Microbial Parameters

The distribution of soil microbial groups is investigated in cropping system (Palaniswami *et al.*, 2007). Crop diversity and level of fertilizer inputs influence the microbial groups in root zone of crops. The bacterial count is low in the root-region of pineapple and control. About two-thirds dose of fertilizer give low number of bacteria than other treatment in coconut. The counts of fungi and actinomycetes are low in root region of banana and coconut. Full dose of fertilizer supports very low count of fungi and actinomycetes. Asymbiotic nitrogen fixers are more in root region of clove and pineapple. The population is maximum in one-third dose of fertilizer treatment. There is an increasing trend from control to one-third dose with decreasing trend. The population of P solubilizers, is maximum in banana among crops with one-third dose of fertilizer. The microbial population decreases with increasing depth. The development of different microbial groups is optimum at moderate doses of mineral fertilizer input, *viz.* one-third and one-fourth when combined with addition of vermicompost produced by recycling of waste biomass.

Higher content of microbial biomass is recorded in root region soils of coconut and component crops at medium levels of fertilizer inputs than full-recommended fertilizer dose or at very low and low fertilizer treatments (Table 28.4) (Palaniswami *et al.*, 2007). The quantity of carbon mineralised, phosphatase and dehydrogenase activities of soil also vary under different fertilizer treatments and a decreasing trend is observed with increasing soil depth. Medium level of fertilizer inputs along with recycling of waste biomass results in development of congenial conditions for higher level of microbial activity in the cropping system. Under root (wilt) affected coconut garden,there is increase in nutrient status of soil and crop over the years with increase in nut yield of coconut (Maheswarappa *et al.*, 2005).

Table 28.4. Microbiological parameters recorded in the coconut and banana hizosphere in HDMSCS

Treatment	*Microbial biomass (Ïg/g soil)*		*C mineralisation (μg C/g soil)*		*Phosphatase activity (μg p-nitrophenol/g soil/hour)*		*Dehydrogenase (μg formazan/g soil/hour)*	
	Coconut	*Banana*	*Coconut*	*Banana*	*Coconut*	*Banana*	*Coconut*	*Banana*
No fertilizer	363.55	231.32	38.25	32.47	32.68	35.98	12.9	15.14
One -fifth	290.35	268.36	37.78	27.79	40.81	39.93	14.87	19.38
One-fourth	447.87	370.36	39.04	37.60	39.54	41.46	18.93	19.48
One-third	393.53	310.07	38.31	32.06	40.74	49.13	14.85	18.86
Two-third	411.34	333.67	36.71	30.36	35.85	39.75	14.71	17.54
Full	421.91	273.96	29.41	30.11	45.82	47.65	12.8	14.89

In Godavari delta, banana is best intercrop with complementary effect on coconut yield (Reddy *et al.,* 1980). There is increase in yield of coconuts by 145.4% and 400% over pre-intercropping period with banana during the first and second year of post intercropping, respectively. The cropping systems also enhance the population of beneficial bacteria and fungi in rhizosphere of coconut including phosphate-solubilising bacteria and *Trichoderma viridae*, a bioagent against basal stem rot. Thus, soil fertility in cropping systems creates a more favourable environment for sustained crop production. Srinivasulu *et al.* (2004) have studied the population dynamics of *Trichoderma* spp. (biocontrol agent) in different crops existed in coconut ecosystem of Andhra Pradesh. Although coconut supports a reasonably good population density of *Trichoderma* spp. in its rhizosphere, it is not free from the most dreaded basal stem rot disease. Mango, ginger, coffee, cocoa, banana and lemongrass supports growth of *Trichoderma* reasonably well and hence can be grown as intercrops in coconut ecosystem.

At Veppankulam (Tamil Nadu), Poovan, Rastali, Karpuravalli and Jurmony performs well as an intercrop (Ramanathan *et al.*,1982). Under high-density multispecies cropping system, Monthan performs better, yielding 17-20kg/plant at Veppankulam (AICRP on Palms, 2006). The banana performs better and provids additional income to coconut farmer in different regions (AICRP on Palms, 2006). Mixed cropping with different crops along with banana has gived in increase in nut yield of coconut, increasing NPK status of soil.

References

AICRP on Palms. 2006. *Annual Report for 2005-06*, Central Plantation Crops Research Institute, Kasaragod, pp. 33-37.

CPCRI. 2008. *Research Highlights 2007-08.* Central Plantation Crops Research Institute, Kasaragod, p.5.

Gopalasundaram, P and Nelliat, E.V. 1979. Intercropping in coconut. *In: Multiple Cropping in Coconut and Arecanut Gardens.* Nelliat, E. V. and Bhat, K. S. (Eds). Central Plantation Crops Research Institute, Kasaragod.

Kushwah, B.L., Nelliat, E.V., Markose, V.T. and Sunny, A.F. 1973. Rooting pattern of coconut. *Indian journal of Agronomy,* **18**:71-74.

Kushwah, B.L., Nelliat, E.V., Markose, V.T. and Sunny, A.F. 1973. Rooting pattern of coconut. *Indian Journal of Agronomy,* **18**:71-74.

Maheswarappa, H.P., Anithakumari, P and Sairam, C.V. 2003. High density multi-species cropping system for root (wilt) affected coconut gardens. Its impact on productivity and economic viability. *Journal of Plantation Crops,* **31**(1):23-27.

Maheswarappa, H.P., Anitha Kumari, P., Kamalakshyamma, P.G. and Shanavas, M. 2005. Influence of integrated nutrient management and high density multi-species cropping system on soil properties, plant nutrition and yield in root (wilt) affected coconut palms. *CORD,* **21**(2):18-29.

Maheswarappa, H.P., Subramanian, P. and Dhanapal, R. 2000. Root distribution pattern of coconut in littoral sandy soil. *Journal of Plantation Crops,* **28**(2):164-66.

Nair, P.K.R. and Balakrishnan, T.K. 1976. Pattern of light interception by canopies in a coconut+cocoa crop combination. *Indian Journal of Agricultural Sciences,* **46**(10):453-462.

Palaniswami, C., George V Thomas, Dhanapal, R., Subramianian, P., Maheswarappa, H.P. and Upadhyay, A.K. 2007. Integrated nutrient management in coconut based cropping sytem. *Technical bulletin No. 49.* CPCRI, Kasaragod, 24 p.

Ramanathan, T., Chandrasekaran, N.K. and Ramachandran, M. 1982. Intercropping in coconut. 448-453. *In: Proceedings of Plantation Crops Symposium* (PLACROSYM V). Indian Society for Plantation Crops, Kasaragod.

Rao, D.V.R., Kalpana, M., Srinivasulu, B. and Khan, H.H. 2004. Performance of coconut based cropping systems for sustainable productivity under coastal ecosystem of Andhra Pradesh. *Journal of Plantation Crops*: (suppl.), 240-44.

Reddy, K.V.S., Sarma, K.N. and Dasarathi, T.B. 1980. Banana as an intercrop in coconut plantations of central Godavari delta (A.P.), *Andhra Agricultural Journal,* **27**(5&6):326-327.

Reddy, D.V.S., Subramanian, P. and Gopalasundaram, P. 2002. Coconut based high density multi-species cropping system under different levels of fertilizers in red sandy loam soils. 106-111. *In: Plantation Crops Research and Development in the New Millennium.* (PLACROSYM XIV), (Eds. P. Rethinam, Khan, H.H., Reddy, V.M., Mandal, P.K., Suresh, K.). (Eds.)Coconut Development Board, Kochi, India.

Srinivasulu, B., Aruna, K., Vijay Krishna Kumar, K. and Rao, D.V.R. 2004. Population status of *Trichoderma* spp. in different coconut based cropping systems of Andhra Pradesh. *Journal of Plantation Crops,* **32**(2):35-38.

Suma, A., Pushkaran, K., Babylatha, A.K. and Nybe, E.V., 1989, Comparative evaluation of banana cultivars in coconut garden. *Indian Coconut Journal,* **19**(7):6-9.

Thamban, C. and Arulraj, S. 2007. Coconut based cropping system for sustainable income. *Indian Horticulture,* **52**(4):22.

29

Improved Technologies for Vegetable Banana Production

Vishal Nath, Dinesh Kumar and V. Pandey

Banana, a wonder berry forms staple food for millions of people across the globe, providing more balanced diet than any other fruit or vegetable (Singh, 2000). It is nearly fat-free but rich in carbohydrates with a calorific value of 67/100g edible portion. In India, mention in the Ramayana (2020BC) and Arthasastra (300-400BC) and presence in paintings of Ajantha and Ellora (600BC) strengthens the claim for banana to be the native of India. Simmonds (1962) mentioned that banana, particularly acuminata or dessert banana, is native of South-East Asia and India forms the major centre of origin for bi-specific clones such as AB, AAB and ABB genomes.

Cultivated banana is mainly of two types. Banana in which ripe fruits are consumed as fresh is referred as desert banana, whereas matured green banana which become palatable after cooking, is popularly known as plantain or cooking banana. It is a staple food in coastal regions of the country, particularly in Kerala where it is consumed as culinary and utilized for various products preparation. Culinary banana has 'ABB' genomic constitution and contains black lines of seeds which are edible. The fruits are large, angular and dark green at maturity. Vegetable banana is characterized by high vigour, drought hardiness and resistance to leaf spot diseases which is inherited from *Musa balbisiana* (Menon and Peter, 2002). The dominance of 'B' genome in vegetable banana imparts starchiness and acidity to fruits which confer specific cooking quality. It is also used for preparation of various value-added products like snacks, chips *etc*. Vegetable banana is also utilized for culinary purpose in various forms like curries, *bhujia*, fried pieces *etc*.

CONSTRAINTS IN VEGETABLE BANANA PRODUCTION IN ORISSA

In spite of the vast potential and uses, vegetable banana is not gaining a status of commercial crop in India in general and Orissa in particular. Keeping in view, a study for constraint analysis in vegetable banana was carried out in 4 districts of Orissa (Khurda, Puri, Cuttack and Jagatsinghpur) leading in vegetable banana production. To identify, prioritize and quantity constrains in vegetable banana production Rank Based Quotient (RBQ) value technique was used (George *et al.,* 2003). The results of study have been summarized and ranks of constraints have been fixed in Table 29.1.

Table 29.1. Effect of NPK fertigation on growth, fruit yield and yield attributes in vegetable banana Monthan (Banthal-ABB)

Treatment	*Days to shooting*	*Days to maturity*	*Bunch weight (kg)*	*No. of hands/ bunch*	*No. of fingers /bunch*
Recommended dose of fertilizers (RDF)	238.66	322.66	13.54	7.50	90.42
75% RDF + option-I	252.14	344.37	12.40	7.00	85.00
75% RDF+ option-II	217.00	320.00	14.67	8.33	105.33
50% RDF+ option-I	246.50	330.50	9.38	6.00	75.66
50% RDF+ option-II	242.33	326.66	9.40	6.00	76.50

Following inference can be drawn which can be helpful in developing strategies for improved vegetable banana production in coastal parts of Orissa and other potential areas of the country.

- Non-availability of disease-free, genuine and quality planting material at a reasonable cost
- Poor irrigation and infrastructural facilities during crucial period of plant growth and fruiting
- Lacking of specific and improved production technologies for cooking banana and their awareness among potential growers
- High cost of planting material, labour, fertilizer, water etc.
- Vegetable banana has poor marketing network and therefore, the produce is sold at very low price in local hots.
- Non-availability of high-yielding standard variety in cooking types. Local land races and ecotypes are ruling varieties in different parts which have regional preference
- Vegetable banana is generally free from diseases and some pests but due to tall plant height it is prone to high wind velocity in costal regions
- Vegetable banana has high product diversification possibility but unfortunately, processing industries in these areas are generally lacking
- Eating preference of vegetable banana is limited to only coastal regions and some other parts, particularly in lean vegetable season. Therefore, round-the-year production and consumption of cooking banana remains a limiting factor for its production.

Beside, vegetable banana has great potential and therefore strategies for improving production and productivity of vegetable banana have to be developed and implemented in potential areas to ensure food and nutritional requirement.

STRATEGIES FOR IMPROVING PRODUCTION OF VEGETABLE BANANA

Overcoming the major constraints in vegetable banana with the appropriate strategy for improving the production of vegetable banana in the potential region. These are:

Ensuring Disease-free and Genuine Planting Material

Presently, sward suckers are used as a major planting material for vegetable banana which are limited in number and lacking in uniform vigour. Such suckers also serve as a source of inoculum, particularly when they are not being taken from well-standardized, disease-free and well-cared plants. To avoid variation in vigour of commercial vegetable banana crop, tissue cultured plants has to be used. The standard protocols for commercial production of tissue-culture plant have been developed. Presently, commercial scale plant production is being done in different public and private sectors but cost and quality of plants are major concerns. Therefore, cost of plants has to be uniform and affordable with minimum variation (somaclonal variation) in tissue cultured plants.

Assured Irrigation in Growing Pockets

Assured irrigation sources and infrastructure must be ensured in vegetable banana for higher production. It requires well-distributed 20-25 irrigations, amounting to 1,620 mm water at weekly intervals for main crop and 15-20 irrigations for each ratoon crop. Owing to beneficial effect of drip irrigation in banana, an optimum irrigations schedule of banana should be used (Hegde and Srinivas, 1991). The most crucial period of irrigation after plant establishment is before shooting and during fruit development. Srinivas *et al.* (2001) reported that sufficient soil moisture at the time of formation of functional pairs of leaves is very much essential. Similarly, during fruit development irrigation helps in accumulation of starch and conversion of starch into sugars. As general practice, flooding or furrow irrigation is being done in vegetable banana. Hegde and Srinivas (1991) reported that drip irrigation is operated on daily basis, replenishing 80% of USWB class 'A' pan evaporation losses. Two emitters /plant are provided one on either side of plant at a distance of 30 cm. The discharge rate of emitters is 4 litre/hour capacity helps in better plant growth and yield in Banthal banana at Bhubaneswar.

Development and Dissemination of Improved Technology

Vegetable banana has not been a focus area for development of agro-techniques. Recently, initiatives have been taken to develop information on planting system and plant geometry, time and methods of planting, nutrition schedule and foliar feeding and fertigation schedule for vegetable banana. Basically, being a homestead garden crop, vegetable banana is being planted at one or two places in backyards where it continues to grow for several years with the help of kitchen waste water. Now, planting techniques, spacing and time of planting have been standardized for different regions based on climatic conditions, which needs to be adopted by growers for commercial orcharding. While planting vegetable banana, care should be taken to avoid the period of low temperature, particularly at shooting/flowering stage.

Balance nutrition plays a crucial role in production of vegetable banana because banana as such is a heavy feeder crop. Application of nitrogenous fertilizer improve the initial plant vigour, whereas the phosphatic and potashic fertilizer improves the bunch weight and finger quality. A fertilizer dose of 100-125g N, 50-70g P_2O_5 and 200-250g K_2O has been found optimum under Orissa condition (Lenka, 2005, personal communication). Doses of fertilizers are split into 3 equal parts and applied before initiation of inflorescence, 40-50, 70-80 and 120-130 days after planting. The soil application of manures and particularly fertilizers gives poor use efficiency and most of nutrients go waste. To improve fertilizer-use efficiency in vegetable banana, fertigation system should be used. IIHR (2006) mentioned that 75% of recommended dose of fertilizer (RDF) applied in a 3:2:1 ratio (option-II) produce highest bunch weight (14.67kg) in Banthal at Bhubaneswar (Table 29.2).

Table 29.2. Cultivars of vegetable banana prevalent in India

State	*Cultivars of vegetable banana*
Andhra Pradesh	Monthan, Yenugu Bontha
Assam	Manohar, Kanchkal, Bharatmoni, Bimkal, Attikal
Bihar	Muthia, Gauria, Kanthali
Jharkhand	Muthia, Hathua
Karnataka	Karibale
Kerala	Monthan, Nendran, Elavazhai
Orissa	Monthan-Banthal, Bateesa
Tamil Nadu	Karpuravalli, Sakkai, Monthan, Ney Poovan, Peyan
West Bengal	Monthan, Kanthali, Bateesa

Foliar feeding with Foliar Spray Formulation (FSF) containing zinc-3%, boran-1.5%, manganese 1% and iron 1.5%, is called as Banana Special-A (Annonymous, 2004) has been found to enhance yield and quality of vegetable banana cv. Banthal at Bhubaneswar. The method of use is as follows:

- Mix 75g banana special + one packet shampoo + juice of two lemon fruit in 15 litre water
- Spraying on plants from fifth month to tenth month (5-6 sprays) at monthly interval.

Vegetable banana, especially Monthan group, is susceptible to Panama wilt, whereas Nendran is susceptible to borers. Effective techniques for their management are available which needs to be disseminated among vegetable banana growers. Mondal and Singh (2003) suggested following measures to manage for major pests and diseases in vegetable banana.

Panama wilt	• Erradication of infected plants • Use of lime in infected pits • Planting tolerant varieties like Nendran • Use of bioagent, *P. fluorescence*
Borers	• Removal of infested plants • Clean cultivation and adoption of phytosanitary measures • Spraying of 0.05% Endosulpan or 0.1% Carbaryl.

Developing Market Network

At present, vegetable banana has producer-retailer-consumer or producer-consumer or producer and consumer chain of market which provides limited scope for marketing of produce beyond a territory. For efficient marketing some points needs to be looked into:

- Establishment of cooperative marketing system in banana growing region.
- System of pre-harvest contact selling to avoid glut and loss during peak harvesting.
- Strengthening infrastructural facility like cold storage, transportation etc. for vegetable banana.
- Market intelligence system to ensure supply as per demand in other parts.

- Fixing the price index for vegetable banana.

Developing Suitable Varieties

Vegetable banana has location-specific variety in our country. Major varieties of vegetable banana in different regions have been recommended (Singh, 2000; Sathiamoorthy and Mustaffa, 2002). Efforts are also being made to identify some superior clones of vegetable banana . A large number of germplasm of ABB, AB or BB are being maintained in the field Gene Banks at different research stations. Vegetable banana varieties available in India needs to be popularized and some new and high-yielding and quality banana varieties needs to be acclimatized and developed. The major crop improvement thrust in vegetable banana should be:

- Introduction and popularization of exotic varieties like Saba
- Selection of superior clone in Monthan and Nendran for high yield and resistance to diseases
- Development of dwarf-statured hybrids of vegetable banana to avoid abiotic stress
- Development of a variety vegetable banana with high quality pulp similar to potato.

Providing Processing Back-up

To assure vegetable banana production and enhance its area in suitable regions, there must be some industrial back-up for processing of excess production. This may include:

- Establishment of smaller/bigger processing unit
- Development of self-help-groups for undertaking processing as cottage industry
- Development of commercial products and its marketing.

References

IIHR, Bangalore. 1980. *Banana cultivation in India.* Extension Bulletin No.21, IIHR, Bangalore.

IIHR, Bangalore. 2004. *Banana Special* (Nutrient Formulation for Foliar Spray for Banana) . *IIHR News* (2):8.

IIHR, Bangalore. 2005. *Annual Report,* IIHR, Bangalore.

IIHR, Bangalore. 2006. *Annual Report,* IIHR, Bangalore.

George, S., Pandey, V., Singh, H.S. and Kumar, Dinesh 2003. Constraints analysis in banana cultivation. *Souvenir,* State level Citrus and Banana Show-Cum-Seminar, Directorate of Horticulture, Government of Orissa.

Hegde, D.M. and Srinivas, K. 1991. Growth, yield, nutrient uptake and water use of banana crop under drip irrigation and basin irrigation with nitrogen and potash fertilization. *Tropical Agriculture,* **68**:268-331.

Mandal, S. and Singh, H.S. 2003. Diseases and insect pest management in banana. *Souvenier,* State level citrus and Banana Show-Cum-Seminar, Directorate of Horticulture, Government of Orissa.

Menon, R. and Peter, K.V. 2002. Vegetable Banana. *Souvenir*-Global Eonference on Banana and Plantain.

NHB Data Base 2006. *Horticultural Data Base -2006.* National Horticultural Board, Gurgoan, Haryana.

Singh, H.P. 2000. Perspective of banana production and utilization in India. *In: Banana – Improvement Production and Utilization.* Singh, H.P. and Chadha, K.L. (Eds). Proceedings of the Conference on ëChallenges for Banana Production and Utilizationí in 21st Centuary, AIPUB, NRC on Banana, Trichy.

Sathiamoorthy, S. and Mustaffa, M.M. 2002. Technological advances in banana Production. *In: Approaches for Sustainable Development of Horticulture:* Singh, H.P. *et al,* DAC, MOA: 164-169.

Simmonds, N.W.C. 1962. *Evaluation of the Bananas.*Tropical Science Series: 170 Longman.

30

Banana Production in Thailand

Pitswat Buara

Bananas are widely cultivated as small farms, backyard trees and commercial plantations in Thailand. There are numerous cultivars of Thai banana. The fruits of different cultivars are available all year round. More than 50 varieties are found throughout the country. They can be divided into 7 groups of different genomes, i.e. AA (Kluai-Khai, Kluai Leb-Mue-Nang), AAA (Kluai-Hom Thong), AAB (Kluai-Nak), ABB (Kluai-Hak-Mook, Kluai-Namwa), ABBB (Kluai-Teparos), BB (Kluai-Tanee) and BBB (Kluai-Leb-Chang-Kood). There are only three varieties mainly grown for economic fruits, also having high potential for export and processing. These are *Musa* (AAA group) "Kluai Hom", *Musa* (AA group) "Kluai Khai" and *Musa* (ABB group) "Kluai Namwa". Kluai Hom, meaning 'aromatic banana' which has a special flavour and fragrance superior to most commercial cultivars of 'Gros Michel' type, grown in other banana-exporting countries. 'Kluai Khai' is meaning 'egg banana' of Sucrier type, although similar to Pisang Amas of Indonesia and Malaysia, is very sweet and has special texture, flavour and aroma. Kluai Namwa is most common banana in Thailand, which is widely grown in the backyard of houses in villages. It is similar to "Pisang Awak' type. It has firm, sticky fresh and can be eaten as fresh, cooked and processed into different type of snacks. It is very nutritious and used for feeding infants. The flowers and immature fruits of this variety can be used for cooking in various dishes. The leaves are used to wrap up foods and made into containers. The inside young stalk can be used as hog feed. The tougher outside or stalk sheet is torn into small strips and made into a durable type of string.

BANANA PRODUCTION IN THAILAND

Area and Production

In 2007, total area under banana was 113,165 ha with a total production of 1,565,817 tonnes, distributed into areas for Kluai-Namwa, totaling to 85,660 ha having a total production of 1,151,937 tonnes. The area planted to Kluai-Hom was 15,856 ha with a total production of 246,834 tonnes, while planted area for Kluai-Khai was 11,649 ha with a total production of 167,046 tonnes (Table 30.1).

Table 30.1. Planted area and yield of banana in Thailand in 2007

Variety	Area (ha)			Total yield (tonnes)	Average yield (tonnes/ha)
	Bearing	Non-bearing	Total		
Kluai-Namwa	83,210	2,450	85,660	1,151,937	13.84
Kluai-Hom	13,642	2,214	15,856	246,834	18.09
Kluai-Khai	9,329	2,320	11,649	167,046	17.91
Total	106,181	6,984	113,165	1,565,817	14.75

Source: Office of Agricultural Economics, 2007

Major banana production areas of Kluai-Hom are: Saraburi, Nonthaburi, Chumphon, Chanthaburi and Phetchaburi. Phetchaburi is an important planted area of Kluai-Hom good quality for exporting to Japan. The area planted are Kluai-Kai is Tak, Kamphaengphet, Nakhonsawan, Chanthaburi and Phetchaburi. Kamphaengplet is an important growing area supply produces for both local and exported markets (Fig.30.1).

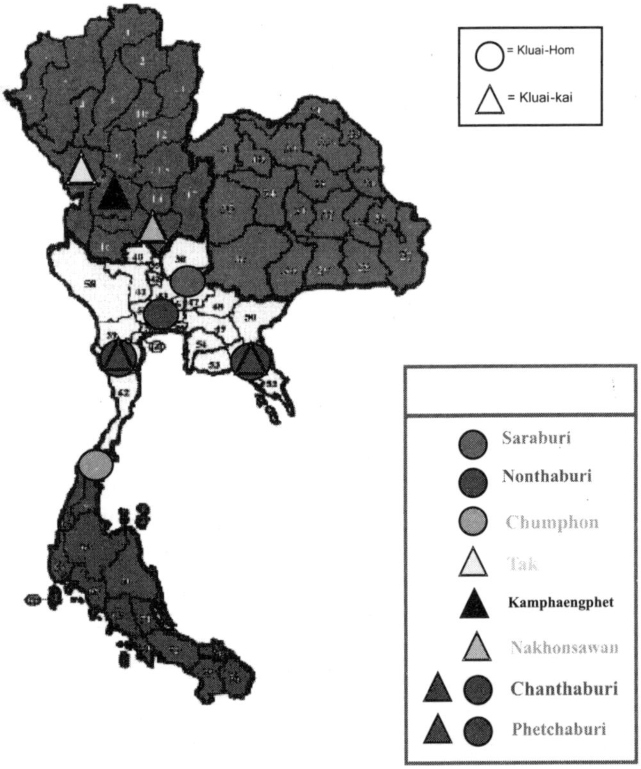

Fig. 30.1. Important planting area of banana in Thailand

Market Situation

The fruits of banana are mostly produced for domestic consumption. Only a small volume is exported. The supply chain for banana in Thailand is divided into 2 ways. The fruits are produced for exporting countries, has to certify in farm and travel directly to importing countries, so the fruit is generally of better quality. The fruits produced for local market have to travel long distance from the production site to the wholesale market through collector agents and then sold to supermarket or open market.

The export quantity of fresh bananas of Kluai-Hom, Kluai-Khai, Kluai-Namwa and other varieties during 2006-2007 ranged between 19,662 and 20,110 tonnes with a value of 193.68 million baht and 170.62 million baht, respectively. The export quantity of banana products (dried) are 1,426 and 2116 tonnes with value of 91.85 and 105.13 million baht, respectively (Table 30.2). The major export market of fresh Kluai-Hom, including processed products are Japan and China. The main market export of Kluai-Khai is Hongkong, China, Loas and Japan.

Importing Country for Fresh Banana

Bahrain, China, Hong Kong, Korea Republic, Malaysia, Australia, Switzerland, Saudi Arabia, Taiwan, Japan, Myanmar, United States, Laos Republic and Netherlands import fresh bananas, while Bahrain, China, Ireland, New Zealand, Singapore, Turkey, USA, Australia, France, United Kingdom, Japan, Netherlands, Denmark, Malaysia, Qatar, Hong Kong, Germany, Guam, Israel, Korea Republic, Switzerland, Philippines, Taiwan, India, Norway and Portugal dried bananas.

Table 30.2. Quantity and value of banana exported during 2006-07

Categories	*2006*		*2007*	
	Quantity (tonnes)	*Value (million baht)*	*Quantity (tonnes)*	*Value (million baht)*
Bananas, (fresh)	19,662	193.68	20,110	170.62
Sweet banana	2,100	55.0	2,960	46.00
Dainty banana	13,837	114.0	13,723	91.00
Kluai-Nam-wa and other bananas	3,725	24.68	3,427	33.62
Bananas (dried)	1,426	91.85	2,116	105.13

Source: Department of Customs, 2008; Sweet banana : Kluai-Khai; Dainty banana : Kluai-Hom

R&D in Banana

The research and development on banana in Thailand is mostly undertaken by the government agencies. There are many institutes involved such as Department of Agriculture, Department of Agriculture Extension, colleges, universities and other government organizations.

As banana is not the main export crop of the country, there are no immense R&D plans. Moreover, financial support on research is very limited. Nevertheless, over the past few years, there are a number of projects concerned banana in various aspects including varietal improvement, propagation techniques, germplasm collection and conservation, plant protection, production technology, fruit quality improvement for export, post-harvest handling and processing technology for new products. The followings are example of R&D activities in Thailand.

Germplasm Collection and Conservation

About 175 accessions have been collected and conserved under the Department of Agriculture, Department of Agriculture Extension and Kasetsart University. Most of these have been characterized on morphology and taxonomy. Some have been used for breeding programmes.

Varietal Improvement

There is only few research works concerning the improvement of banana variety. However, in the past few years, Kasetsart University has released the new banana variety called Kasetsart Banana 2. This new variety occurred from the mutation of callus that is initiated by radiation.

Production System

The R&D on production systems are aimed to improve yield and qualities. The followings are the projects that have been done in the past few years.

- Development of production technology for improving growth, yield and qualities of Grande Naine.
- Effect of sucker size on growth of Grande Naine.
- Sucker-dividing effect on growth of Kluai-Hom, Kluai-Khai and Kluai-Nam-Wa.
- Effect of media on growth of Grande Naine seedlings.
- Effect of some foliar fertilizers on growth of Grande Naine seedlings.
- Study on growth of inflorescence and fruit of Kluai-Hom-Kom.
- Selection of saline-tolerant varieties through tissue culture techniques.
- Effect of number of genomes on propagation via tissue culture methods.
- Increasing of number of chromosomes of Kluai -Leb-Mue-Nang using colchicines under aseptic condition.
- Study on possibility of conservation of banana germplasm with liquid nitrogen.
- Study on causal agent of leaf speckle symptoms and its control measures.
- Study on effectiveness of some fungicides on the control of sigatoka disease of Kluai-Hom, Taiwan variety.

Production Constraints

Kluai Namwa is the most popular variety in Thailand. It is grown and widely adapted to several areas in the country, i.e. southern region, central plain and northeast region. In Thailand, it is found that this variety is very susceptible to Fusarium wilt. The causal organism has been identified as *Fusarium oxysporum* f.sp. *cubense*. The outbreak of this pathogen in some banana planting areas is due to disease-free suckers are not used as new planting materials and lack of quarantine and sanitation to prevent the spread of the pathogen to uninfected areas.

At present, Thailand has not yet serious problems of this wilt disease infection. Good agricultural practice (GAP) for banana is established by Department of Agriculture to evaluate all registered banana farm, particularly in commercial production of large-scale farmers. The selection of disease-free suckers as new planting materials is a major step towards preventing the spread of Fusarium wilt to disease-free fields. There is the use of banana tissue culture plants as new planting materials to prevent Fusarium wilt in commercial production of Kluai

Namwa. Acid soil is more conductive to Fusarium wilt. Amendment of soil with lime is recommended to lower the spread of disease in the field. Severity of Fusarium wilt is also more with the application of nitrate form of nitrogenous fertilizer. Furthermore, sanitation of equipment management and tools and selection of diseased-free planting material play an important role in practices to prevent the introduction of *Fusarium* wilt into clean banana fields.

References

Anupunt, P. 2002. Banana in Thailand. *Advancing Banana and Plantain R&D in Asia and Pacific*, Vol. 11, pp.149-54.

Bansiddhi, K. 2003. Current status and prospects of banana R&D in Thailand. *Advancing Banana and Plantain R&D in Asia and Pacific*, Vol. 12. pp.111-15.

Chandraparnik, S., Ditchaiwong, C., and Bansiddhi, K. 2004. Comparison of Musa germplasm in Thailand. *Advancing Banana and Plantain R&D in Asia and Pacific*, Vol. 13. pp.131-34.

Kooariyakul, S. 2008. Survey, collection and characterization of banana *Fusarium* wilt pathogen in Thailand, 13p.

OAE, Thailand..2007.Statistical data of agricultural economics. Office of agricultural economics. Ministry of agricultural and Cooperative. Thailand. 97p.

Silayoi, B. 2006. New variety, Kasetsart Banana 2. Workshop on mass production of Kasetsart banana 2 by tissue culture, 29 May ñ 2 June 2006 at Cetral Laboratory, Kasetsart University, 112p.

Vichitrananda, S.2007. Current banana research and development activities in Thailand, 8p.

31

Production Standards for Banana for Export in Taiwan

Chih-Ping Chao

PEI-CHIAO, a Giant Cavendish banana clone, cultivated mainly under subtropical region in Taiwan and characterized by its good eating quality, has been oriented for export to Japan since 1960. Taiwan dominated Japan's banana market in the mid 1960s and had exported a record high volume of 27 million cartons (12.6 kg/carton) of banana, to Japan in 1966. However, due to the challenge of establishing intensive plantation of Cavendish banana in Mindano island of Philippines by the multiple international companies, their advantages of low labour cost, no typhoon damage, and high percentage of appealing banana in that production area lead the export business of Taiwan banana declines drastically, sharing less than 10% banana market in Japan annually at the end of the last century. In 2008, probably only 0.8 million cartons of Taiwan banana are available for export.

Nevertheless, banana industry is still considered by the Taiwan Government to be the flagship among all fruit crops with potential to export on condition that banana produce can be supplied stably with high quality standard in the future.

QUALITY STANDARD OF EXPORTED BANANA

To assure that Taiwan banana exported to Japan can be evaluated as high quality banana and offered with a stable and lucrative price, a quality standard chart, which was learnt from countries with strong banana industries in central and south America, has been applied to bananas assembled for export in Taiwan since 1970. The criteria of high-quality banana for export is listed as follows:

Packing Specification

About 4-6 hands per carton (12.6 kg/carton) are packed. Each hand should be free of blemish as much as possible, with good outlook and proper maturity. Hand size ranges 1.9 to 4.6 kg/hand. Finger length should be > 15 cm.

Maturity (age-days needed from shooting to harvesting) Control

Spring banana (harvested from mid-March to mid-June): 75-85% fullness, 135-80 days. Summer banana (mid-June to late-September): 70% fullness, 60-80days. Autumn banana (late-September to mid-January): 80% fullness, 75-100 days. Winter banana (mid-January to mid-March): 80-85% fullness, 100-135days.

Cleanness

Finger without flower or contaminants such as latex, dust, ash, pesticide residue, bird, ant, bee or spider web.

Smooth Crown Surface

Make hand crown smooth with a small knife.

No Insect Attachment

Hand with insect such as moth, stem weevil or mealybug should be rejected.

Defect Tolerance

When more than 10% hand fingers affected by a defect like pulp bruising, sun burn, insect boring or scarring, chilling, or pesticide residue, reject the defected hand. When 20% fingers per hand showing corky scab symptom on the peel, caused by flower thrips, or when >10% fingers with 1/6 peel surface appearing 80 freckle spots, the sampled hand is considered as a defect too.

Good Formation of a Hand

Reject the hand with one layer only, extra large, double finger or malformed shape. When a overweight hand with edge cut or a hand with 2-3 fingers just detached is observed, it should not be accepted.

Reject Banana Hand with Risk of over Ripening or Turning

Only hand with proper age and maturity can be packed.

GUIDELINES FOR PRODUCTION OF HIGH-QUALITY BANANAS

Since outbreak of mad cow disease in England in 1986, food safety becomes a big concern for consumers worldwide. Affected by consumer's opinions in Japan, banana exported to Japan has to document what kind of pesticides applied in the production system since 2000. To fulfill the biosafety demands by Japan's importer of Taiwan banana, the standardized operation procedures for banana production, or the so called Taiwan Good Agriculture Practice (TGAP), was initiated in 2004. Currently, all bananas packed for Japan's market can be traced for all components related to field management, for example, packing station, location of plantation, cultivar and plating material, type and source of fertilizer or pesticide applied, outcomes of phytotoxity or heavy metal analysis, through barcode traceability system.

Banana Mosaic

- Always use certified virus-free tissue culture plantlets as planting material.
- Healthy suckers to replace plantlets could control spread of this disease in good condition.
- Tissue culture plantlets should not be intercropped with or cultivated near by the intermediate hosts of CMV such as legume or solanaceous crops including cowpea,

cucumber, green or hot pepper, tomato etc.

- When climate is cool and dry, lay out silver mulching along soil grounds of plantlets in a row at planting stage could repel numbers of intruding alate aphids and the population density of wingless aphids propagated from the alate aphids. Covering the plantlet with an aphid-proof, 32-mesh white net can also effectively prevent the incidence of banana mosaic.
- Regular spray, at an interval of 5-7days, of a systemic insecticide such as Confidor mixed with non-phytotoxic mineral oil over foliages of plantlets may decrease the population of infested aphids, especially at the first two months after planting.
- Eradicate the CMV-infected plantlets as soon as possible in the field.
- Dry soil and overdose of N-fertilizer at early planting stage of plantlets may conduce to the susceptibility of plantlets to banana mosaic.
- Weeds in cut leaf ground cherry (*Physalis angulata*) and black nightshade (*Solanum nigrum*), can be source of CMV too (Tsai *et al.*, 1986). A good weed management is also helpful to reduce chance of plantlets infected by CMV.

Bunchy Top

- Use healthy, virus-indexed plantlets as planting materials.
- Location for plantlets should be free of BBTV inoculum .
- Besides the ability of banana aphid to host on abaca, taro and *Heliconia* spp., it can also persist in ginger lily (*Alpinia speciosa*), shell flower (*Hedychium coconarium*), and goose grass (*Eleusine indica*). Not only the BBTV-infected banana plants should be eradicated, all the mentioned ornamental plants and weeds surrounded should also be removed simultaneously.
- Insecticides such as Dimethoate, Furadan, and Tamaron gave good control of banana aphid.

Heart Rot

- Growth medium amended with 1% calcium sulfate and 0.2 ppm sodium borate (Ko *et al.*, 1997) could avoid the occurrence of heart rot for plantlets in nursery.
- When larger pots are used for planting plantlets, heart rot can be well prevented .
- If the microclimate such as temperature and soil moisture could be well adjusted for growth of plantlets in nursery, incidence of heart rot should be lower.

Fusarium Wilt

- Always avoid the risk by assurance of new planting site away from *Foc* infested area. Soil conditions of a proper planting site should have a pH range of 6.5-7.5, good drainage, and sufficient organic contents. Use fusarium wilt free tissue culture plantlets or other planting material as source of seedling (Hwang and Ko, 1984).
- Deployment of resistant cultivars, such as the recent release of 'Formosana' in Taiwan, is the most effective way to curb spread of Fusarium wilt. Screening resistant clone with good horticultural traits and resistance to Fusarium wilt by somaclonal variation approach is promising and less time consuming, compared to traditional hybridization programme (Hwang and Ko, 2003).

- Abide by quarantine regulation that Foc infected suckers or other uncertified planting material should not be used as the source of seedlings in new banana plantation confined the disease occurred only in localized areas (Jones, 2000).
- Once incidence of Fusarium wilt is recorded in a banana plantation where Fusarium wilt did not exit in the past, it would be better just keeping the infected mats and the surrounding soil in intact situation and terminating of infected plants by injecting 10X glyphosate into center of pseudostem or spraying kerosene to enhance the dry out of the infected plants (Hwang and Tsai, 1986).
- When a resistant cultivar is not available, rotating the land with paddy at least for 6 months each for two consecutive years, could significantly reduce the amount of chlamydospores of Foc survived in the soil and delay the return of Fusarium wilt significantly (Su *et al.*, 1986).
- Soil amended with some microbial organic manures with *Streptomyces* spp. and hyperphosphate may reduce disease incidence to certain degree and delay the spread of Fusarium wilt in a banana orchard (Chao, 1998).

Black Sigatoka and Leaf Freckle

Both foliar diseases of banana plants could be managed in an economically feasible way by justifying the on-time chemical spray recommendation suggested by disease forecast system. Disease intensity and number of functional leaves per prediction site in banana planting areas are calculated by the international rating scale for Black sigatoka (Simmond and Stover, 1987).

Alternation of broad-spectrum Triazole fungicides such as Tilt in the rainy season with protectant Dithane M-45 as an example in dry season could give a satisfactory control, minimizing the risk of generating resistant strains of pathogen to chemicals at the same time.

Removing the severely infected leaves before application of fungicide could obtain a better management of black sigatoka and leaf freckle. The earlier covering of fully developed fruit bunch by either perforated blue polyethylene bag or kraft paper bag could protect fruit from being infected by conidia of freckle disease spread from infected leaves.

Role of Planting Material

Before 1980, sucker and ratoon crop are the major source for farmers to grow bananas in the southern and central Taiwan, respectively (Hwang *et al.,* 1984). Even though planting suckers or follow ratoon cropping system is usually money-saving, and a high yield potential and earlier harvesting could be expected, chance of these kinds of traditional planting material carrying diseases or pests to new areas is also increased. Through the utilization of virus-indexed programme to produce healthy tissue culture seedlings of banana, less labour is required for establishing a banana orchard with the small, and easy carried plantlets, harvesting time can be well adjusted, cost of managing diseases and pests of banana is reduced. Furthermore, fruit quality of plants derived from plantlets is generally more homogenous than those from suckers or ratoon crop. This explains why 2-3 million plantlets each year had to be produced for banana growers during the last five years.

CONCLUSION

Whether Taiwan banana export industry can be sustainable or not, the future strongly depends on the quality control of banana packed for export. To assure that 85-95% of packed banana with good quality, we believe that it is a must to construct a good and efficient extension network. Only when banana growers can really follow the good management approaches suggested and have the vision that high quality banana is their only future and Taiwan banana export industry hopefully be sustained.

Disease are always important limiting factors affecting growth condition and quality of crops. Just like most of farmers in Taiwan, banana grower tends to solve problems by using pesticides. Actually, they realize that some diseases cannot be effectively handled only by chemical control. By continuous R&D, and the extension of different know-how to farmers, the construction of an effective, multi-disciplined IPM programme should not be far.

The well established foliar disease forecast system, deployment of virus-indexing technique to produce healthy banana seedlings, and unconventional way to screen clones with good traits and stable resistance to Fusarium wilt stand for a good beginning for build-up of an effective IPM for banana industry in Taiwan. By sharing the experience of a mature IPM for certain disease of banana in other country, target of managing a harmful organism under the economic threshold and maintaining the sustainability of producing high quality banana in good shape should be hopeful to achieve.

References

CA, Taiwan. 2005. Safety management system for fruit exported to Japan. 2005. Bananas, 4-6 pp. Council of Agriculture, Taiwan.

Chao, C.P. 1998. Disease and pest control of banana in 1998. *TBRI Annual Report*, 73pp.

Chao, C.P. 1999. Disease and pest control of banana in 1998. *TBRI Annual Report*, 65pp.

Hwang, S.C and H.J. Su. 1980. Disease forecast for black sigatoka in banana plantation in Taiwan and its technique improvement trial. *TBRI Special Report*. 1:1-12.

Hwang, S.C., C.L. Chen and F.L. Wu. 1984. Susceptibility of banana clones to Fusarial Wilt, Freckle and Marginal scorch disease in Taiwan. *Plant Prot. Bull.* (Taiwan, R.O.C). **26**:155-66.

Hwang, S.C., C.L.Chen., J.C.Lin and H.L. Lin. 1984. Cultivation of banana using plantlets from meristem culture. *Hort. Sci.* **19**:231-33.

Hwang, S.C and H.J. Su.1998. Production of virus-free banana plantlets in Taiwan. *FFTC Extension Bull.* No.460, 7pp.

Hwang, S.C and W.H.Ko. 2003. Cavendish banana cultivars resistant to Fusarium wilt acquired through somaclonal variation in Taiwan. *Plant Diseases* **88:**580-87.

Liu, H. C. 2000. Strains of Banana Bunchy-top, Virus and Transmission Natures of Vector Aphids. MS thesis, NTU, Taipei, Taiwan. 104pp.

Su, H.J., Hwang S.C. and Ko W.H. 1986. Fusarial wilt of Cavendish banana in Taiwan. *Plant Diseases* **70:**814-18.

TBRI, Taiwan. 2004. Guidelines for production of high-quality banana, Taiwan Banana Research Institute. Taiwan 20pp.

TBRI, Taiwan. 2006. Taiwan Good Agricultural Practice (TGAP)-Banana, Taiwan Banana Research Institute. Taiwan, 27pp.

Tsai, Y.P., Hwang M.T., Chen S.P. and Liu S.S. 1986. Population dynamics and chemical control of banana aphid Pentalonia nigronervosa. *Plant Prot. Bull.* (Taiwan, R.O.C.) **28:**147-53.

Tsai, Y.P., Hwang M.T., Chen S.P. and Liu S.S. 1986. An ecological study of banana mosaic. *Plant Prot. Bull.* (Taiwan, R.O.C.) **28:**383-87.

Tsai, P.L. 1997. From entrance to prosperityñBriefing history of Taiwan Banana (in Chinese). *Historial Monthly* **119:**34-44.

32

Cultivation of Banana in *Terai* Zone of West Bengal

C.P. SURESH, S.K. GHOSH, P. DEB AND M.A. HASAN

BANANA or 'Adam's Fig' is believed to be one of the oldest fruits, probably originated in the warm moist tropical Asia. It is widely grown in India with great socio-economic significance and cultural heritage of the country. North-East India including the North Bengal rears the wide genetic base of this fruit. The delicious, soft, flavored fruit is a rich source of sugars, minerals and vitamins. It is also dessert fruit to millions of people apart from staple food owing to its rich and easily digestible carbohydrates. A large number of local cultivars are under cultivation in West Bengal along with commercial cultivars. However, information on performance of promising local cultivars in this zone is scanty. Therefore, performance of banana cultivars in *Terai* zone of West Bengal is discussed.

The experiment was cconducted in a randomized block design with three replications at Uttar Banga Krishi Viswavidyalaya, Pundibari, Cooch Behar (West Bengal) during 2006. Four banana cultivars, Amritsagar (AAA), Rasthali or Malbhog (AAB), Champa (AAB) and Karpuravalli (ABB) were taken for study. The recommended cultural practices were followed to raise uniform and successful crop. Observations on growth, yield per plant, yield-attributing characters and biochemical parameters were recorded. Biochemical parameters were estimated as per Ranganna (1979). Statistical analysis of the data was carried out according to Panse and Sukhatme (1978).

These were significant variation in characters among cultivars. The data reveal that highest pseudostem height (281.33cm) and girth (57.79cm) were found in Karpuravalli. Amritsagar produced highest number of functional leaves (12.33) and lowest being in Malbhog (8.23). Leaf area was maximum in Kapuravalli (1.14m^2). The lowest phyllochron was observed in Amritsagar (7.62 days). Maximum number of suckers was produced in Malbhog (7.60). Amritsagar was found most precocious (264.63 days taken for shooting). Karpuravalli took maximum days for shooting (344.93). Similarly, significant variations in growth characters of several varieties of banana have been refunded by Shaikh *et al.* (1985). In yield-attributing characters, highest finger length was found in Amritsagar (20.05cm), followed by Karpuravalli (19.21cm). Finger girth was also maximum in Karpuravalli (12.19cm) which was close to Amritsagar (11.73cm). Amritsagar possessed maximum volume of fruit (182.47 cc), followed by Karpuravalli (173.70cc). But Malbhog was considered as highest in pulp: peel ratio (3.24) which was statistically similar with Karpuravalli (3.06). Whereas maximum finger weight (188.93g), number of hands (8.69), number of fingers/bunch (201.85) and bunch weight or yield/plant bunch (30.61kg) are recorded in Karpuravalli. These findings are more or less similar to those reported by Baruah *et al.* (2007). The findings of Chattopadhyay *et al.* (2001) corroborate with our data on growth, yield and yield-attributing characters. Arulmozhian and Ayyanperumal (1996) reported closer findings on Karpuravalli and Champa (Poovan). Rajamanickam and Rajmohan (2005) also found significant variation in fruit characters of different banana cultivars. The highest TSS was recorded by Karpuravalli (22.60°Brix) and lowest by Champa (16.13°Brix). Maximum amount of total sugar was recorded in Karpuravalli (21.48%) and minimum in Champa (15.86%). Again reducing sugar was highest in Karpuravalli (8.17%) and lowest in Amritsagar (5.58%). Karpuravalli was found to be lowest in acidity (0.42%), while Champa was highest (0.67%). Baruah *et al.* (2007) also reported similar findings in Champa and Malbhog. Thus, it is concluded that Karpuravalli was best among four cultivars in yield, yield-attributing character, and biochemical quality. It may be due to the richness of Karpuravalli (ABB) cultivar with Balbisiana (B) genome which enables it to perform better naturally.

Table 32.1. Growth, fruit and quality characters of different banana cultivars

Cultivar	*Growth characters*						
	Pseudostem height (cm)	*Pseudostem girth (cm)*	*No. of functional leaves*	*Leaf area (m^2)*	*Phyllochron (days)*	*No. of suckers*	*Days taken to shooting*
Amritsagar	174.85	52.94	12.33	1.10	7.62	5.18	264.63
Malbhog	165.41	49.34	8.23	1.06	9.34	7.60	308.34
Champa	251.74	52.80	8.67	0.97	8.27	6.75	272.28
Karpuravalli	281.33	57.79	10.67	1.14	8.81	6.13	344.93
CD (0.05%)	16.03	4.35	0.77	0.09	0.17	0.19	10.97

	Fruit characters							
	Finger length (cm)	*Finger girth (cm)*	*Finger volume (cc)*	*Pulp: peel ratio*	*Finger weight (g)*	*No. of hands/ bunch*	*No. of fingers/ bunch*	*Bunch weight (kg)*
Amritsagar	20.05	11.73	173.70	2.27	184.81	8.33	181.10	23.49
Malbhog	15.19	10.71	126.81	3.24	135.95	7.46	97.67	25.44
Champa	13.42	9.23	83.61	2.54	94.96	6.12	85.34	17.53
Karpuravalli	19.21	12.19	182.47	3.06	188.93	8.69	201.85	30.61
CD (0.05)	0.43	0.42	16.90	0.58	16.83	15.98	0.52	0.72

	Quality characters			
	TSS (°Brix)	Total sugar(%)	Reducing sugar (%)	Acidity (%)
Amritsagar	19.20	16.32	5.58	0.47
Malbhog	20.80	20.27	7.83	0.53
Champa	16.13	15.86	6.30	0.67
Karpuravalli	22.60	21.48	8.17	0.42
CD (0.05%)	1.91	1.97	0.27	0.07+

SUMMARY

An experiment was conducted at Uttar Banga Krishi Viswavidyalaya, Pundibari, Cooch Behar, to evaluate the performance of different cultivars of banana. It was observed that Karpuravalli gave highest yield (30.6 kg/plant), followed by Rasthali or Malbhog (25.4 kg/plant) and other yield-attributing characters. The TSS was maximum in Karpuravalli (22.6 °Brix) followed by Rasthali (19.2 °Brix). Total sugar content was also highest in Karpuravalli (21.48%) with minimum acidity (0.42%). Karpuravalli was found to be suitable large-scale plantation in *Terai* Zone of West Bengal.

References

Arulmozhiyan, R. and Ayyamperumal, A. 1996. *In: Conference on Challenges for Banana Production and Utilization in 21 st Century.* NRC, Banana, Trichy, 24-25 September, pp. 19-20.

Baruah, K., Sharma, B. and Sut, D. 2007. Genetic Variability in Banana Cultivars in Assam Conditions. *Indian J. Hort.*, **64(3)**:282-85.

Chattopadhyay, P.K., Abdul Khader, J.B.M. and Hasan, M.A. 2001. Banana. *In:Fruits: Tropical and Subtropical*, Vol-1,3rd rev. edn. Bose, T. K., Mitra S. K. and Sanyal, D., pp. 226-330.

Panse, V.G. and Sukhatme, P.V. 1978. *Statistical Methods for Agricultural Workers*, 3rd edn., ICAR., New Delhi.

Rajamanickam, C. and Rajmohan, K. 2005. Studies on diploid varieties (AA, AB and BB) of banana (*Musa* spp.). *South Indian Hort.*, **53(1-6)**:1-7.

Ranganna, S. 1979. *Manual of Analysis of Fruit and Vegetable Products*, Tata Mc.Graw Hill Publshers, New Delhi.

Shaikh, M.R., Rane, S. and Razvi, S.I. A. 1985. Varietal trial on banana in Sind. *Pakistan J. Agric. Res.* **6(2)**: 189-91.

33

Production and Utilization of Banana in Karnataka

K. Ramakrishanappa

Karnataka is a major banana producer in southern peninsula. The tropical conditions prevailing in Karnataka are congenial for grand growth and production of banana. The hills of Western Ghats are natural home of banana and still several wild types occur in this area. The vast diversity of varieties in Western Ghats gives an impression that several local varieties might have originated in this part. The wild varieties also exist in this area.

In recent years, banana has emerged as a commercially viable crop. Therefore, its cultivation is making headways in all over Karnataka. There are many instances in which farmers have made huge profits out of banana cultivation. Another important factor that is promoting its cultivation is related to low risk involved in the cultivation. Under tube well irrigated conditions, it is a major competitive crop to even most profitable vegetable crops. Because of these reasons, many farmers in the state are coming forward to cultivate banana. The state government gives subsidies and other assistances under different state sector and other schemes, including the National Horticulture Mission Scheme.

PRODUCTION STATISTICS

Area and Production

In Karnataka, banana is grown in 54,788 ha with a total production of 13.45 lakh tonnes annually. More than half a dozen varieties are under commercial cultivation in Karnataka. Cavendish group (Robusta, Dwarf Cavendish, Grand Naine (G 9) and Red Banana occupies 40% of the total area, followed by Yalakki (Ney Poovan) (25%), Mysore (Salem) (15%), Javari (5%), Budu Bale (2%), and others (Nendran, Nanjangud Rasabale and native types) (13%). Important banana-growing regions/districts are: Bangalore (Urban), Bangalore (Rural), Mandya, Tumkur, Mysore and Chamarajanagar, (South) Shimoga, Davanagere, Chitradurga and Bellary (Central), Dakshina Kannada, Udupi, Uttara Kannada, Kodagu, Hassan and Chikkamagalur (Coastal) and Western Ghats, Belgaum, Bagalkot, Bijapur, Koppal, Raichur, Gulbarga, Dharwad and Haveri, in North Karnataka.

Scope for Banana Cultivation

Banana cultivation is increasing in all banana-growing areas because of its remunerative cultivation. There is a growth rate of about 5% every year. The subsidy factor to banana cultivation under NHM is quite instrumental in its area expansion. It is expected that the same trend will continue in the near future also. The easy availability of tissue culture plants is helping the farmers to extend its cultivation throughout the state. There is a tremendous scope for area expansion in command areas of major irrigation projects. It is estimated that the area can go up to 1 lakh ha.

Planting Material

In banana, propagation through tissue culture plants is picking up fast. Everywhere, farmers have realized its advantages though tissue culture plants are costlier than suckers. The tissue cultured plants are free from disease. The Department of Horticulture is also educating farmers on advantages of tissue cultured plants. There are 20 tissue culture labs in the state, 19 being in private sector and 1 in public sector. The total annual production of tissue culture plants is around 30 million. Biocenter, Hulimavu, is the only lab under public sector under Department of Horticulture. This centre, on an average produces and supplies about 1 million plants per annum. Further, Department of Horticulture is planning to expand its capacity to 2 million plants. To meet the requirement of farmers in northern Karnataka, Department of Horticulture has initiated steps to set up another Lab at Belgaum, with an installed production capacity of 0.5 million plants per annum.

Methods of Irrigation

About 50% of banana plantations is irrigated through drip system. This is a very convenient mode of irrigation, especially when the plantations are taken up in vast areas, the drip system proves very handy. Application of nutrients through drip (fertigation) is also gaining up. The Department of Horticulture is extending subsidies from 75 to 100 % to banana growers adopting drip irrigation system under Microirrigation Mission Scheme.

Subsidy for Banana Cultivation

Banana growers need to invest heavily on commercial crop production. On an average, farmers spend Rs 50,000-60,000/acre. This is a heavy amount for most of the small and marginal farmers. Realizing this problem, subsidies are given to banana growers. The farmers are eligible for a subsidy of Rs 15, 000/ha, up to 4 ha, in 2 installments of Rs10,500 during first year (70%) and Rs 4,500 during second year (30%). This subsidy is given to newly started banana orchards. Preference is given to small and marginal farmers while extending the subsidy.

Problems

- Fast decline in groundwater reserve and consequent failure of bore-wells.
- Poor electricity supply and frequent load sheddings in rural areas, affecting uninterrupted operation of irrigation pumpsets.
- High cost of tissue cultured plants and transportation.
- High cost of chemical fertilizers especially of water-soluble fertilizers for fertigation, pesticides, *etc.*

- High cost of drip irrigation systems and staking materials.
- Serious threat caused by Panama wilt and Bunchy Top diseases.
- Natural calamities such as hailstorms, stormy winds, floods, unusual torrential rains, *etc.*
- Disorganized and improper marketing facilities with poor marketing infrastructures.
- Lack of awareness among farmers and traders of scientific handling, grading, packaging, transporting, storage, *etc.*
- Poor scope for processing and value-addition in banana, leading to heavy losses in fresh fruit market.
- Unstable and inconsistent prices for banana in market.
- Frequent gluts leading to fall in prices.

Organic Banana Production

With increasing consciousness about food safety, many consumers opt for organically-grown bananas. By and large, problem of residual toxicity in banana fruits is not as alarming as in grape, pomegranate, apple, fig, etc. Some farmers resort to unnecessary sprayings of pesticides as precautionary measures against possible attacks of pests and diseases. This has caused some concerned about the safety of bananas.

Adoption of scientific organic farming techniques is gaining popularity among many banana growers in Karnataka. With proper preparedness and technology, it is not difficult to grow bananas organically. The Jaivik Krishik Society, an organization exclusively working for promotion of organic cultivation and popularization of consumption of organic food, is helping the farmers to produce organic bananas and also providing facilities for certification and marketing through its outlets. There are many other similar organizations promoting organic cultivation of bananas and other horticultural crops in the state. There are many success stories on record. Many farmers are venturing into organic cultivation of banana. Finding the organically grown bananas is not very uncommon in many super markets in Bangalore and it is likely that the trend will continue further in the days to come. For export purpose in future, only organically grown bananas will be preferred, and this factor may become the driving force for switching over to organic cultivation of banana.

Processing and Value-Addition

In Karnataka, at present, only a few stray attempts have been made for processing and value-addition of bananas. Fresh banana fruits are available throughout the year. The scope for processing and value-addition has not gained real momentum. Yet, because of the urban pressures and changed life-styles, many people are forced to consume processed foods and hence banana may find better future with regard to its processing and value-addition. There is a very good demand for some of value-added products of bananas such as dehydrated banana figs, banana powder (banana custard), banana chips and banana squash. Hence, venturesome entrepreneurs can make use of such opportunities of processing banana into various products that have ready demand among consumers. The Department of Horticulture is extending a subsidy of up to 40 % on the capital investment made for processing of horticultural produce, including bananas, under scheme "Processing in Horticulture". Similarly, Ministry of Food Processing Industries,

Government of India, is also extending a subsidy of 25% for fruit processing industries. Under NHM scheme also, 25% subsidy is given. These subsidies will undoubtedly prove as very good incentives to those who can enter into the field of processing.

34

Production and Utilization of Banana in Maharashtra

B.V. Gopal Reddy and P.V. Firke

In Maharashtra, banana is grown in 72,000 ha, with an annual production of 45.35 lakh tonnes and productivity of 63 tonnes/ha, which is highest in country. Maharashtra occupies 14% area of the country under banana, contributing 28% in production. Maharashtra ranks second in country in area and first in production. Commercial production of banana is being done in Maharashtra and the same is marketed all over the country.

PRODUCTION SYSTEMS OF BANANA

Banana crop is raised under two distinct production systems in Maharashtra. In general, banana is grown with the planting material maintained by farmer for himself or taken from other banana-growing farmer. It is mainly raised with surface irrigation. The other system is tissue cultured planting material. This is necessarily cultivated using micro-(drip) irrigation, along with fertigation. The cultivars are either selections of locally popular cultivars, i.e. Basarai, Ardhapuri (Dwarf Cavendish) or imported cultivars such as Grand Naine. In general, vegetables are grown as intercrop in initial months. Many a times, turmeric, ginger and wheat are grown as intercrop. There is a small plantation of tall cultivars in Vasai area of Thane district. Ratooning is a general practice. Tissue cultured plants give three crops in two years, giving more returns to farmers. Basrai, Shrimanti, Shendurni, Ardhapuri, Padalse, Grand Naine, Harisal, Lal Velchi, Safed Velchi, Rajeli, Valha etc. are popular banana cultivars. Jalgaon, Dhule, Nandurbar, Parbhani, Hingoli, Nanded, Thane, Pune, Raigad, Ratnagiri, Solapur and Kolhapur are major banana growing areas.

Tissue Culture

Tissue culture in banana has many advantages. They are: large number of uniform, true-to-type plants produced at any given time, availability of disease-free and virus-indexed planting material, guaranteed yield, and easy transportation of material. Yield from tissue cultured plants is three times higher than that propagated through suckers. Maturity period is 11-12 months. In 30 months, one main and two ratoon crops are harvested. In Maharashtra, there are 21 tissue culture laboratories are producing different types of plants. However, major ones producing tissue cultured plants are: Jain Irrigation Systems, Jalgaon; H.U.Gugale Biotech, Jamkhed, District Ahmadngar; Kumar Florists, Pune; Indrayani Bio-tech, Pune; Nirmiti Bio-tech, Pune; EPC, Nasik; Gussif, Mumbai; Marathwada Agricultural University, Parbhani; Maharashtra State Seeds Corporation, Akola.

Banana has a lot of promise for *in vitro* propagation but this shall come true only if the relevant problems are addressed at an early stage in a scientific manner. For obtaining better yields and quality production, efforts needs to be made to bring about 80% of banana area under tissue culture plantlets.

Production Technology

The soils of Jalgaon region where maximum area is located are well-drained, sandy clay loam to deep black with a pH of 7.5-8.0. Banana is strictly a tropical crop. The optimum climatic conditions are tropical climate with high humidity with a mean temperature of not more than 33 °C during summer and not below 12 °C in winter. However in Jalgaon, major banana-growing district, maximum summer temperatures are as high as 40-46 °C and minimum temperature (in winter) being 10 °C. Despite these adverse climatic conditions, high yield is realized by farmers, which could be attributed to cultivar (Dwarf Cavendish-Basrai), crop management system (high-density planting), use of windbreaks especially on south-west sides, use of microirrigation, ingenuity and enterprise of banana growers.Quite often paired row system is seen where plants are spaced at 0.6-1.0 m in paired rows on raised beds and at 2.0 m along rows. The distance between two laterals is usually 2.75 - 3.0 m. This reduces the cost of laterals. For planting pits of 0.45m × 0.45m × 0.45 m size are dug and each pit is filled with 20-30 kg of well-decomposed farmyard manure. The fertilizer dose is 440 kg N, 338 kg P_2O_5 and 778 kg K_2O/ha. Nitrogen and potash are applied in four equal split doses 30, 75, 120 and 165 days after planting. The farmers use nitrogen 700 kg, phosphorus 500 kg, potash 1000kg, micronutrients 50 kg (each of Mg, Zn and Boron) and farmyard manure 30 tonnes/ha with mulching with dried leaves and water suckers.

Microirrigation

Banana is a water-intensive crop. Irrigation management therefore plays an important role in is cultivation. With ever-increasing water scarcity problem in Maharashtra, it is very essential to use the available water more economically and scientifically. Use of drip irrigation, therefore, is being popularized. At present, area under drip irrigation is 3.83 lakh ha. Of which, area under banana is 0.80 lakh ha. Thus, area under drip in banana is quite high.

Fertigation

The use of balanced fertilizers at appropriate time plays a vital role in enhancing the productivity. At present, farmers in Maharashtra use conventional fertilizers either, straight or complex fertilizers of different grades. These fertilizers are not completely soluble in water and hence they are not suitable for delivery through drip irrigation. The high salt index of most conventional fertilizers restrict their use in sensitive fruit and vegetable crops. The fertilizers are generally applied to banana manually by ring method which is often labour-intensive. With the introduction of microirrigation system, ring method of fertilizer application has become old. The annual application of crop dose needs splitting in about 3 doses.

Many a times fertilizers are placed in areas beyond the active root zone and are no longer useful to plants. Normally, 50-60 % fertilizers are utilized and rest are wasted either by leaching or volatilization. Considering the advantages of fertigation through drip system, water-soluble fertilizer grades are approved for this purpose. They are:

Solid (N : P : K)	13 : 13 : 13	10 : 36 : 10	
	12 : 61 : 0	13 : 0 : 46	
	5 : 15 : 30	16 : 8 : 24	
	19 : 19 : 19	20 : 10 : 10	
	13 : 00 : 45	0 : 52 : 34	
	Calcium nitrate Potassium magnesium Sulphate		
Liquid (N : P : K)	8 : 8 : 8	6 : 12 : 6	6 : 6 : 12

The annual production of solid water-soluble fertilizers is about 1,000 million tonnes, while import of these fertilizers is about 400 million tonnes annually.

MAJOR PROBLEMS

Transportation of Banana

The harvesting of banana starts from mid-May with its peak period during August-December. The banana is transported mainly by railways and trucks. The transportation by trucks is quick but costly as compared to railways. However, there are many problems in railway transportation. They are:

- The railway wagons supplied at present are not sufficient in number.
- There is no regular and timely supply of wagons.
- The charges of freight are high.

The experience shows that when wagons are demanded they are supplied after 2-3 days. Moreover, when rake load of banana at a time is available for transport, i.e. around 70 wagons, then only a few wagons are supplied at a time. This create either delay in transportation of banana to marketing centres or creates glut in the market.

Banana is a perishable fruit crop which requires urgent transport to consumer market centre so as to avoid losses due to spoilage. It is, therefore, desirable that supply of wagons for banana transport should be timely, regular and in sufficient number at affordable rates.

Benefit of Banana Transport by Railways

If railway wagons in required number are provided quickly, it would be in the interest of farmers and of the railways. The rake carrying railway takes almost 100 hours from Bhusaval to Delhi, whereas a fast passenger train takes hardly 22 hours. Many a times at Jhansi, Itarasi, Agra and such other stations, the wagons carrying banana are shunted for more than 8-10 hours. This has to be avoided. This would help to shorten the time and also make the wagons available for banana transport quite speedily adding to the income of Railways.

The Railway Ministry of India should consider following demands on priority so as to solve the transport problems of banana growers.

- The regular and sufficient number of wagons supply should be ensured by Railways.
- The rates of freight should be reasonable.
- A committee consisting of railway authorities and representatives of fruit sale cooperative societies to monitor this problem is necessary.
- The wagons for banana transport should be white painted with ventilation and wooden flooring. The complete wooden wagons are the best for banana transportation.

Pre- and Post-Harvest Technology

Pre-harvest care includes standard package of practices to get better fruits. In post-harvest, importance is given to maintenance of uniform size of fruits in bunch, harvesting maturity for optimum desirable quality and shelf-life. For this, de- budding is done keeping desirable number of hands per bunch. While transporting, bruising has to be avoided by using padding material. If bunches are cut into hands, latex from cut ends should be dried. Anti-fungal paste can be applied on cut ends to avoid crown rot.

FUTURE NEEDS AND INFRASTRUCTURE

Considering these factors, it is presumed that the demand for this fruit will increase up to 250 lakh tonnes in next 20 years. To achieve this production level under adverse conditions like lowering down of the water table ultimately resulting in decreasing the area. It is necessary to concentrate on evolving improved cultivation practices, irrigation management, IPM etc.

Export

Harvesting continues from July to February. In the remaining months, banana arrivals are low. Our exports do not commensurate with production base. There is a good export market for banana, particularly in UAE, Dubai, Sharajh and Abu Dhabi, Sultanate of Oman, Kuwait, Saudi Arabia, Qatar, Bahrain, Singapore, Nepal, Sri Lanka and Malaysia. Small portion of bananas are exported to these countries from merchant exporters but not from grower exporters. At the same time, a formidable competition is also there from Asian exporters like Philippines. However, Asia imports more than the exports. There is a gap of 1.3 lakh tonnes which at present is being met by imports from Africa and South America. In 2004, Ecuador, Costa Rica, Philippines and Colombia accounted for 63 % of total export. Ecuador having 9% of the total world area alone provided more than 30% of global banana exports. However, main banana producing countries, such as India (23% area of world) or Brazil (9% area of the world), are hardly involved in international trade. Bananas are imported mainly by European Union, USA and Japan, accounting for 67% of the total world imports in 2004 (*see* Box).

Banana exports from Maharashtra

Year	*Quantity(tonnes)*	*Importing countries*	*Districts from which exported*
2002-03	13.62	Dubai	Jalgaon
2003-04	24.00	Dubai	Jalgaon
2004-05	32.50	Dubai	Jalgaon
2005-06	17.57	Dubai	Jalgaon

The quality of our produce in terms of flavor and taste is very good. This would enable us to capture at least a small share of the European markets which annually imports more than 3 million tonnes of bananas.

A farmers' marketing organization, Mahabanana, has been established by MSAMB in 2002. The MSAMB exported 16 tonnes of banana in February 2005 to Dubai through 40 refer container from Jalgaon.

Nowadays, it is possible that with little modification in cultural operations, we will have comparable quality produce available for exports. With proper planning, Maharashtra should be in a position to export 30,000 tonnes to west Asia, 15,000 tonnes to Europe and 15,000 tonnes to South-East Asia.

For this, we will have to gear up our production and fine tune the production technology both pre-and post-harvest. For the same, MSAMB is working by setting of pre-cooling, cold storage and modern packhouse-facility in banana-growing areas. Recently, Government of India has declared Agri-Export Zone for banana-growing districts, *viz.* Jalgaon, Nandurbar, Dhule, Hingoli, Parbhani, Nanded, Buldhana and Wardha. Now, MSAMB has proposed to create modern packhouse for banana at Raver, district Jalgaon and Basmatnagar, district Hingoli which includes mechanical handling system, pre-cooling, cold storage and ripening chamber.

Domestic Market

Banana is highly perishable. At ambient temperature banana has an average market life of 1-10 days. The difficulties associated with the short storage life of plantain are worsened by marketing system. The marketing system in India usually involves several retailers. Buying and selling take time, leading to increased damage. Transport is often delayed and can fail altogether, because of poor conditions of vehicles and roads. A combination of high perishability, high ambient temperatures, slow marketing systems and poor market conditions lead to losses in fruit quality and ultimately to post-harvest losses. Shelf-life could be extended by appropriate storage at 12-13 °C. To maintain product quality throughout the market chain, the whole system, from the farm to consumer, should be considered.

Majority of banana supply and marketing is governed by private sector except in some cases where cooperatives handle its marketing. Unlike in other commodities, cash marketing is most common, where money payment is done immediately after every transaction. Banana markets are specialized markets located in all big cities and most of the marketing functions are followed. Primary wholesale market is major marketing yard for all commercial clones in the country. Irrespective of location of the farm, bunches are transported by growers to nearest towns. In some cases, bunches are sold by farmers to pre-harvest contractors even before harvesting, who pay them 50% of the cost in advance.

Disposal of fresh banana in domestic market is still going to be the primary way of marketing of this important produce from the state. In this case, setting up of modern banana ripening chambers in some key terminal markets such as Delhi and Kolkata which receive banana from the state are warranted.

Processing

Banana fruits are available in plenty in tropical countries like India. A sizeable quantity of banana is wasted due to poor transportation and storage facilities. Therefore, processing and product development is of utmost importance. Various processed products like figs, clarified juice, banana powder, flour, starch,jam, chips, stem candy and fermented products like ethanol, brandy and beer, are prepared and used commonly in India.

Processing of banana into puree, presently carried out by medium-sized modern processing units based at Jalgaon, Mumbai, Wardha and Sangli is going to benefit from availability of processable varieties. Since domestic demand of banana puree is quite small and the present market is primarily export-oriented. To be globally cost-competitive, the right quality banana with higher solid content has to be made available to processors in Maharashtra. Only then the market for this produce can grow in coming years.

35

Technologies for Banana Leaf Production

T.N. Balamohan, V. Jegadeeswari, J. Auxilia and L. Nalina

Banana is fourth important food in world and most celebrated fruit in India of great socio-economic relevance. Apart from its commercial dual utilities as a dessert and culinary, other banana-based industries have not gained much attention. However, banana leaf production has gained business status in Tamil Nadu, Karnataka, Kerala and Andhra Pradesh. The annual turnover of leaf industry is estimated to be Rs128 million (Singh, 1996) approximately equivalent to one-seventh of annual turnover of banana industry. Currently annual turnover is estimated as Rs 250 million. Use of banana leaves as biodegradable dining plates has both cultural and ecological significance. Compared with other banana-based industries such as fibre industry, this has become a source of livelihood for several marginal and small farming communities. Banana leaf production has proved advantageous nowadays due to sustainable demand for leaf throughout the year, year-round sustained source of income for farming families, ability to balance the price fluctuation faced by farmers in fruit industry to a greater extent and applicability to adopt to different banana production systems, including garden, wetland and high-land cultivation.

BANANA LEAF INDUSTRY

Banana-leaf harvesting has been a commercial venture of most banana growers because of continuous demand for leaves. In southern India, serving food on banana leaves is considered very religious and auspicious in traditional households. It has been a practice to have 2-3 clumps of banana plants as a source of leaf for dining purpose along with dessert and cooking bananas in backyards. But this has grown into an industry over the past decade with standardized practices and organized marketing channels. Banana leaves are predominantly used by Hindus and Buddhists as a decorative element for auspicious functions, marriages and ceremonies in India and Southeast Asian countries. The Indians also believe that banana leaf gives a special taste to the food served on it. South Indian food is usually served in banana leaf. Some South Indian and Khmer recipes use banana leaves as a wrapper for frying and removed later to retain flavour. Three banana varieties are used to recover waxes from leaves. Wax yields of *Musa liukiuensis, M. acuminata* and *M. chiliocarpa* are 0.58, 1.05 and 1.41% per dry weight respectively. The melting point of banana waxes is 78-82 °C, which is higher than commercial natural waxes except carnauba wax. The components of *M. chiliocarpa* are similar to carnauba wax by thin layer chromatography analysis. Fatty acids with carbon numbers ranging from C14 to C30 are detected in *M. chiliocarpa* leaf wax after saponification. The C22 fatty acid is found abundant in these fatty acids.

Factors Affecting Leaf Production

Unfurling of leaves takes about six days under favourable tropical conditions (Pillai, 1975 and Balakrishnan, 1980). A new leaf emerges from the heart by the time the previous one is fully opened; therefore emergence of one leaf may be taken as a general rate of leaf production in a week. Turner (1970) is of the view that total number of leaves produced per plant may vary from 23 to 50 depending upon cultivars. It has been observed by Simmonds (1966) that in the initial stages of crop growth, leaf size gradually enlarges and subsequently size reduces before flowering.

The leaf production in banana is dependent upon genome, ploidy, temperature, age of plant and availability of nutrients, particularly nitrogen. Total number of leaves/plant is lowest (32.6) in *Musa acuminata* (AA) and highest (47.4) in *M.balbisiana* (BB). Leaf area, leaf number and leaf area duration are apparently influenced by ploidy. Within the predominant hybrids of bi-specific origin, Ney Poovan (AB) is found to produce a larger number of leaves at shorter interval of 5.49 days as reported by Nambisan and Rao (1980). The rate of leaf production is reduced by low temperature. A monthly average temperature of 27 °C and above is ideal for optimum leaf production in banana.

Packages of Practices

In Tamil Nadu, banana leaf production is also done simultaneously in all the districts where banana is cultivated for fruits. In general, after harvesting under wetland systems, side suckers are maintained for leaf purpose till suckers are used for planting. However, bananas are also planted exclusively for leaf purpose in all banana-growing districts. Though, banana leaf cultivation is confined under coconut grooves, of late, due to its attractive income, farmers started cultivating bananas exclusively for leaf production in Coimbatore, Tanjore, Madurai, Trichy, Tirunelveli, Erode and Tuticorin districts. Out of 85,000 ha cultivated in Tamil Nadu, nearly about one-tenth of area might be under leaf production which is nowadays as lucrative as that of fruit production.

Selection of Variety for Leaf Industry

The primary criterion for selection of an accession for leaf industry is that it should produce a greater number of quality leaves in a given span of time and a bunch with marketable quality. Though leaves are harvested at unfurled stage (before complete opening), its freeness from leaf spot diseases is a desirable quality. Usually younger leaves are preferred than mature and older that are liable to tear either in transit or in use. The accession should possess stability in yield traits, since the crop is maintained as one plant crop with two ratoon. A profuse suckering habit is another good trait to yield a greater number of plants/clump. The leaves need to be soft and flexible rather than brittle.

Elavazhai (BB): It is a wild diploid variety grown in Western Ghats. Although it bears a bunch with seeded fruits, it is highly suitable for marginal lands, showing a high degree of resistance to leaf spot diseases, and is conventionally used as a shade crop in areca plantations.

Kunnan (AB): A popular variety in Kerala and Tamil Nadu, it is a suitable for mixed cropping systems as it tolerates shady conditions. It has a shorter life-cycle, semi-tall in stature, with profuse suckering (producing thick and succulent water suckers) and is tolerant to nematodes and leaf spot diseases.

Poovan (Mysore AAB): It is a leading commercial cultivar grown throughout the country with location-specific ecotypes like Palayankodan in Kerala, Poovan in Tamil Nadu, Karpura Chakkarakeli in Andhra Pradesh and Alpan in North-Eastern Region. Poovan is also commercially cultivated for leaf industry throughout Tamil Nadu and in certain parts of Kerala.

Monthan (ABB): Monthan is also cultivated for production of leaves in Trichy and Tanjore districts of Tamil Nadu. It has many desirable qualities like less incidence of leaf spot disease and to salt tolerance. The cultivar yields normal bunch weight even under marginal condition, but it is susceptible to *Fusarium* wilt disease.

Saba (ABB): An exotic variety introduced from the Philippines and a substitute for Chakkia group (ABB) is suitable for marginal lands, with dual utility and high yielding capacity. The cultivar requires less care exhibits tolerance to leaf spot diseases and field tolerance to Fusarium wilt. This cultivar has a strong stability for yield over generations and hence needs fewer replanting sequences over time.

NRCB Selection 1 (ABB): A single plant selection, it is a high-yielding with 40-45 kg bunch and good marketability. Expresses good phyllochron over others of Pisang Awak subgroup and suited for a 1+3 cropping system (1 plant crop+3 ratoons). Expresses good yield stability and tolerance to leaf spot diseases.

Peyan (ABB): Suitable for backyards and mixed cropping systems. It is a choice variety, suitable for fresh leaf market and is highly prized for its medicinal values. It is slightly longer in duration compared to other backyard varieties. It shows field tolerance to *Fusarium* wilt and a high degree of tolerance to leaf spot diseases.

Kechulepa (ABB): Suitable for fresh leaf market, it produces a robust bunch with good quality and dual-purpose fruits with high TSS (30-31°Brix). It is a choice variety of north-eastern India, possessing tolerance to leaf spot diseases and field tolerance to *Fusarium* wilt.

Initially, the leaf industry is not cultivar-specific but varieties those satisfy both bunch and leaf purposes, such as Poovan (AAB-Mysore), Monthan (ABB-Monthan), Peyan (ABB-Unique), Sakkai (ABB-Bluggoe) and Karpooravalli (ABB-Pisang Awak) are grown. Among them, Poovan is most preferred variety for its leaf quality besides freedom from root diseases. In some areas, plant crop is generally left for bunch production, while first and second ratoon is used for leaf production.

The agro-techniques adopted by farmers vary according to purpose for which crop is raised either for fruit or leaf production. Special package of practices is used by farmers.

Season of Planting

There is no season of planting for leaf production, however in a few areas, planting is restricted to June-July to cultivate intercrops like onion, tomato, okra, radish, *etc.* and other pulses can be raised.

Planting Systems

Under coconut grooves, 400-800 suckers are planted based on spacing between rows and plants of coconuts. However, for pure planting, a spacing of 4½ × 4½' is practised, accommodating 2,150 plants/acre. In those gardens, where bananas are grown exclusively for leaves intercrops, *viz.* onion, tomato, okra, radish, etc. are taken as intercrops during the initial phase.

Nutrient Application

At the time of planting 10 kg of farmyard manure is applied in each pit. Thereafter, fertilizers are applied once in two months till the leaf harvesting is over. Optimum dose of fertilizers for leaf production is 113:113:339 g/plant which enhances number of leaves/plant.

Effect of Nitrogen

Nitrogen is chief promoter of growth. It induces vegetative growth of pseudostems and leaves, giving them desired healthy green colour. A healthy robust vegetative frame is an essential prerequisite for high yield and nitrogen is mainly responsible for such a vegetative frame. Banana plants which have not received nitrogen produce only 7 leaves against 17 leaves produced by banana plants supplied with adequate nitrogen. If N is deficient in bananas, leaves take 23 days for unfolding against 10 days for leaves of banana supplied with nitrogen. Greater the number of healthy and large leaves produced during the first 4-6 months, larger is size of fruit bunch. The N influences the longitudinal growth of petioles. Nitrogen increases bunch grade, and sucker production. Nitrogen deficiency causes slow growth and paler leaves with reduced leaf area and rate of leaf production. Lack of N causes short, thin and compressed leaf petioles, thin profuse roots and lesser number of suckers. Phosphorus uptake is found higher due to N deficiency.

Intercultural operations

After planting, till first harvesting, 3-4 manual weeding are taken up. Meanwhile, 3-4 healthy side suckers are also allowed from five to six months onwards. In each mat, 3-4 suckers along with mother plant are allowed and maintained.

Harvesting

The harvesting of leaves lasts for 15-20 months with an average yield of 40-45 leaves/plant. In an acre, within a span of 24 months, 2.7-3.0 lakhs leaves are harvested (2,500-3,000 bundles of 100 leaves). The bundles are sold at the rate of Rs110-130/bundle, which fetches Rs 2.75 lakh. Hence, a farmer may get around Rs1.37 lakh/acre as a gross income. The cultivation cost comes Rs 50,000-55,000/year. On an average, a farmer may get a net profit of Rs 90,000/year/acre.

There are three predominant stages in leaf production of banana. The first is *"Thady stage"* (looks like a stick) which is not harvestable stage. The second stage is called *"Ponthan stage"* (leaf about to unfurl) is best stage for harvest since the leaf is very flexible and silky. The third stages is called *"Virikuruthu"* (three-fourth leaf unfurled) is also used for leaf purpose. However, this is considered slightly over matured compared to *"Ponthan stage"*. The leaves are normally kept at shady area to avoid any more drying. In each bundle, there are 100 leaves packed in to 10 folders of 10 leaves each. Leaf harvesting is done in the early morning hours and bundled during noon and ready for transportation in the evening (Table 35.1).

Table 35.1. Economics of banana leaf cultivation

Particulars	*Cost per unit (Rs)*	*No./Quantity required*	*Cost (Rs)*
Ploughing	-	-	600.00
Cost of planting material	10.00	2200 plants/acre	11,000.00
Cost for taking pits	3.50	2200.00	7,700.00
Filling of pits	3.00	2200.00	6,600.00
Planting	2.00	2200.00	4,400.00
Fertilizers			
FYM	500/t	4500.00	5,000.00
Urea	4.90/kg	539.00	2,441.00
Single superphosphate	3.48/kg	1553.00	5,404.00
Muriate of potash	4.45/kg	1216.00	5,411.00
Application cost	1.00/pit	2200 plants/acre	1,000.00
Intercultural operations			
Irrigation	100/time	80times	8,000.00
Manual weeding	500/time	6 times	3,000.00
Plant protection			
Pesticides	245.00	6 times	1,477.00
Nematicide	60.00/kg	12 kg	720.00
Spraying cost	250.00/spray	4 times	1,200.00
Harvesting and packaging	Rs.20/bundle	2500-3000 bundles	50,000.00
Expenditure			1,13,953.00 /24 months/acre 56,978.00/year/acre
Yield			2,75,000 leaves/ 24 months 1,37,500 leaves/year
Gross Income	= Rs 110/bundle (2500*110) = Rs 2,75,000/24 months/acre = Rs 1,37,500/year/acre		
Net profit			Rs 80,522/year/acre
Cost: benefit ratio			1:1.40

References

TNAU, Coimbatore 1984. Nutrient requirement for banana. *Reports on Achievements of Banana Research at the Faculty of Horticulture*, Tamil Nadu Agricultural University, Coimbatore, pp.6.

Arunachalam, A. 1972. Studies on the effect of nitrogen on Cavendish clones (*Musa Cavendishii* Lambert). M.Sc.(Ag.) thesis, Annamalai University, Annamalainagar.

Balakrishnan, R. 1980. Studies on growth, development, sucker production and nutrient uptake at different ploidy levels in banana. Ph.D. Thesis, Tamil Nadu Agricultural University, Coimbatore.

Bhakthavatsalu, C.M. and Sathiamoorthy, S. 1973. A note on nutrition with reference to leaf, fruit and sucker production in Poovan banana in Thiruvaiyaru area. *Monthly Reporter*, Madras Agricultural Department.

Chattopadhyay, P.K., Holder, N.C., Maiti, S.C. and Bose, T.K. 1980. Effect of nitrogen on growth, yield and quality of Giant Governor banana. *National Seminar on Banana Production Technology*, Tamil Nadu Agricultural University, Coimbatore, pp.109-12.

Geetha, D., Suharban, M., Vijayan, M. and Pramod, R. 2005. Coir pith compost for increasing banana crop production efficiency. *Indian Coconut Journal,* **35**:7-9.

Kumar, N. and P. Jeyakumar. 2002. Developments in plant and soil sciences. *Plant Nutrition,* **92**:354-55.

Nambisan, K.M and V.N.M. Rao. 1980. The influence of specific origin on leaf production and associated growth characters in South India banana. *National Seminar on Banana Production Technology*, Tamil Nadu Agricultural University, Coimbatore.

Pillai, O.A. 1975. Studies on the effect of number of functional leaves maintained on the growth and development of banana. M.Sc. (Ag.) thesis, Tamil Nadu Agricultural University, Coimbatore.

Prabhuram, R. 1992. Effect of organic manures and urea on the growth and development of banana cv. Rasthali and changes in the soil. M.Sc.(Hort) Thesis, Tamil Nadu Agricultural University, Coimbatore.

Ramaswamy, N. and C.R. Muthukrishnan. 1974. Effect of application of different levels of nitrogen on Robusta banana. *Prog. Hort.,* **5** (4): 5-16.

Shanmugam, K.S. and K.S. Velayutham. 1972. Better manure your bananas. *Agrl. Dig.*, **3**:17-29.

Sharma, A.K. and A.R. Roy. 1973. Fertilizer cum spacing trial on banana (*Musa paradisica* L.). *Indian J. Agric. Sci.*, **41**(6):493-98.

Simmonds, N.W. 1966. *Bananas*, 2nd edn. Longmans, London.

Summerville, W.A.T. 1944. Studies on nutrition as qualified by development of *Musa cavendishii* Lambert. *Queensl. J. Agric. Sci.,* **1**:1-127.

Turner, D.W. 1970. The growth of the banana. *J. Aust. Inst. Agric. Sci.,* **36**:102-10.

36

Using Bunch Covers to Get Quality Banana

M.A. Hasan, S. Sarkar, R. Ray, Chowdhury and D. Majumder

Production and crop duration of banana mainly depends on climatic factors like temperature (Haque, 1984). A sigmoid relationship exists between bunch maturation period and temperature with maximum responses at 29 °C and minimum at 11 °C (Green and Kunhe, 1975). Hence the fruits that grow and develop during cold season are inferior in all aspects to those develop during other seasons (Dhua *et al.*, 1988). Plant growth regulation using synthetic materials has opened new vistas in plantation crop production and has helped horticulturists in mitigating certain complicated problems regarding banana production round the year. Low temperature causes physico-chemical changes in chlorophyll content (Taylor and Craig, 1971) and brings variations in quality of their relative fractions ultimately reducing photosynthesis rate (Khantaniar, 1989). Bunch cover with synthetic materials may alter this situation by increasing the temperature up to 10 ºC in a modified microclimatic condition around the bunch (Reddy, 1989). This microclimate helps in proper finger growth and development as well as reduction in shooting-harvest interval irrespective of season. Bunch covers are reported to reduce infestation of scarring beetle on fruits. Thus, it helps to reduce application of chemical pesticides and their pollution. Therefore, to improve productivity and fruit quality by bagging bunches and to correlate fruit growth with microclimate inside bunch covers was of utmost improtance in banana Grand Naine.

The experiment was conducted at the Horticultural Research Station, Mondouri, Bidhan Chandra Krishi Viswavidyalaya, Mohanpur, Nadia, West Bengal. For standardizing the perfect combination of polythene cover colour, thickness and its perforation to increase yield and quality: 1, colour of polythene (C_1 = transparent and C_2 = blue); 2, levels of polythene thickness (G_1 = 20µ and G_2 = 40µ) and 3, levels of polythene perforation (P_1 =10%, P_2 =20% and P_3 =30%) was evaluated for both plant and ratoon crops. The experiment was laid out in augmented 3-factor factorial CRD followed by principal component analysis (PCA) based on correlation matrix for all the characters to extract performing treatment (s) due to specific set of character (s)/parameter (s). Regression analysis was done for abiotic factors inside bunch covers to find out their contribution on physio-chemical properties of banana. The data were recorded on bunch weight, finger weight, finger length, finger circumference, finger volume, finger density, pulp and peel weight, pulp: peel ratio, pulp thickness, shooting-harvest interval, harvest index, total soluble solids, total, reducing and non-reducing sugars, ascorbic acid, temperature inside and outside bunch cover, humidity inside and outside bunch cover, light interception inside and outside bunch cover and photosynthetically active radiation (PAR) inside and outside bunch cover.

The results revealed that transparent polythene bunch cover (C_1), appreciably increased finger circumference, pulp thickness, total soluble solids (TSS), total and reducing sugars in fingers (Tables 36.1, 36.2 and 36.4). Transparent polythene bunch covers reduce shooting-harvest interval by around 18 days as compared to the control (Table 36.3). Covering the bunches with blue, colourless or white polythene bag after differentiation of male buds is also found to increase bunch weight by 32% (Weerasinghe and Ruwanpathirana, 2002).

Table 36.1. Effect of colour, thickness and perforation of polythene bunch cover on bunch weight and finger weight

Treatment	*Bunch weight (kg)*			*Yield (tonnes/ha)*		
	Plant crop	*Ratoon crop*	*Pooled*	*Plant crop*	*Ratoon crop*	*Pooled*
Polythene colour × polythene thickness × polythene perforation (C × G × P)						
$C_1G_1P_1$	16.43	16.33	16.38	50.72	50.41	50.57
$C_1G_1P_2$	18.70	18.63	18.67	57.72	57.51	57.61
$C_1G_1P_3$	17.13	16.73	16.93	52.88	51.65	52.26
$C_1G_2P_1$	16.17	15.90	16.03	49.90	49.07	49.49
$C_1G_2P_2$	18.73	18.40	18.57	57.82	56.79	57.31
$C_1G_2P_3$	17.07	16.73	16.90	52.68	51.65	52.16
$C_2G_1P_1$	16.13	16.03	16.08	49.79	49.49	49.64
$C_2G_1P_2$	17.07	16.90	16.98	52.68	52.16	52.42
$C_2G_1P_3$	16.27	16.03	16.15	50.21	49.49	49.85
$C_2G_2P_1$	16.80	16.53	16.67	51.85	51.03	51.44
$C_2G_2P_2$	17.70	17.17	17.43	54.63	52.98	53.81
$C_2G_2P_3$	16.30	16.13	16.22	50.41	49.79	50.10
SEm (±)	0.777	0.629	0.687	2.399	1.942	2.121
CD(P=0.05)	2.259	1.828	1.997	7.002	5.668	6.191
Control	14.83	15.20	15.02	45.78	46.91	46.35
Control vs. rest						
SEm (±)	1.529	1.235	1.348	4.72	3.814	4.16
CD(P=0.05)	3.143	2.539	2.771	9.742	7.872	8.586

Polythene bunch covers having 20% perforation significantly improve finger circumference, peel thickness as well as bunch weight. Covering bunches with blue, colourless or white polythene bags after differentiation of male buds also increases bunch weight by 32% (Weerasinghe and Ruwanpathirana, 2002).

The thickness of polythene bunch covers did not produce any considerable change on physico-chemical properties of fingers. Interception of light and photosynthetically active radiation (PAR) inside polythene bunch covers is markedly enhanced by transparent (C_1) and thinner thickness (G_1) and highest perforation percentage (P_3) of polythene bunch covers. Higher perforation percentage (P_2) proves to be cost effective as if produces highest benefit: cost ratio.

Table 36.2. Effect of colour, thickness and perforation of polythene bunch covers on finger weight and finger length

Treatment	*Finger weight (g)*			*Finger length (cm)*		
	Plant crop	*Ratoon crop*	*Pooled*	*Plant crop*	*Ratoon crop*	*Pooled*
Polythene colour × polythene thickness × polythene perforation (C × G × P)						
$C_1G_1P_1$	125.67	117.73	121.70	16.63	15.64	16.13
$C_1G_1P_2$	127.80	118.93	123.37	17.17	16.27	16.72
$C_1G_1P_3$	127.50	118.87	123.18	16.30	15.24	15.77
$C_1G_2P_1$	125.33	117.03	121.18	16.20	15.17	15.69
$C_1G_2P_2$	126.57	117.40	121.98	16.97	16.25	16.61
$C_1G_2P_3$	125.30	116.73	121.02	16.50	15.51	16.01
$C_2G_1P_1$	123.33	115.23	119.28	15.90	14.95	15.42
$C_2G_1P_2$	124.80	115.83	120.32	16.00	15.04	15.52
$C_2G_1P_3$	124.67	116.43	120.55	16.53	15.54	16.04
$C_2G_2P_1$	124.40	116.17	120.28	16.07	15.10	15.58
$C_2G_2P_2$	126.70	119.47	123.08	17.03	15.72	16.38
$C_2G_2P_3$	125.67	117.37	121.52	16.20	15.23	15.71
SEm (±)	2.829	3.293	3.011	0.486	0.449	0.463
CD(P=0.05)	NS	NS	NS	NS	NS	NS
Control	115.07	107.4	111.23	14.17	13.64	13.91
Control vs. rest						
SEm (±)	5.590	6.470	5.924	0.961	0.883	0.913
CD(P=0.05)	11.490	13.299	12.177	1.975	1.815	1.877

Table 36.3. Effect of colour, thickness and perforation of polythene bunch covers and their interaction on shooting-harvest interval and harvest index

Treatment	*Shooting-harvest interval (days)*			*Harvest index*		
	Plant crop	*Ratoon crop*	*Pooled*	*Plant crop*	*Ratoon crop*	*Pooled*
Polythene colour × Polythene thickness × polythene perforation (C × G × P)						
$C_1G_1P_1$	132.00	124.33	128.17	0.435	0.481	0.458
$C_1G_1P_2$	131.33	123.67	127.50	0.441	0.482	0.461
$C_1G_1P_3$	134.33	126.33	130.33	0.437	0.482	0.460
$C_1G_2P_1$	132.00	124.33	128.17	0.435	0.482	0.458
$C_1G_2P_2$	133.00	125.33	129.17	0.441	0.482	0.461
$C_1G_2P_3$	133.00	125.33	129.17	0.433	0.483	0.458
$C_2G_1P_1$	141.00	132.67	136.83	0.435	0.482	0.459
$C_2G_1P_2$	142.00	133.33	137.67	0.437	0.484	0.460
$C_2G_1P_3$	140.33	132.00	136.17	0.435	0.482	0.458
$C_2G_2P_1$	142.00	133.67	137.83	0.439	0.482	0.460
$C_2G_2P_2$	140.33	132.00	136.17	0.434	0.480	0.457
$C_2G_2P_3$	141.67	133.00	137.33	0.436	0.482	0.459
SEm (±)	1.934	3.598	2.719	0.003	0.001	0.001
CD(P=0.05)	5.622	NS	7.900	NS	NS	NS
Control	151.00	141.33	146.17	0.430	0.481	0.455
Control vs. rest						
SEm (±)	3.880	7.115	5.400	0.006	0.002	0.003
CD(P=0.05)	7.975	14.625	11.100	NS	NS	NS

Table 36.4. Effect of colour, thickness and perforation of polythene bunch covers and their effect on total soluble solids (TSS) and total sugar

Treatment	*TSS (Brix)*			*Total sugar (%)*		
	Plant crop	*Ratoon crop*	*Pooled*	*Plant crop*	*Ratoon crop*	*Pooled*
Polythene colour × polythene thickness × polythene perforation (C × G × P)						
$C_1G_1P_1$	21.63	21.57	21.60	19.56	19.53	19.54
$C_1G_1P_2$	21.50	22.03	21.77	19.70	19.70	19.70
$C_1G_1P_3$	21.43	21.40	21.42	19.34	19.33	19.34
$C_1G_2P_1$	21.23	21.27	21.25	19.30	19.27	19.28
$C_1G_2P_2$	21.43	21.50	21.47	19.64	19.67	19.66
$C_1G_2P_3$	21.40	21.33	21.37	19.24	19.21	19.22
$C_2G_1P_1$	19.20	19.27	19.23	17.36	17.38	17.37
$C_2G_1P_2$	19.60	19.63	19.62	17.64	17.66	17.65
$C_2G_1P_3$	19.00	19.07	19.03	17.28	17.28	17.28
$C_2G_2P_1$	19.07	19.03	19.05	17.16	17.17	17.16
$C_2G_2P_2$	19.73	19.77	19.75	17.92	17.94	17.93
$C_2G_2P_3$	18.93	18.90	18.92	17.20	17.23	17.22
SEm (±)	0.431	0.359	0.389	0.429	0.417	0.423
CD(P=0.05)	1.253	1.044	1.131	1.247	1.212	1.230
Control	16.33	16.23	16.28	14.76	14.57	14.66
Control vs. rest						
SEm (±)	0.857	0.717	0.774	0.865	0.840	0.852
CD(P=0.05)	1.762	1.474	1.591	1.778	1.727	1.751

Note: C_1, Transparent polythene cover and C_2, blue polythene cover; G_1,20ì thick polythene cover and G_2, 40ì thick polythene cover; P_1, 10% perforation of polythene cover, P_2,20% perforation of polythene cover and P_3,30% perforation of polythene cover; NS, non-significant.

The first order interaction among colour, thickness and perforation of polythene bunch covers revealed that treatments like transparent polythene cover having 20% perforation (C_1P_2) and 20μ-thick polythene covers having 20% perforation (P_2G_1) significantly improved finger circumference. Reduction in shooting-harvest interval is markedly noticed in 40μ-thick transparent polythene covers (C_1G_2), closely followed by 20μ thick transparent polythene covers (C_1G_1) and transparent polythene covers having 20% perforation (C_1P_2) closely followed by transparent polythene covers having 10% perforation (C_1P_1).

The colour, thickness and perforation of polythene covers, bunch weight, yield, finger weight, finger length and finger circumference were significantly higher with 20μ-thick transparent polythene covers having 20% perforation ($C_1G_1P_2$) (Table 36.1 and 36.2). The 40μ thick transparent polythene covers having 20% perforation ($C_1G_2P_2$) and 40μ thick blue polythene covers having 20% perforation ($C_2G_2P_2$) also produced higher yield as well yield attributing parameters. They were more or less statistically at par with $C_1G_1P_2$. The treatment 20μ-thick transparent polythene covers having 20% perforation ($C_1G_1P_2$) produce bunch weight, finger weight, finger length and finger circumference of 18.67 kg, 123.37 g, 16.72 cm and 15.53 cm about 24, 11, 20 and 18% respectively more than the control. The bunch cover colour on yield and yield-attributing characters, transparent polythene bunch cover performed better over blue polythene covers because of more light penetration through and light is essentially required for nitrate reductage enzyme synthesis (Careli and Magalhaes, 1979). Thus, higher nitrate reductage activity and chlorophyll of finger skin tissues under transparent bunch covers (Hasan *et al.*, 2000) ultimately helpes significantly to promote yield-attributing characters and obviously finger yield.

Maximum pulp weight (86.07g) was recorded with 40μ-thick blue polythene covers having 20% perforation ($C_2G_2P_2$), followed by 40μ-thick transparent polythene covers having 20% perforation ($C_1G_2P_2$) and 20μ-thick transparent polythene covers having 20% perforation ($C_1G_1P_2$). No significant variation was observed in finger density, pulp: peel ratio and peel thickness among different treatment combinations compared with the control.

The different treatment combinations of colour, thickness and perforation of polythene bunch covers were found to decrease shooting-harvest inverval from about 8-19 days as compared to the control. The 20μ-thick transparent polythene covers having 20% perforation ($C_1G_1P_2$) caused minimum shooting-harvest interval of 127.5 days compared with the control (146.17 days) (Table 36.3). This treatment also produced significantly higher TSS, total sugar, reducing sugar and non-reducing sugars in ripe fingers as compared to the control (Table 36.4). Whereas, the treatment of bunch covers could not cause any significant change in acidity and ascorbic acid content in ripe fingers.

Table 36.5. Effect of colour, thickness and perforation of polythene bunch covers and their effect on benefit: cost ratio

	Plant crop	*Ratoon crop*	*Pooled*
Polythene colour × polythene thickness × polythene perforation (C × G × P)			
$C_1G_1P_1$	2.983	2.965	2.974
$C_1G_1P_2$	3.394	3.382	3.388
$C_1G_1P_3$	3.110	3.037	3.074
$C_1G_2P_1$	2.922	2.874	2.898
$C_1G_2P_2$	3.386	3.326	3.356
$C_1G_2P_3$	3.085	3.025	3.055
$C_2G_1P_1$	2.928	2.910	2.919
$C_2G_1P_2$	3.098	3.068	3.083
$C_2G_1P_3$	2.953	2.910	2.931
$C_2G_2P_1$	3.037	2.989	3.013
$C_2G_2P_2$	3.200	3.103	3.151
$C_2G_2P_3$	2.953	2.916	2.934
SEm (±)	0.141	0.114	0.124
CD(P=0.05)	0.410	0.331	0.360
Control	2.390	2.449	2.42
Control vs. rest			
SEm (±)	0.277	0.224	0.244
CD(P=0.05)	0.569	0.460	0.502

Note: C_1, Transparent polythene cover and C_2, blue polythene cover; G_1, 20μ-thick polythene cover and G_2, 40μ- thick polythene cover; P_1,10% perforation of polythene cover, P_2, 20% perforation of polythene cover and P_3, 30% perforation of polythene cover; NS, non-significant.

Considerably highest light interception (1668.65 lux) and photosynthetically active radiation (725.93 μ mol/m²/second) were found in 20μ-thick transparent polythene covers having 30% perforation ($C_1G_1P_3$). The treatment 20μ-thick transparent polythene covers having 20% perforation ($C_1G_1P_2$) produced significantly highest benefit: cost ratio of 3.388.

Inside bunch covers, physical and biochemical properties of banana showed differential responses to combined effect of meteorological parameters on multiple regression analysis. The influence of four meteorological observations (light interception and photosynthetically active radiation inside bunch covers) varied from 2.5% in finger pulp: peel ratio to as much as 79.6% in TSS. Significantly negative association with temperature was in shooting-harvest interval at 5% level of significance, whereas significantly positive association with temperature was observed in TSS, total sugar and non-reducing sugars. By and large other physical and biochemical properties were not influenced solely by any of the four meteorological parameters but on their cumulative effect.

To identify major parameters responsible for almost similar treatments principal component analysis (PCA) was used on the basis of variance explained more 5% of total variance and Eigen value more than one (Table 36.6). Thus, three factors were derived and total variance

Table 36.6. Component loading, Eigen value and accounted for variance based on correlation matrix input for bunch and finger quality characters of banana.

Parameter	*Factors*		
	1	*2*	*3*
Bunch weight	0.838	-0.424	0.143
Finger weight	0.920	0.037	-0.083
Finger length	0.947	-0.081	0.049
Finger circumference	0.896	-0.273	-0.013
Finger volume	0.933	0.074	-0.073
Peel weight	0.740	0.367	-0.303
Pulp weight	0.865	-0.204	0.060
Pulp thickness	0.896	-0.273	-0.025
Peel thickness	-0.358	0.103	0.859
Shooting-harvest	-0.871	-0.416	-0.144
Harvest index	0.778	-0.365	-0.241
TSS	0.910	0.317	0.150
Total sugar	0.920	0.303	0.157
Reducing sugar	0.919	0.214	0.055
Acidity	-0.308	0.747	-0.281
Ascorbic acid	0.840	0.160	-0.123
Benefit: cost ratio	0.941	-0.319	0.038
Non reducing sugar	0.870	0.362	0.235
Eigen value	12.673	1.891	1.149
Variance (%)	70.403	10.506	6.384
Cumulative (%)	70.403	80.909	87.294

explained was confined to over 87%. Bunch weight, finger weight, finger length, finger circumference, finger volume, peel weight, pulp weight, pulp thickness, TSS, total sugar, reducing sugar, non-reducing sugar, ascorbic acid and benefit: cost ratio dominate factor 1 with positive loadings as expressed by principal component (Factor 1) and Factor 1 explained over 70% of total variance. Similarly, Factor 2 which explained further 11% of total variance was dominated by positively loaded characters like finger acidity. From the scatter diagram of regression factor scores for Factor 1 vs Factor 2 it was concluded that the treatment $C_1G_1P_2$ followed by $C_1G_2P_2$, $C_1G_1P_3$, $C_1G_1P_1$, $C_1G_2P_1$, $C_1G_2P_3$, and $C_2G_2P_2$ are influential treatment combinations for characters representing Factor 1. But second factor was headed by $C_1G_2P_1$, $C_1G_1P_1$, $C_1G_2P_3$ and $C_1G_1P_3$. So, considering both factors the performing treatment combination were $C_1G_1P_1$ (20µ-thick transparent polythene covers having 10% perforation), $C_1G_2P_1$ (40µ thick transparent polythene covers having 10% perforation), $C_1G_2P_3$ (40µ thick transparent polythene covers having 30% perforation) and $C_1G_1P_3$ (20µ-thick transparent polythene covers having 30% perforation) where characters dominant in both such factors are present.

Factor 3 is dominated by peel thickness with highest positive loading and explained above 6% of total variance (Table 36.7). The treatment $C_2G_2P_1$, i.e.40µ thick blue polythene covers having 10% perforation and $C_2G_2P_3$ i.e.40µ thick blue polythene cover having 30% perforation were the treatment combinations. Considering all these three factors, it was concluded that among the bunch covers, the best performing treatment was transparent polythene covers of 20µ thick having 20% perforation in respect of bunch weight, finger weight, finger length, finger circumference, finger volume, peel weight, pulp weight, pulp thickness, TSS, total sugar, reducing sugar, non reducing sugar, ascorbic acid and benefit: cost ratio, followed by 40µ-transparent polythene covers having 20% perforation (Tables 36.1-36.5).

Table 36.7. Factor-wise distribution of performing treatments due to positively and negatively loaded dominant characters

Factor	*Dominant character with negative loading*	*Dominant character with positive loading*	*Treatments dominated positively by loaded characters*	*Treatments dominated by negatively loaded characters*	*Variance explained (%)*
1.	Bunch weight, finger weight, finger length, finger-circumference, finger volume, peel weight, pulp weight, pulp thickness, TSS, total sugar, reducing sugar, non-reducing sugar, ascorbic acid, benefit-cost ratio	Shooting-harvest interval	$C_1G_1P_2$, $C_1G_2P_2$, $C_1G_1P_3$, $C_1G_1P_1$, $C_1G_2P_1$, $C_1G_2P_3$, $C_2G_2P_2$	Control	70.40
2.	Acidity	-	$C_1G_2P_1$, $C_1G_1P_1$, $C_1G_2P_3$, $C_1G_1P_3$	-	10.50
3.	Peel thickness	-	$C_1G_2P_3$, $C_1G_2P_2$	-	6.38

So, it can be suggested that use of 20-40µ-thick transparent polythene covers having 20% perforation would be beneficial for improving yield and finger quality in banana.

SUMMARY

Transparent polythene bunch covers appreciably increase finger circumference, pulp thickness, total soluble solids, total and reducing sugars in fingers. Transparent polythene bunch covers reduce shooting-harvest interval by around 18 days as compared to the control. Polythene bunch covers having 20% perforation significantly improve finger circumference, peel thickness as well as bunch weight. Whereas, thickness of polythene bunch covers do not produce any considerable change on physico-chemical properties of fingers. Interception of light and photosynthetically active radiation inside polythene bunch covers is markedly enhanced by transparent and thickness and highest perforation of polythene bunch covers. Higher perforation proved to be cost-effective having maximum benefit: cost ratio.

References

Anonymous 2006 a. History. http://www.nal.usda.gov/fmc/foodcomp

Anonymous 2006 b. Market. www.banana.com

Careli, M. A. and Magalhaes, A. 1979. Partitioning of nitrate reductage activity in chlorophylls tissues of soyabean seedlings. *Pl. Physiol.*, **63** (suppl.)**:**25.

Dhua, R.S., Sen, S.K. and Bose, T.K. 1988. Physico chemical development in banana CV. Giant Governor during fruit growth and maturity. *Prog. Hort.,* **20**:128-32.

Green, G.C. and Kunhe, F.A. 1975. Physiological changes of banana with temperature. *Agrochemophysica,* **7**:27-32.

Haque, M.A. 1984. Effect of planting time on the production of banana in Bangladesh. *Bangladesh J. Agric.,* **9**:23-32.

Hasan, M.A., Bhattacharjee, S. and Debnath, U. 2000. Studies on the chlorophyll content of fruit peel during growth and development of banana (*Musa* AAA) subgroup Cavendish, CV. Giant Governor) under different bunch. *Orissa. J. Hort.,* **28**:36-40.

Khantaniar, L. 1989. Effect of time of planting on banana cv. Jahaji (Cavendish subgroup AAA). M.Sc. (Agric.) Thesis, Assam Agricultural University, Jorhat.

Reddy, S.A. 1989. Effect of bunch covers on bunch maturity and fruit size in high density Robusta banana orchards. *J. Res., APAU,* **17**:81-82.

Rodriguez, M.G.V., Souto, R.F. and Menegucci, J.L.P. 2001. Influence of polyethylene banana bunch cover for irrigated banana tree in the north of Minas Gerais state. *Revista Brasileira de Fruticultura,* **23**(3):559-62.

Taylor, A.D. and Craig, A.S. 1971. Plants under climatic stress. II. Low temperature, high light effect on chloroplast ultrastructure. *Pl. Physiol.,* **47**:719-25.

Weerasinghe, S.S. amd Ruwanpathirana, K.H. 2002. Influence of bagging material on bunch development of banana (*Musa* spp.) under high density planting system. *Annals of the Srilanka Department of Agriculture,* **4:**47-53.

37

Development of Flavour in Ratoon Crop of Tissue Cultured Banana

L.Nalina, N.Kumar, P. Jeyakumar, J.Auxcilia, P.S.Kavitha and T.N.Balamohan

Banana is a tropical fruit with a pleasant flavour, widely consumed throughout the world. The primary objective in any production system is to obtain maximum fruit yield per unit area without affecting fruit quality. The fruit quality in banana is mainly judged by sugar content and acidity in the pulp. Flavour and aroma are properties of food that strongly influence its acceptance by consumers. The characteristic aroma of banana arises from a complex mixture of compounds, esters of short-chain fatty acids such as acetates, butanoates and 3-methylbutyl esters (Macku and Jennings, 1987; Perez *et al.*, 1993).

DEVELOPMENT OF FLAVOUR IN RATOON BANANA

The experiment was laid out with tissue cultured banana Robusta (AAA) as a test crop in a randomized block design with six treatments and four replications. The treatments consisted of application of 100, 150 and 200% of recommended NPK (110:35:330g/plant) in three splits 3, 5, 7 months after planting and four splits 2, 4, 6 and 8 months after planting. Fruit quality parameters, viz. TSS, acidity, ascorbic acid and sugar content were analysed as per standard procedures. The flavour compounds were analysed using gas chromatography (Varian Analytical Instruments, GC model CP-3800).

A marked effect on fruit quality was observed with the application of adequate amount of nutrients. Higher levels of TSS, ascorbic acid, reducing, non-reducing and total sugars and high sugar: acid ratio with a lower acidity were recorded in treatment received165:52.5:495g of NPK/plant (Table 37.1). Higher fruit quality especially sugar content could be due to better role of nutrients particularly potassium which is involved in carbohydrate synthesis, breakdown and translocation of starch, synthesis of protein and neutralization of physiologically important organic acids (Tisdale and Nelson, 1966).

Table 37.1. Effect of different fertilizer treatments on fruit quality traits and flavour compounds of tissue culture Robusta banana

Treatment	*TSS (%)*	*Acidity (%)*	*Ascorbic acid (mg/100g)*	*Total sugar (%)*	*Acetaaldehyde (%)*	*Ethanol (%)*	*Iso amylacetate(%)*
T_1	18.46	0.23	14.10	17.32	12.23	11.67	0.20
T_2	19.00	0.17	14.85	18.00	13.85	11.57	0.81
T_3	19.50	0.15	15.42	18.88	14.20	4.66	0.57
T_4	20.63	0.12	16.23	19.44	15.37	4.11	0.94
T_5	17.40	0.24	12.55	16.00	10.74	19.09	0.12
T_6	18.21	0.22	13.88	17.18	10.77	12.35	0.16
S Ed	0.13	0.004	0.14	0.18	0.77	2.79	0.14
CD (p=0.05)	0.28	0.010	0.33	0.40	1.58	6.20	0.32

THE RESULT

Increased level of K_2O application reduced the acidity of pulp. This may be due to fact that under low K_2O regime, phosphoenol pyruvate (PEP) is apparently shunted into alternate pathways, resulting in shortage of acetyl Co-A (Patee and Teel, 1967). Hence, oxalo acetate appeared to be preferentially formed from PEP in plants with low levels of K_2O in as much as organic acid derivatives accumulate. Neutralization of organic acids due to high K^+ level in tissues also could be attributed to reduction in acidity (Tisdale and Nelson, 1966).

The fruit quality parameters, TSS, ascorbic acid and sugars are significantly influenced by split application of fertilizers. The highest value for these is observed in T_4 (165:52.5:495 g NPK/plant) in four splits at two months interval. The effect of split doses of fertilizers on fruit quality of bananas has been reported (Mustaffa, 1987, 1988; Ram and Prasad, 1988; Agrawal *et al.*, 1997; Tirkey *et al.*, 1998).

Presence of flavour compounds are determined as the flavour of banana fruit depends on quantity of esters of short chain fatty acids such as acetates, butanates and 3-methyl butyl esters (Perez *et al.*, 1993) present in fruits. Flavour compounds such as isoamyl acetate, acetaldehyde and ethanol were estimated in banana fruits in different treatments. The results reveal maximum isoamyl acetate and acetaldehyde and minimum ethanol in treatment T_4(NPK @ 165:52.5:495 g/plant in four splits). Fruits also had higher amount of carbohydrates, which could provide significant quantity of precursors for flavour compounds during ripening. The highest quality and flavour compounds are observed in the treatment T_4, i.e. application of NPK @ 165:52.5:495g/plant (150% of recommended dose of fertilizers) in four splits.

SUMMARY

Higher levels of TSS, ascorbic acid, reducing, non-reducing and total sugars and high sugar: acid ratio with a lower acidity were recorded in treatment received165:52.5:495g of NPK/plant. Estimation of flavour compounds revealed that maximum isoamyl acetate, acetaldehyde and minimum ethanol was found in the same treatment.

References

Agrawal, S., Pandey, S.D. and. Tiwari, B.L. 1997. Studies on the effect of high status of nitrogen and potassium on qualitative characters of *in vitro* banana fruit cv.Robusta. *The Orissa J. Hort.,* **25 (1):**67-72.

Macku, C., and Jennings, W.G.1987. Production of volatiles by ripening bananas. *J. Agric. Food Chem.,* **35:**845-48.

Mustaffa, M.M. 1988. Effect of spacing and nitrogen on growth, fruit and yield of Robusta banana grown under rainfed conditions. *South Indian Hort.,* **36**(5):228-31

Mustaffa, M.M. 1987. Growth and yield of Robusta banana in relation to potassium\ nutrition. *J. Potassium Res.,* **3 (3):**129-32.

Perez, A.G., Sanz, C, Rios, J.J. and Olias, J.M. 1993. Comparison of main aroma constituents of apple, banana and strawberry. *Rev. Esp. Cienc. Tecnol. Aliment.,* **33**(6):665-77.

Ram, R.A. and Prasad, J. 1988. Effect of nitrogen, phosphorus and potassium on fruit quality of banana cv. Campierganji local. *South Indian Hort.,* **36**(6):290-92.

Tisdale, S.L. and. Nelson, W.N. 1966. *Soil Fertility and Fertilizers.* Macmillan Co., London, 81 p.

Tirkey, T., Pandey, S.D .and Sharma, G.L. 1998. Effect of nitrogen levels and time of application of NPK on yield and quality of *in vitro* raised banana cv. Dwarf Cavendish. *South Indian Hort.,* **46**(1-2):65-67.

38

Banana for Better Nutritional and Livelihood Security

C. K. Narayana, S. D. Sivakumar and M. M. Mustaffa

BANANA, rich in carbohydrates, is a major nutritional component of daily diet. It is free from fat and rich in minerals and vitamins, hence known as a wholesome fruit. With changing lifestyle, low calorie diets are preferred more with abundant roughages. The recommended dietary allowances (RDA) indicate that an adolescent requires 3,000 k calories of energy, 0.5-0.6 g calcium, 25 mg iron, 750 µg retinol, 1.5 mg thiamin, 1.7 mg riboflavin, 50-100 µg folic acid, 30-50 mg vitamin C, 3,000 µg carotene and 200 IU vitamin D. Because of their impressive potassium content, bananas are highly recommended by doctors for patients whose potassium is low. One large banana packs 602 mg of potassium and 140 calories. It has 2g of protein and 4g of fibres. A large banana weighing 150 g has less than 2 mg of sodium and is good for people suffering from hypertension. Vitamins and minerals are abundant in banana offering 123 IU of vitamin A, 0.07 mg thiamin, 0.15 mg riboflavin, 0.82 mg of niacin, 0.88 mg vitamin B6 and 29 mcg of folic acid. Thus, banana is among the healthiest of fruits. The plantain, when cooked, rates slightly higher on nutritional scale in vitamins and minerals but similar to banana in proteins and fibre content. It is good for many of the common ailments like anaemia, blood pressure, constipation, depression, stress, peptic ulcers and heartburns (*see* Box).

Nutrition value of banana and plantain

Parameters		*Banana (raw)*	*Plantain (raw)*
		Nutrients (per 100 g)	
Carbohydrates	-	23 g	32 g
Fats	-	0 g	0 g
Proteins	-	1 g	1 g
Dietary fibre	-	3 g	2 g
Energy	-	89 calories	122 calories
		Minerals (per 100 g)	
Calcium	-	5 mg	3 mg magnesium
	-	27 mg	37 mg
Phosphorus	-	22 mg	34 mg
Potassium	-	358 mg	499 mg
Sodium	-	1 mg	4 mg
Iron	-	0.26 mg	0.6 mg
Manganese	-	0.27 mg	0 mg
Zinc	-	0.15 mg	0.14 mg
		Vitamins (per 100 g)	
Vitamin A	-	64 IU	1127 IU
Vitamin C	-	8.7 mg	18.4 mg
Vitamin E	-	0.1 mg	0.14 mg
Vitamin K	-	0.5 mcg	0.7 mcg
Thiamin		0.031 mg	0.052 mg
Riboflavin	-	0.073 mg	0.054 mg
Niacin	-	0.665 mg	0.686 mg
Vitamin B6	-	0.367 mg	0.299 mg
Folate	-	205.6 mcg	225.6 mcg
Pantothenic acid	-	0.334 mcg	0.260 mcg

BANANA FOR NUTRITIONAL SECURITY

Banana provides a source of livelihood and a source of nutritional security to millions across the globe. With the spiraling costs of inputs, increase in production, falling market prices and tough competition in international trade, the net income from banana farming has started showing dwindling sign. Therefore, where we have to think about diversification for both livelihood and nutritional security of rural households in particular and country's population in general. In the next 10-20 years, scientists will breed many more cultivars with high production capability. Will more banana and plantain mean more money and greater well being for farm households and rural communities? This is a question to be answered by all the concerned with the development of programme and policy for banana farmers. In developing countries, 55% of the economic value of agricultural sector derives from post-production activities, while in more developed economies over 80% of the value of agriculture is from post- production. Therefore, more income and greater well being from more banana and plantain is more likely to come from post-production activities. However, post-production is controlled by urban-based businesses and traders. Then how can the increasing value of banana and plantain that comes from processing and marketing benefit rural households and communities? There are two elements which are important in this aspect. One is competitive small and medium rural enterprises for processing and marketing and another is a dynamic information and services support system which supports innovation and competitiveness among rural enterprises.

Livelihood Security

Though India is largest producer of banana in the world, the per capita availability is about 1.5 kg/person/year. It is a source of livelihood for millions of people in several states both in production and post production stages. The livelihood can be defined as the capabilities, assets (including material and social resources) and activities required for a means of living. A recently conducted study by INIBAP has shown that thousands of micro-enterprises which process banana and plantain are extremely important to marginal households as a livelihood strategy for survival (see Box). More than 96% of the bananas produced in country are sold as fresh fruits domestically, while 0.1% is processed and 0.8% is exported in processed form. Among plantains, 45% are sold as fresh fruits and 53% are processed into chips and other products, and the rest is exported.

Micro-enterprises in banana

Type of business	*Products/countries*
Micro-enterprises: single product prepared and sold on the street (100's to 1000's in each country)	Philippines – Banana cue Cameroon – Banana fritters, roasted plantain Nigeria – dodo ikere, roasted plantain Malawi – Banana pancakes, muffins, snack foods
Micro-enterprises: single product prepared at home for sale on street or from home (100's to 1000's in each country)	Nicaragua – chips, sweets, baked ripe plantain, chocolate covered ripe banana, vinegar India – Chips, sweets, handicrafts, fibres Costa Rica – Chips Nigeria – Chips, Beer Tanzania – Beer, Brandy Malawi – Banana pancakes, muffins, snack foods Cameroon – Chips
Small to medium enterprises: mixed snack or food production with company label and license (few to 50 per country)	Chips – Nigeria, Malaysia, India, Costa Rica, Nicaragua, Philippines, Cameroon Banana relish, juice, jam, figs, sweets – India Flour – Nigeria, Cameroon Banana Catsup – Philippines Alcohol – Cameroon
Micro and small enterprises promoted by small NGOs (few to 10 per country)	Figs – Cameroon, TanzaniaNicaragua, Costa Rica Wine – Malawi, Tanzania Paper products and handicrafts – Nicaragua, Tanzania, Malawi Pre-frozen food for export – Costa Rica Yogurt – Nicargaua
Medium to large businesses with modern processing technology (few per country)	Frozen pre-cooked yogurt – Costa Rica Dairy products – Cameroon, Costa Rica, Nicaragua, India, Malaysia, Malawi, Puree – Costa Rica, India, Philippines, Chips – Costa Rica, Nicaragua, India, Flour – Nigeria, Wine – Tanzania

The initiative of INIBAP has helped in bringing out a clear picture about the global status of banana processing and its support systems available in major banana-growing countries across the globe. Of the countries surveyed by INIBAP, about 5% of the total production of dessert banana and 24% of plantain processed. Most of these processed products are made fresh on streets everyday, thereby providing livelihood to several marginal households, particularly in Costa Rica, Nigeria, Cameroon and India. The most common products in these countries are chips, sweets and roasted plantain. In Philippines, 30-40% of cooking bananas produced are processed into beer, alcohol, chips, catch-up and street foods like 'banana-que'. Even in India, considerable quantity of cooking bananas go for making street foods.

Banana figs produced by sun-drying of ripe bananas is one of the oldest processed banana products. It is widely produced in the Philippines, Thailand, Brazil and Honduras. It is exported to USA and European markets. Banana puree or mashed bananas are highest volume processed product in the world. Other processed banana products are dehydrated flakes, powder, chips and catch-up. The Philippines exported 36,538 tonnes of banana chips, valued at US$ 36.86 million in 2004. Banana flour, meal and powder were exported to a tune of 2.2 tonnes valued at US$ 14,771. In India, most popular processed banana product is chips, while pickles, flour, fig and juice have started picking up. Most of these producers are in the unorganized sector. The trade in chips in India is estimated to be to a tune of Rs 500 crore annually with a volume of production of 1.0 lakh tonnes per annum. It provides employment to more than one million people directly and indirectly. About 3-4% of total banana produced is processed in India, while Philippines process 16% and Latin America 25%. Therefore, there is a tremendous scope for diversifying into non-farming activities like processing of banana. Among non-edible products, banana fibre is prominent. It is being produced and utilized in all banana-growing countries. However, highest commercial exploitation of banana fibre for economic development is being done in Philippines. The yield of banana fibre varies from 1 to 1.5% depending on variety and method of extraction. It is estimated that we can obtain 28 million tonnes of banana fibre annually.

The survey conducted by INIBAP involving 9 banana-producing countries across the globe summarized that processing and product diversification have potential to contribute to rural development and to the reduction in poverty, but there is a need for systematic and sustained investment in business services environment and in the management capacity of potential small business operators. The strategies for Costa Rica and India to promote added value in rural communities indicate that progress is possible.

MUSA PROCESSING TO REDUCE RURAL POVERTY — A CASE STUDY

Livelihood Security

By providing employment opportunities, *Musa* processing industries create means of living for labourers. Hence the government has placed greater emphasis on promoting agro-processing industries, especially in rural areas. *Musa* processing industries have made an impact in the lives of labourers employed in these industries. From the set of industries in which in-depth data collection and analysis was made, three industries were purposively selected for the analysis. The industries selected are given in Table 38.1. The labourers in each unit were selected by simple random sampling method. The sample size was 60 labourers. An interview schedule was prepared for collecting data from labourers.

Table 38.1. Industries selected for study

Selected Industry	*No. of employees in the unit*	*Sample size*	*Products made*	*Remarks*
Dorven	30	20	Small scale processing unit that produces juice, pickle, fig, etc.	Adopts modern technology and new products
Branded Banana Chips	30	20	Chips	Traditional technology, labour intensive, branded traditional product
Kolvel Cooperative Fibre Society	400	20	Fibre products	Traditional technology, labour intensive, traditional product

There were two major branded banana chips making units at Coimbatore, which accounted for 80% of the sales of the branded segment. The sum of employees in both units was 30, hence both units were considered to draw an adequate sample of 20 employees. The data were analyzed and the results are:

General Characteristics of Sample

Age, education, gender, average family size, employment and income of sample households are discussed in this section to understand the profile of workers in selected *Musa* processing industries. The age distribution of sample households is given in Table 38.2. The results reveal that majority (83%) of employees were young (16-35 years). This trend could be observed in all industries.

Table 38.2. Age distribution of sample labourers (number of labourers)

Age groups (years)	*Dorven*	*Banana slice*	*Cooperative society*	*Overall*
16 ñ 35	17	15	18	50(83)
36 ñ 55	3	5	2	10(17)
Total	20	20	20	60(100)

The educational status of employees is given in Table 38.3. About 95% of the employees were literate. Employees with diploma or graduates accounted for only 7% of the sample. It could be concluded that majority of employees had only higher secondary (12 years of education at school) level of education. The gender of employees was analyzed and the results are presented in Table 38.3. On the whole, women accounted for major (73%) share of employees. It could be seen that fibre making was entirely done by women. In chips industry, it was male dominated. Since Dorven was located in biotechnology park initiated to promote role of women, all employees were women.

Table 38.3. Educational status of sample labourers (number of labourers)

Education	*Dorven*	*Banana slice*	*Cooperative society*	*Overall*
Illiterate	3	—	—	3 (5)
Primary	6	6	—	12(20)
SSLC	9	10	11	30(50)
H.Sc.	—	2	9	11(18)
Diploma	—	1	—	1(2)
Graduate	2	1	—	3(5)

Table 38.4. Gender of employees in sample industries (number of labourers)

Gender	Dorven	Banana slice	Cooperative society	Overall
Men		16	—	16 (27)
Women	20	4	20	44(73)
Total	20	20	20	60(100)

The family size, employees and income of sample households are presented in Table 38.5. The average family size of sample households ranged between 4 and 6, indicating that sample households were of small size. The average number of workers in sample households did not vary much and it ranged between 2 and 2.5. The average income of households was maximum among workers in chips industry (Rs 5351.25) and it was lowest among workers in fibre units. The average income was found to be below poverty line among households in fibre and Dorven industries and it was only marginally higher in the households in chips industry.

Table 38.5. Family size, employees and income of labour households

Particulars	Dorven	Banana slice	Cooperative society	Overall
Average family size (number of persons)	6	5	4	5
Average of working family members in the households (number of persons)	2.5	2	2	2.16
Average family income (Rs/month)	3,780.00	5,351.25	2,985.50	4,038.51

Thus, it could be concluded that majority of employees were young and moderately educated (without any special skill) and gender orientation could be observed in the type of products made. They came from small families with one-third to 50% workers and yet majority of them were under poverty.

Employment and Income Contribution

The duration of employment, number of days of employment and income of employees in *Musa* industries are discussed in this section. The duration of employment of employees is presented in Table 38.6. Majority of the employees (68%) had been working in these industries only during the last five years, while 27% of the employees had been working for the past 6-10 years. Dorven was started during 2004, hence all the employees had one year or less service. Fibre unit was started during 1970s and branded banana chips units were started 10 years ago. Considering the age distribution and duration of employment it could be concluded that only young and middle-aged persons were interested to work in these industries. The number of days of employment in these industries was analyzed and the results are presented in Table 38.7. Majority of employees (83%) had 25-26 days employment (except on Sundays, when the unit was closed), which could be rated as very good.

Table 38.6. Duration of employment in sample industry (number of labourers)

Duration (years)	*Dorven*	*Banana slice*	*Cooperative society*	*Overall*
dî 5	20	10	11	41(68)
6–10	—	9	7	16(27)
11–15	—	—	—	—
16–20	—	—	—	—
> 20	—	1	2	3(5)
Total	20	20	20	60(100)

Figures in parentheses indicate percentage to total

Table 38.7. Number of days of employment in musa processing industries (number of labourers)

Days employed per month	*Dorven*	*Banana slice*	*Cooperative society*	*Overall*
20–22	—	—	2	2(4)
23–24	—	—	6	6(20)
25–26	20	18	12	5082)
27–28	—	—	—	—
29–30	—	2	—	2(4)
Total	20	20	20	60(100)
Average days employed	25	26	24	25

The income of sample households from *Musa* industries and other sources were analyzed and the results are presented in Table 38.8. The share of income from *Musa* industries accounted for about 50% in households of chips and dorven industries, while it was only 44% fibre industries. However, it could be concluded that income from *Musa* industries was substantial in households. The distribution of the share of income from *Musa* industries is presented in Table 38.9. In about 50% of households in all industries, the share of income from *Musa* industries to that of the total family income was 41-60%. In about 8% households the *Musa* industries were sole source of income. In 30% of the households, share of income from *Musa* industries was 21-40%. Thus, it could be concluded that income contribution of Musa industries in majority of households was substantial.

Table 38.8. Income of sample households from various sources (Rs per month)

Average monthly income (Rs)	*Dorven*	*Banana slice*	*Cooperative society*	*Overall*
Musa Processing Industries	1,970.00 (52)	2,751.25 (51)	1,340.50 (44)	2,020.58 (50)
Other sources	1,810.00 (48)	2,600.00 (49)	1,645.00 (56)	2,018.33 (50)
Total	3,780.00 (100)	5,351.25 (100)	2,985.50 (100)	4,038.91 (100)

Table 38.9. Share of income earned from musa processing industries to total family income (number of labourers)

Share of income from Musa processing industries (%)	*Dorven*	*Banana slice*	*Cooperative society*	*Overall*
d" 20	—	—	—	—
21–40	3	6	9	18 (30)
41–60	10	8	11	29(48)
61–80	6	2	—	8(14)
81–100	1	4	—	5(8)
Total	20	20	20	60 (100)

Figures in the parentheses indicate percentage to total

Table 38.10. Mode of expenditure of income from Musa processing industries (number of households)

Particulars	*Dorven*	*Banana slice*	*Cooperative society*	*Overall (n= 60)*
Meet food expenditure	19	16	8	43(72)
Repay loan obtained	16	9	12	37(62)
Education of children	14	8	10	32(53)
Health expenses	19	11	—	30(50)
Savings	9	4	6	19(32)
Purchase jewels	—	—	3	3(5)
Materials	—	—	1	1(2)

Figures in parentheses indicate percentage to total

Expenditure Pattern of Income from Musa Industries

Income from any activity could used to meet a wide range of needs and the allocation of money for various purposes would reflect the status of household. Hence, manner in which the income earned from *Musa* industries was spent by sample households was analyzed to ascertain whether the income was used for creating assets, meet basic needs of family or used for investment in other ventures. The results of analysis are presented in Table 38.10. In about 72% of households income was spent on food. The other major use of the income was to repay loan obtained, followed by education expenses of children and health care expenses. Only 32% of households could save something out of the income from *Musa* industries. Asset creation (purchase of materials and jewels) could be observed in only 8.3% of households. Thus, it could be concluded that majority of households used the income to meet basic needs of family, which is characteristic of households in poverty.

39

Shelf-Life and Quality Development in Banana

K.B. SHEELA, M. RENI, V.K RAJU AND P.B. PUSHPALATHA

BANANA has gained popularity in international fruit trade. India is largest producer of banana and plantain in the world. The post-harvest losses in banana are 20-30% which can be minimized by systematic and scientific management. Hence an investigation was undertaken at the Department of Processing Technology, College of Horticulture for studying the effect of post-harvest treatments on quality and shelf-life of banana Njalipoovan.

DEVELOPMENT OF SHELF-LIFE AND QUALITY IN BANANA

Hands from matured branches were subjected to different post-harvest treatments and then kept under ambient conditions for ripening. The different treatments were: hot water treatment at 45±5 °C for 2 minutes(T_1), precooling with tap water for 10 minutes(T_2), precooling with cold water at 10 °C for 15 minutes (T_3), precooling with ice flakes for 10minutes (T_4), coating with edible waxes (waxol 2%) at cut end of hands (T_5), vacuum infiltration with 1% $Cacl_2$ for 3 minutes (T_6), dipping in 1% $Cacl_2$ for 3 minutes (T_7), and control (T_8). Observations were recorded on shelf-life, physiological loss in weight, days to 50 and 100% yellowing and chemical composition on ripening.

THE RESULTS

Physiological loss in weight registered an increasing trend during storage (Table 39.1). Throughout the storage period, there was minimum physiological loss in weight (T_6), followed by T_2, whereas the control samples recorded maximum value. The T_6 was found to be superior to other post-harvest treatments in extending the days to reach 50% yellowing, 100% yellowing and shelf-life (17.5 days) (Table 39.2).

Precooling with tap water and cold water were the next best treatments in extending the days to reach ripening and shelf-life. Longer shelf-life was obtained for precooling with tap water (17.0 days), followed by cold water (16.0 days), $Cacl_2$ dip (15.5 days) as compared to the control fruits (11.0 days). High calcium fruits stores well due to less metabolic breakdown as explained by Miklos (1980). Yuen (1994) reported that ethylene production and respiration rates were immensely related to calcium concentration in flesh of fruits. At higher levels, calcium counteracted the effect of nitrogen and kept the respiration low. Reduction in physiological loss in weight by $CaCl_2$ infiltration may be due to retardation of respiration and transpiration by calcium salt. Similar reduction in physiological loss in weight by calcium was reported by Singh (1988) in guava, Chukwu *et al.* (1995) in plantain, (Pathak *et al.*, 1999).

Table 39.1. Effect of post-harvest treatments on PLW during ripening in Njalipoovan banana under ambient conditions

Treatment	3 DAS (%)	6 DAS (%)	9 DAS (%)	12 DAS (%)	15 DAS (%)	18 DAS (%)
T_1	15.600	22.750	29.500	34.600	ND	ND
T_2	10.100	15.950	19.750	22.400	26.00	ND
T_3	10.450	16.600	19.350	23.200	25.05	ND
T_4	12.050	18.600	20.350	31.500	ND	ND
T_5	12.150	16.250	19.100	22.550	ND	ND
T_6	3.100	4.600	7.100	9.350	15.85	23.15
T_7	12.000	21.150	23.550	27.900	30.20	ND
T_8	15.450	28.600	33.900	42.900	ND	ND
CD (0.05)	3.234	3.191	NS	NS	NA	NA

ND, not determined due to spoilage of fruits; NA, not analysed statistically DAS: Days after scope

Table 39.2. Effect of post-harvest treatments on days to 50 and 100% yellowing and shelf-life in Njalipoovan banana

Treatment	Days to 50% yellowing	Days to 100% yellowing	Shelf-life
T_1	6.0 (2.550)[c]	8.0 (2.910)[bc]	12.5 (3.605)[ij]
T_2	8.5 (2.999)[ab]	10.5 (3.316)[ab]	17.0 (4.183)[de]
T_3	9.0 (3.078)[a]	10.5 (3.309)[ab]	16.0 (4.060)[ef]
T_4	7.0 (2.732)[abc]	7.0 (2.732)[c]	12.0 (3.536)[jk]
T_5	6.5 (2.644)[bc]	8.5 (2.999)[abc]	12.5 (3.605)[ij]
T_6	9.0 (3.082)[a]	11.0 (3.391)[a]	17.5 (2.242)[cd]
T_7	7.0 (2.732)[abc]	9.5 (3.161)[abc]	15.5 (4.000)[fg]
T_8	5.5 (2.447)[c]	8.5 (2.999)[abc]	11.0 (3.391)[k]

Figures in parentheses are transformed values; values with same superscripts do not differ significantly

Post-harvest treatments did not have any significant effect on chemical constituents of fruits, viz. TSS, total sugars and reducing sugars on ripening (Table 39.3) The TSS was highest in T_4 and the control samples (30.5° Brix). Acidity varied significantly among various treatments, maximum (0.64%) being in T_8 (control) and minimum (0.39 %) in T_3 (precooling with cold water). The values for total and reducing sugars ranged from 19.5 to 18.06% in T_2 to 23.8 to 22.85% in T_8.

Table 39.3. Effect of post-harvest treatments on chemical composition of ripe fruits of Njalipoovan

Treatment	TSS (Brix)	Acidity (%)	Total sugars (%)	Reducing sugars (%)	Non-reducing sugars (%)
T_1	27.500	0.545	19.880	18.550	1.330
T_2	27.500	0.590	19.600	18.060	1.540
T_3	27.500	0.390	20.930	19.410	1.520
T_4	30.500	0.420	22.450	19.925	2.525
T_5	30.000	0.530	21.990	19.235	2.755
T_6	29.000	0.600	21.000	19.140	1.860
T_7	28.500	0.570	22.150	20.725	1.425
T_8	30.500	0.640	23.800	22.845	0.955
CD (0.05)	NS	0.41*	NS	NS	1.21*

* - Significant at 5% level; NS, non-significant

SUMMARY

Minimum physiological loss in weight throughout the storage period was registered by vacuum infiltration with 1% $Cacl_2$, followed by pre-cooling with tap water. Vacuum infiltration with 1% $Cacl_2$ was effective in delaying 50% yellowing and 100% yellowing, extending shelf-life by 17.5 days. Longer shelf-life was also obtained for pre-cooling with tap water (17.0 days), cold water (16.0 days) and $Cacl_2$ dip (15.5 days).

References

Dhoot, L.R., Desai, U.T. and Rane, D.A. 1984. Studies on shelf-life of guava with polythene packaging and chemical treatment. *J. Maharashtra Agric. Univ.,* **9**:185-88.

Chukwu, E.U., Olorunda, A.O. and Ferris, R.S.B. 1995. Extension of ripening period of *Musa* fruit using calcium chloride infiltration, Semperfresh and Brilloshine-1 coating. *Musa Africa,* **8**:14-15.

Miklos, F. 1980. Modern concepts in fruit nutrition. *Mineral Nutrition of Fruit Trees.* Atkinson, D., Jackson, J.E., Sharples, R.D. and Waller, W.M. (Eds). Butter Worths and Co. Ltd, London.

Pathak, N., Sanwal, G.G. and Pathak, N. 1999. Regulation of ripening of banana (*Musa acuminata*) fruits by chemicals. *Indian J. Agric. Sci.,* **69**(1):17-20.

Singh, G. 1988. Effect of calcium nitrate and plant growth regulators on the storage of Allahabad Safed guava. *Indian J. Hort.,* **45**:45-50.

Yuen, C.M.C. 1994. Calcium and fruit storage potential. *Postharvest Handling of Tropical Fruits.* Champ, B.R., Highley, E. and Johnson, G.I. (Eds). *Proceedings of the International Conference on Postharvest Handling of Topical Fruits,* Chiang Mai, Thailand, pp.218-25.

40

Osmotic Dehydration in Banana Slices

Dattatreya M. Kadam, Devinder Dhingra, Mathhew Prasad and R.T. Patil

Dehydration in fruits is an alternative to reduce post-harvest losses in fruits and also a process to produce dried fruits, which can be directly consumed or become part of foodstuffs like cakes, pastries and many others. Osmotic dehydration is a method for partial removal of water from plant tissues by immersion in a hypertonic (osmotic) solution. The driving force for diffusion of water from tissue into solution is provided by higher osmotic pressure of hypertonic solution (Rastogi and Raghavarao, 2004). The diffusion of water is accompanied by simultaneous counter diffusion of solute from osmotic solution into tissue. Since membrane responsible for osmotic transport is not perfectly selective, other solutes present in cells can also be leached into osmotic solution. The rate of diffusion of water from any material made up of such tissues depends upon factors such as: temperature and concentration of osmotic solution, size and geometry of material, solution to material mass ratio and level of agitation of solution. (Bongirwar and Sreenivasan, 1977; Pokharkar and Prasad, 1998). Osmotic dehydration processes are normally designed with the aim of maximizing water removal meanwhile restraining solid uptake, to obtain a product whose taste and flavor have little changes in respect to that of the fresh food. The ratio of water loss to solid uptake is a good index of the efficiency of the process. Therefore, to investigate osmotic dehydration kinetics of banana slices in sugar syrup solution having different process conditions and to optimize combination of banana osmosis for weight loss (water removed) and solid gain are necessary.

OSMOTIC DEHYDRATION IN BANANA SLICES

The fruits of banana cavendish were procured from the local market of Ludhiana (India). The fruits of similar ripeness, having uniform visual quality and size were selected. There were peeled from three-fourth side and one-fourth unpeeled side of banana fruit was kept on stainless steel table, to avoid contamination, during manual cutting with stainless steel knife having uniform slice thickness of 4, 8 and 12 mm.

Food grade sugar was procured from local market and known quantity of weighed sugar (kg) was poured in known quantity of tap water (kg) to prepare known concentration of sugar syrup. Sugar mixed in water is stirred till complete sugar is dissolved in water to form sugar syrup. Sugar syrup was sieved before pouring it in to a plastic container.

As per predetermined banana sample-syrup ratio, sugar syrup and banana slices were weighed and poured in plastic containers. In each experiment and time interval, samples were taken out of sugar syrup, surface syrup was removed using blotting paper, weighed and removed from experimental studies. Each experiment was replicated thrice.

Two quantities represent adequately the osmotic process: water loss (WL), indicating the water that diffuses from the fruit to solution and solid gain (SG). The solid gain represents the amount of solids that diffuses from the solution to the fruit.

The water loss (WL) and solid gain (SG) of fruit after time 't' of osmotic treatment are defined as (Singh *et al.*, 2006)

$$WL\ (\%) = \frac{(M_0 - m_0) - (M - m)}{M_0} \times 100 \qquad \text{.................................. (1)}$$

$$SG\ (\%) = \frac{(m - m_0)}{M_0} \times 100 \qquad \text{..................................(2)}$$

where, M_0 is initial mass of fresh fruit before osmotic treatment; M is mass of banana after time 't' of osmotic treatment; m is dry mass of banana after time 't' of osmotic treatment and m_0 is dry mass of fresh fruit.

Seven experiments were conducted using a factorial completely randomized design (CRD) with 3 sugar syrup concentration, 3 slice thickness, 3 sample to syrup ratios and 7 time intervals and 3 replications each. Data were analysed using Ag Res statistical package for single and two factor ANOVA.

THE RESULTS

Weight loss and solid gain increased with increase in sugar syrup concentrations (Figs 40.1 and 40.2). The increase in weight loss and solid gain may be due to synergistic effect of sugar syrup to develop high osmotic potential. Ade–Omowaye *et al.* (2002); Marani *et al.* (2006); Barat *et al.* (2001); Sutar and Gupta, (2007) and Chenlo *et al.* (2007) also observed similar effect of concentration of osmotic solution. The time and concentrations are highly significant during weight loss and combination of both are non-significant. The time, concentrations and combination of both are non-significant. Hence, there is no effect of time and concentration on banana osmosis in terms of solid gain.

The weight loss and solid gain increased with decrease in banana slice thickness during osmosis process (Figs 40.3 and 40.4). It was also noted that banana slice thickness had a negative effect on weight loss and solid gain. Similar effects was observed by Agnelli *et al.* (2005) where water loss and solute gain were more in apple cubes of 1.5 cm as compared to 2 cm during osmotic dehydration. In case of thinner slices, distance to be travelled by water to reach the fruit surface is smaller; this could be the reason for higher water loss for thinner slices. An increased flow of sugar syrup solution to fruit slice will result in higher solute gain. Weight loss indicates that time interval and syrup concentrations are highly significant and combination of both are non-significant. Effect of slice thickness on solid gain found that the time and concentrations are highly significant and combinations of both are non-significant.

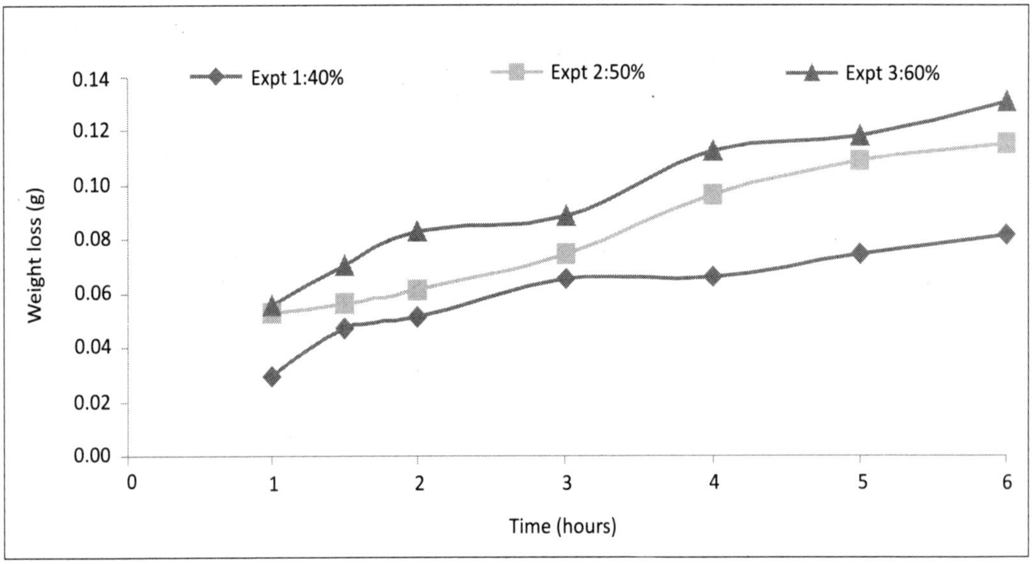

Fig 40.1. Effect of sugar syrup concentration on weight loss of banana due to osmosis

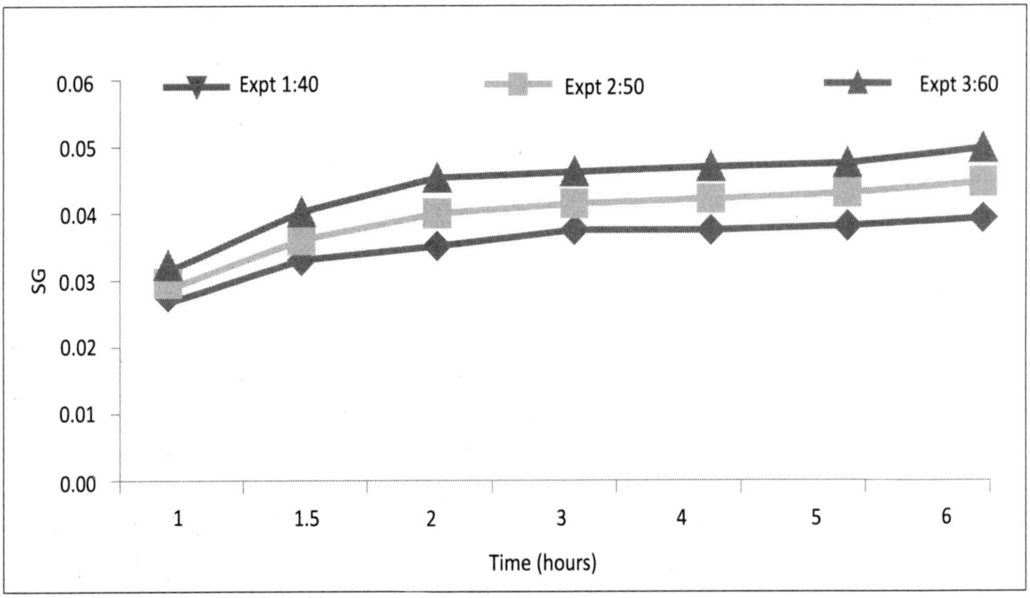

Fig 40.2. Solid gain (SG) due to osmosis of banana at different sugar syrup concentration levels (40, 50 and 60%)

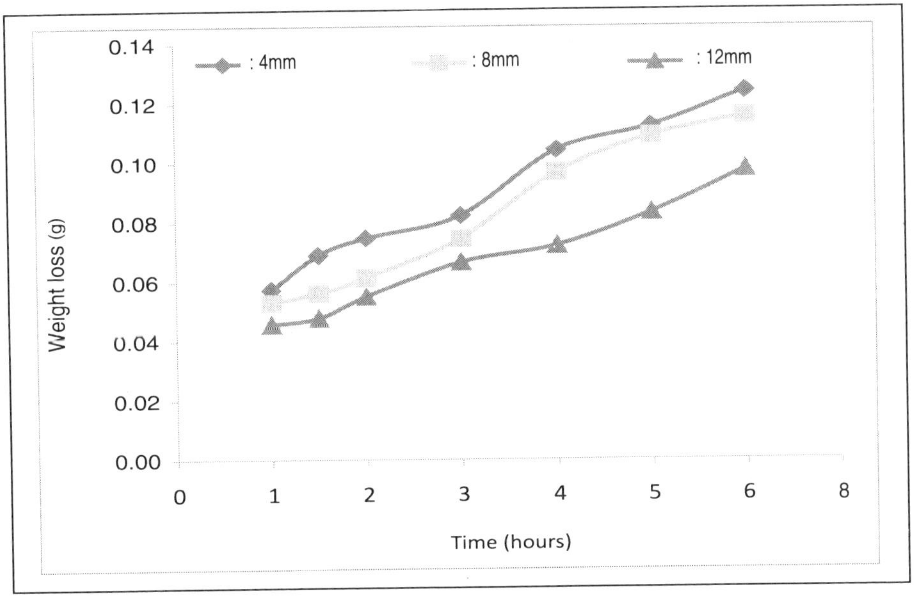

Fig 40.3. Effect of slice thickness on weight loss of banana due to osmosis

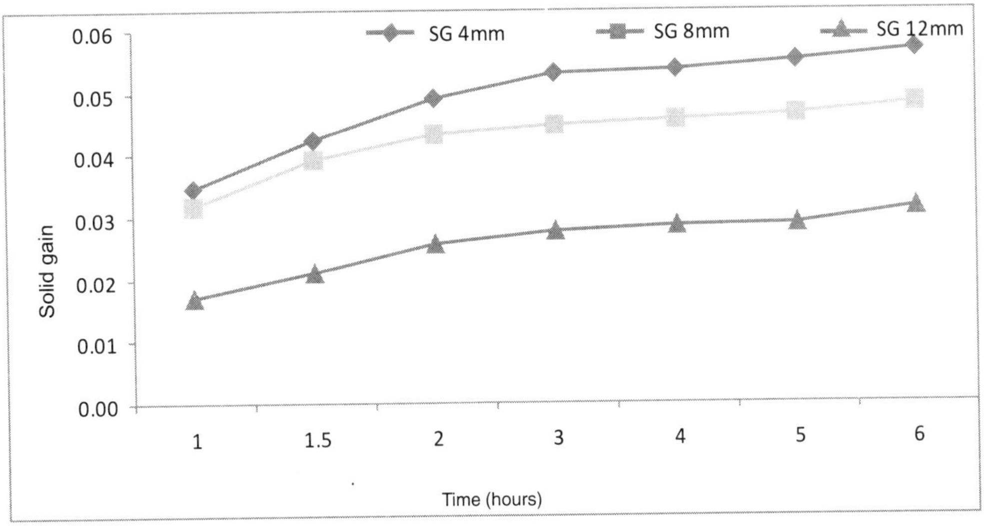

Fig 40.4. Solid gain due to osmosis of banana at different slice thickness (4, 8, 12mm)

There is an increase in weight loss and solid gain with increase of banana sample to sugar syrup solution ratio (Figs 40.5 and 40.7). Singh *et al.* (2006) also studied the effect of different sample to solution ratios (1:2, 1:4 and 1:6) on mass transfer kinetics of carrots and observed that there was an increase in water loss and solute gain with increase of fruit to solution ratio. This behaviour could be explained on the basis that water loss and solute gain took longer time to acquire equilibrium during osmotic dehydration. Pokharkar and Prasad (1998) also reported the increased rate of water loss and solute gain with increase of solution to fruit ratio. The time and concentrations are highly significant and combination of both are non-significant (Table 40.5). It is clear that time and concentrations are highly significant and combinations of both are non-significant during solid gain.

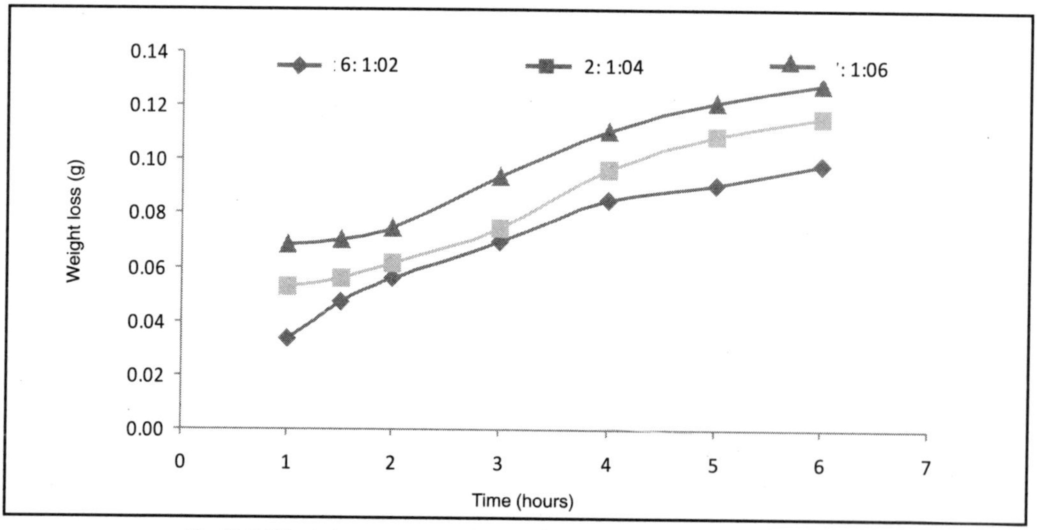

Fig 40.5. Effect of sugar syrup ratio on weight loss of banana due to osmosis

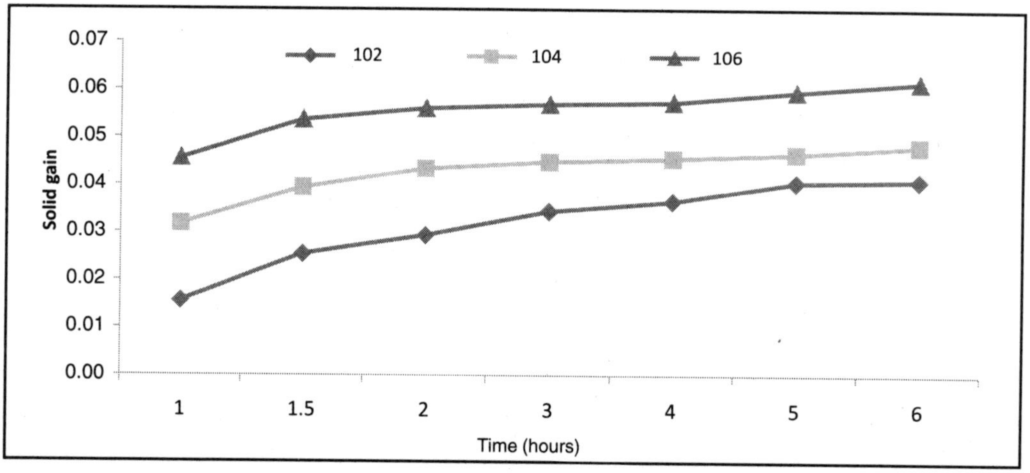

Fig 40.6. Sold gain due to osmosis of banana at different sugar syrup ratio

Mean comparison of weight loss and solid gain due to osmosis of banana was analysed statistically using AgRes by LSD with descending order. Almost all treatment combinations were significantly different. The treatment combinations ranked in terms of water loss and solid gain due to effect of time during osmosis of banana slices. Osmotic dehydrated banana slices obtained from treatment combination of 8 mm banana slices dipped in 1: 4 sample to sugar ratio having 60% sugar syrup concentration and combination of 8 mm banana slices dipped in 1: 6 sample to sugar ratio having 50% sugar syrup concentration level ranked first in terms of weight loss values which can be obtained in 5-6 hours of osmosis. On other hand, osmotic dehydrated banana slices obtained from treatment combination of 8 mm banana slices dipped in 1: 6 sample to sugar ratio having 50% sugar syrup concentration level ranked first in terms of solid gain values which can be obtained in 3 hours of osmosis.

SUMMARY

Weight loss and solid gain increased with increase in sugar syrup concentrations, weight loss and solid gain increased with decrease in banana slice thickness. There was an increase in weight loss and solid gain with increase on banana slice sample to sugar syrup solution ratio. The experimental design of osmotic dehydration of banana-slices was prepared and their combinations of different parameters were studied. The optimum combination of banana osmosis for weight loss (water removed) and solid gain were found to be: 60% sugar syrup solution + 4 mm banana slice thickness + 1: 6 sample to sugar syrup solution ratio.

References

Ade-Omowaye, B.I.O., Rastogi, N.K., Angersbach, A., Knorr, D. 2002. Osmotic dehydration of bell peppers: influence of high intensity electric field pulses and elevated temperature treatment. *J. Food Eng.,* **54**:35-43.

Agnelli, M.E., Marani, C.M., Mascheroni, R.H. 2005. Modeling of heat and mass transfer during (osmo) dehydrofreezing of fruits. *J. Food Eng.* **69**:415-24.

Barat, J.M., Fito, P., Chiralt, A. 2001. Modeling of simultaneous mass transfer and structural changes in fruit tissues. *J. Food Eng.,* **49**:77-85.

Bongirwar, D.R., Sreenivasan, A. 1977. Studies on osmotic dehydration of banana. *J. Food Sci. Technol.,* **14**:104-13.

Chenlo, F., Moreira, R., Fernandez-Herrero, C., Vazquez, G. 2007. Osmotic dehydration of chestnut with sucrose: mass transfer processes and global kinetics modeling. *J. Food Eng.,* **78**(3):765-74

Kar, A., Gupta, D.K. 2001. Osmotic dehydration charactersistics of button mushrooms. *J. Food Sci. Technol.,* **38**:352-57.

Marani, C.M., Agnelli, M.E., Mascheroni, R.H. 2006. Osmo-frozen fruits: mass transfer and quality evaluation. *J. Food Eng.,* (DOI 10.1016/j.jfoodeng.2006.03.022).

Pokharkar, S.M., Prasad, S. 1998. Mass transfer during osmotic dehydration of banana slices. *J. Food Sci. Technol.,* **35**(4):336-38.

Rastogi, N.K., Raghavarao, K.S.M.S. 2004. Mass transfer during osmotic dehydration of pineapple: considering Fickian diffusion in cubical configuration. *LWT-Food Sci Technol,* **37**(1):43-47.

Singh, B., Kumar, A., Gupta, A.K. 2006. Study of mass transfer kinetics and effective diffusivity during osmotic dehydration of carrot cubes. *J Food Eng* (DOI 10.1016 /j.jfoodeng. 2006.01.074).

Sutar, P.P., Gupta, D.K. 2007. Mathematical modeling of mass transfer in osmotic dehydration of onion slices. *J. Food Eng.,* **78**(1):90-97.

41

Modern Packaging Techniques for Banana

R.T. Patil and D.R. Rai

Packaging provides convenience of handling and extension of shelf-life by keeping the products fresh as well as free from contamination. Most food products after harvesting receive artificial treatments for preservation. The main purpose of the treatments is essentially to lengthen the safe handling time from harvesting to consumption. The longer the time, the farther the food can be distributed from the site of harvesting to consumers. Modern science has developed drying, cooking, heating, refrigerating, freezing, pasteurizing, sterilizing, irradiating, and packaging. Packaging is extremely important and can maximize the time of handling from harvesting to consumption and therefore warrants the longest shelf-life of any food product. Food packaging has become an increasingly important area in modern society. The purpose of food packaging is to safely preserve foods and distribute them in near to fresh, nice, sweet, tasty, and delicious form to consumer, preventing deterioration under various environmental conditions. Different packaging materials and their application technologies have been developed for food packaging. They include:

- Retortable packaging
- Aseptic packaging
- Free-oxygen scavenging packaging
- Vacuum packaging
- Controlled atmosphere packaging
- Modified atmosphere packaging
- Active packaging
- Anti-microbial packaging
- Nano-technological packaging
- Smart or intelligent packaging
- Micro-perforated packaging.

These packaging techniques have been further divided into different sub-techniques depending upon the requirement.

PRINCIPLES OF PACKAGING

Packaging is defined as techniques and states for applying suitable materials or containers to goods to protect and maintain product quality during shipping and storage. Since packaging includes all operations necessary before the product reaches the consumer, it should be planned to meet distribution processes and the external environment. The principles of packaging include:

- Containment for handling, transportation and use
- Preservation and protection of contents for required shelf-life
- Protection from external environmental and mechanical hazards
- Identification of contents, quantity, quality and manufacturer
- Food safety and quality retention
- Facilitation of dispensing and use
- Ease of opening, re-closure, portioning application, unit of use, multipacks, second use or re-use
- Working features; such as found in the cook-in pouches.

Packages must be designed to meet internal or external difficulties that may be encountered at respective stages of distribution, to protect goods safely from such difficulties, and to send goods to the consumers satisfactorily. The packaging engineer and researcher must know that the food packer's main aim is not to sell a package but its contents.

Primary, Secondary and Tertiary Packaging

There are three kinds of packaging. First is primary or item packaging, which is defined as packaging of individual goods and states for applying suitable materials or containers in order to attract the consumer or to protect product quality or in other words primary packaging is the main package that holds the fresh or processed food. The primary packaging is mostly carried out at the producer's level and depicts the manufacturing date, status, date of expiry etc. on each package; e.g. each milk pouch contains the information about its date, place of manufacturing, selling price and date of expiry, batch number etc. Secondary packaging combines the primary packages. It is defined as the inner packaging of packaged freight to protect goods; e.g. the individual milk pouches are either bagged or taken in plastic containers for transportation.Tertiary packaging combines all of the secondary packages into one pallet and is defined as outer pack of packaged freight for the purpose of shipping.

Harvesting, Handling and Packaging

The bananas are harvested at various stages of its maturity depending upon the purpose for which it is cultivated, such as culinary, table purpose etc. and distance to the market (three-fourths full maturity in Robusta for distant markets, while full maturity for local market etc.). The assessment of maturity is usually done by visual judging. The banana bunch weighs on an average 30-35 kg. The yield of Dwarf Cavendish in Tamil Nadu and Maharashtra is 30-40 tonnes/ha, for Poovan perennial plantations 15-20 tonnes/ha, wetland or gardenland 20-25 tonnes/ha and Hill bananas (perennial) 15-17.5 tonnes/ha.

As much as possible care is taken that it does not hit the ground. For packing and transportation, the bunch is padded with banana leaves. A fungicidal treatment is applied to cut ends to prevent stem end rot. After the paste dries, bunches are put into polythene tubes (175-gauge, 55 cm wide with 1.8 cm with holes 10 cm apart). The bunches are then stacked vertically or horizontally on the ground added with wilted or dry banana leaves in case there is delay in transportation. For exports, bananas are removed from stem and hands and clusters of bananas are packed in corrugated boxes with perforated polyethylene liners. The curved side of hands is kept facing upwards making sure that the crown of the upper hands do not damage the banana underneath.

Banana needs artificial ripening as green bunches are generally transported from field to the place of consumption. Smoke treatment is commonest method to induce ripening in Maharashtra and Tamil Nadu. Smoking is done with straw, leaves and cowdung in a closed chamber with bunches arranged in a heap for 18-24 hours in summer and 48 hours in winter. Bunches are then transferred to a well-ventilated room for development of colour. Smoke treatment cause ripening of bunches within 3 days. Ethrel, (ethylene) and temperature induced respiration is used for artificial ripening. High concentration (1000 ppm) and/or a prolonged treatment with ethrel (5 minutes) and ethylene at a concentration of 1 part per thousand (48 hours) causes intensive respiration and initiate ripening of banana. In commercial trade, ripening is initiated by using various chemical substances. Growth regulating chemicals, such as 2, 4-D, 2, 4, 5-T, IAA and TBZ have been tried to hasten ripening. Use of acetylene gas generated from CaC_2 for ripening banana was started as early as 1932. Since then CaC_2 is used on a large scale in case of ripening mature green bananas. However, due to health hazards this methods is done away with the ethylene induced ripening.

The hands are graded based on number and size of fingers in each hand. Overripe and injured fruits are discarded at this stage. Banana is sent to local market as bunches.

Bananas can be stored at a temperature slightly above 13 °C and a relative humidity of 85-95% for about three weeks, and is ripen in a week or two at 16.5-21 °C. Banana fruit becomes blackened at lower temperatures and should not be placed in a refrigerator. Under controlled atmospheric conditions, 2-5% oxygen and 2-5% carbon dioxide should be used to supplement temperature and humidity management during transportation and storage. Maintaining ethylene concentration below 1 ppm can extend post-harvest life of mature green bananas. Mature-green bananas can be stored for up to 3 weeks in ethylene-free air or up to 6 weeks in a controlled atmosphere at 14 °C. Internally, banana is carried either by rail or by road in un-refrigerated carriage. Dipping of bananas at 200 ppm TBZ has been approved and recommended as a post-harvest treatment. A double coating of 12% wax emulsion prolonged the storage life of Dwarf Cavendish banana by 10-12 days at 14.5 °C.

Banana is most important fruit in international trade. The value of world trade in bananas is higher than all other tropical fruit combined. For supermarket buyers, fruit should be packed in one layer in corrugated fibre boxes. The final packages should be tight to reduce fruit movement in transit but not be forced so as to cause damage during the packing process. All cartons should be weighed and be of standard weight and count. Handwritten labeling should be minimized early on by printing information during carton manufacture or through the use of adhesive labels.

Minimizing, simplifying and encouraging careful handling of this product throughout the chain is important for its safer post-harvest life. Strapping of cartons in twos or threes encourages upright stacking, discourages throwing and prevents pilferage. More expensively, but more reliably, the use of smaller 2-3 kg packs placed inside a 'mother' carton with a total weight of 10-15kg can also achieve same results. Strapping is recommended initially. It is essential with this product that cartons are handled horizontally and not tipped. Tipping cartons sideways will damage fruit.

Quality Considerations

At Harvesting

- Shape/length
- Maturity
- Skin appearance
- Skin blemish

After Ripening

- Uniformity of coloration
- Flesh colour
- Absence of progressive defects

Precautions at Packhouse

- Very careful handling and transportation.
- Fruit of uniform age and maturity are to be stem-suspended, in source group, under shade with good air circulation and are not dumped on the ground. For this purpose fibre sheets can be used which can be spread on ground and fruits can be placed over them. These sheets are tough and have better life as compared to other materials.
- Cooling in shortest possible time of arrival.
- The packing station should be clean, hygienic and well-managed.
- Only approved cleaning agents may be used.
- Ensure before packaging that no flower-ends remain on fruit tips.
- Bore-hole or mains supply water should be used in washing tanks.
- Use sharpened knives or proper tool for uniform cutting and minimum cell damage.
- Careful checks at every stage of handling clusters for insects, amphibians, spiders etc.

Packaging

Individual bagging of clusters in perforated polythene bags is recommended. Polypropylene bags add visibility so those must be used. Bag packaging reduces weight loss, promotes good peel colour on ripening and reduces abrasion of one fruit on the next.

- Weigh to specified weight to avoid under- or overweight as quality is impaired under both conditions.
- Correct choice of bag: perforated or non-perforated modified atmosphere packaging.
- Well-trained staff for handling and packaging.
- High quality, recyclable packaging material, glued/sealed instead of stapled.
- Full audit-trail information on every carton, plus all legally required information.

- Palletization at the point of packing is highly desirable to ensure maintenance quality and avoid excessive handling/mishandling.

RECENT ADVANCES IN BANANA PACKAGING

Modified Atmosphere Packaging of Banana

Bananas in trays and held them at 29-30 °C with covering of polyvinyl chloride film. They found that it prevents the early senescent peel spotting, which is typical for that cultivar. Carbon dioxide and ethylene concentrations within the packages increased, but inclusion of carbon dioxide scrubbers or ethylene absorbents, which considerably affect gas composition, has no effect on spotting. Experiments with continuous low oxygen concentrations confirmed that the effect of the package is mainly due to low oxygen. Relative humidity is higher in the packages but this had no effect on spotting. The positive effect of modified atmosphere packaging on peel spotting is accompanied by reduced *in vitro* phenylalanine ammonia lyase (PAL) activity in the peel, and by an increase of *in vitro* polyphenol oxidase (PPO; catechol oxidase) activity. They concluded that senescent spotting of banana peel required high oxygen levels. The reason for low spotting at low oxygen levels may be due to increase of *in vitro* PPO activity apparently showing an increase in potentially active protein.

The modified atmospehere reduced the chilling injury symptoms. Phenylalanine ammonia lyase and polyphenol oxidase activities in the peel of control bananas are considerably higher than in MA-packed fruit which may be related to CI-induced peel browning. The pulp softness, sweetness and flavour of MA-packed fruit are better than in control fruits.

Amcor P-Plus Modified Atmosphere Packaging (MAP) Technology

Amcor P-Plus MAP technology works by tailoring the gas permeability of the flexible film to extend the life of fresh produce by naturally creating a modified atmosphere. In this case, the Amcor P-Plus MAP technology gives an extra 2 days shelf-life on bananas over standard non MAP bags. The specialised wicketted bags are produced from a recyclable material and an antimist system is utilised so that full visibility of the product is achieved. The bags are Amcor P-Plus treated and flexo printed.

Temperature-Activated Intelimer Polymer-Based Packaging Technology

Landec Corporation, has developed a banana packaging technology for extended shelf-life of Banana. Their intelimer polymer based packaging technology for storing and ripening bananas the polymer customizes abruptly to change its physical characteristics when heated or cooled through a pre-set temperature switch. The bag like package is made of a customised membrane that controls the rate at which oxygen enters and carbon dioxide leaves, maintaining the optimal gas mix. The success of Intelimer represents a marketing breakthrough for membrane packaging. The active, controlled and intelligent packagings which are being developed for oxygen scavengers, antimicrobial films and gas permeable packages has led effective modified atmosphere packaging, moisture absorbers and other hybrid forms of packages to keep foods fresh. M/s Chiquita has evaluated the Intelimer packaging technology for bananas for two years and, based on the results, is confident that product can extend the shelf-life of fruit. Chiquita Brands International is one of the largest banana producers in the world and a major supplier of bananas in Europe and North America. The intelimer-based food packaging

Modern packaging of banana

Individual fruit packing of banana

Packaging of banana boxes in CFB

Modified atmosphere packaging of banana

Natural rest handling of banana for longer shelf-life

Nicely washed and cooled bananas store well

(membranes and bags) regulates the levels of oxygen and carbon dioxide within a package to maintain the optimum atmosphere for a particular vegetable in order to extend the shelf-life of the produce. Package atmospheres can be uniquely tailored to accommodate high respiration rates and compensate for modest ranges of temperature abuse, resulting in extended shelf-life and higher quality taste, smell, and appearance of fresh produce. Landec manufactures and sells temperature-activated and other speciality polymer products for a variety of food and agricultural companies. The company's temperature-activated products are based on its proprietary polymers. The company's products differ from other polymers as they can be customised to abruptly change their physical characteristics when heated or cooled through a pre-set temperature switch.

Vacuum Packaging of Banana

Various banana grown in South India generally ripen 3-5 days after harvesting. Ripening during the long distance transport and export, results in huge post-harvest losses. To delay ripening of fruits, vacuum packaging is one of the methods, where matured banana hands are packaged and prevented from contact with air/oxygen. Using a simple gadget for vacuum packaging, banana can be stored under vacuum. This delays ripening up to 21 days and further ripening is completed within one week after opening the package. The process requires capital of Rs 1,00,000 and cost operation is only Rs 15/hour with 0.5 hp power requirement. Recently, Woon Ho Bak (2008) has filed an international patent for vacuum packaging and freezing of bananas. The method prevents discoloration of bananas during storage, and thus does not require use of artificial colorants and preservatives. It involves sterilization, washing and peeling of bananas, then sequential vacuum packaging, cold storage (ageing), quick-freezing and frozen storage. Banana flavour is retained by regulating the degree of vacuum applied during vacuum-packaging and uniform dispersion into the banana flesh of juice produced during cold storage.

42

Management of Optimum Harvesting Stage in Banana

S.N. GOHIL, B.T. GAIKWAD AND M. D. MALI

BANANA is known as a poor man's fruit. It fulfills the requirement of full diet when consumed with milk. Maharashtra ranks first in productivity (60 tonnes/ha) in the country. Though its production is maximum in Maharashtra, particularly in Jalgaon district, banana growers do not get profitable share. Most of the produce is directly harvested by commission agents at an early stage of maturity to get maximum profit. The commission agents purchase it on the basis of weight while they sell it to retailers on the basis of number, hence they harvest the produce at early stages. Secondly, at the time of shortage of banana fruits in market or by considering high demand they harvest the produce by paying little bit more price to growers. Therefore, study was undertaken to know weight loss in produce when harvested in between 50 and 130 days from emergence of bunches.

MANAGEMENT OF OPTIMUM HARVESTING STAGE IN BANANA

The study was undertaken at Mahatma Phule Agricultural University, Rahuri District Ahmednagar, Maharashtra, during 2005-06. Banana Grand Naine was utilized for study. The variety fruits were harvested 50, 65, 85, 100, 115 and 130 days after full emergence of hands. The fingers from middle of second to fifth hand were randomly selected for observations. The statistical design used was RBD with 6 treatments and replications. The results were statistically analysed as per Panse and Suckhatme (1978).

THE RESULTS

There were maximum concave and convex fruit length, (27.40 cm and 19.60 cm respectively) when fruits were harvested at 130 days maturity stage, while it was lowest (21.23 cm and 14.53 cm) when fruits were harvested at 50 days maturity stage. The curvature of fruits decreases as maturity period increases.

The fruit curvature is minimum (2.00cm) when fruits are harvested at 130 days maturity, while it is maximum (3.50 cm) when fruits are harvested at 65 days maturity stage. The fruit diameter increase as maturity increases. Fruit diameter is maximum (3.43 cm) when fruits are harvested at 130 days maturity, while it is minimum (2.63cm) when harvested at an early maturity, i.e. 50 days.

The fruit circumference is minimum at 50 days of maturity, while it is maximum (14.10 cm) when harvested at 130 days maturity. There is maximum fruit weight (242g) when fruits are harvested at 130 days after bunch initiation stage, while minimum fruit weight (78g) is recorded at 50 days after initiation of bunch. The fruit weight increases as maturity stage enhances (Table 42.1, Figs. 42.1 and 42.2). Fruit takes more time for ripening (20 days), when harvested at 50 days stage, while it takes minimum ripening time (7 days) when harvested at 130 days maturity stage. The ripening time required is in descending order as per maturity stage of fruits. The weight loss is maximum (25 %) from fresh fruits to ripening stage when fruits are harvested at 50 days maturity stage, while it is minimum (16 %) when fruits are harvested at 100 days maturity stage followed by 130 days maturity stage.

The pulp content is maximum (86.52 %) when fruits are harvested at 115 days maturity stage followed by 130 days maturity stage. The pulp content is minimum (74. 14%) when fruit are harvested at early stage of maturity, i.e. 50 days . The skin content is maximum (25.86 %) when fruits are harvested at an early stage, 50 days, while minimum skin content (13.48%) is recorded when fruits are harvested at 115 days of maturity, followed by 130 days maturity.

Pulp: skin ratio is maximum (6.42)when fruits are harvested at 115 days maturity stage, while it is minimum (2.87) when fruits are harvested at 50 days maturity stage. The total soluble solid content is maximum (24.2 °Brix) when fruits are harvested at an early stage, 50 days, while it is minimum (18.6 °Brix), when fruits are harvested at late stage, 130 days. The TSS content decreases as maturity stage increases.

The economical loss in fruit weight is maximum (164 g/fruit) when fruits are harvested at 50 days maturity stage, while it is 15g/ fruit, when fruits are harvested at 115 days maturity stage. The economical loss in fruit weight is maximum (210 %) when fruits are harvested at an early stage of maturity, 50 days, while it was 7 % when fruits are harvested at 115 days maturity stage.

It is concluded that fruits should be harvested at an early stages to extend natural ripening in transit for long distance transportation. The fruit length, diameter, circumference, weight and pulp increase as maturity stage increases, i.e. from 50 to 130 days from initiation of all hands in hunch. There is continuous decrease in skin weight, TSS and days to ripening as maturity stage increases. There is maximum loss (210 %) when fruits are harvested at 50 days maturity stage and it decreases as maturity stage enhances. These findings are supported by Laylieam and Kosittrakun (1998), Mustaffa *et al.* (1998), Ramma *et al.* (1999), Trejo *et al.* (2000) and Ahmad (2006).

Table 42.1. Effect of harvesting at different maturity times in banana Grand Naine

Treatment	*Days to harvesting from full emergence of hands*	*Fruit length (cm)* Concave	Convex	*Fruit curvature (cm)*	*Fruit diameter (cm)*	*Fruit circumference (cm)*	*Fresh weight (g)*	*Ripening weight (g)*	*Pulp weight (g)*	*Skin weight (g)*	*TSS (°Brix)*	*Days to riping*
1.	50	21.23	14.53	3.00	2.63	9.77	78	58	43	15	24.2	20
2.	65	22.03	15.68	3.50	2.97	12.30	132	103	97	16	23.8	15
3.	85	23.70	16.83	2.90	3.23	13.33	169	137	114	23	21.6	13
4.	100	25.28	17.20	2.77	3.36	13.40	174	146	119	27	20.4	11
5.	115	25.60	18.45	2.47	3.40	14.00	227	178	154	24	19.0	10
6.	130	27.40	19.60	2.00	3.43	14.10	242	198	169	29	18.6	07
SE		0.61	0.58	0.41	0.06	0.09	1.43	1.07	1.38	0.75	0.06	0.61
CD		1.83	1.73	1.23	0.17	0.26	4.32	3.21	4.16	2.26	0.18	1.84

Treatment	*Loss in weight up to ripening*	*Weight loss (%)*	*Pulp (%)*	*Skin (%)*	*Pulp: skin ratio*	*Loss in weight up to ripening (g/fruit)*	*Loss (g/ fruits)*	*Loss (%)*
1.	20	25.64	74.14	25.86	2.87	20 (25%)	164	210
2.	29	21.97	84.47	15.53	5.44	29 (22%)	110	83
3.	32	18.93	83.21	16.79	4.96	32(19%)	073	43
4.	28	16.09	81.51	18.49	4.41	28 (16%)	68	39
5.	49	21.59	86.52	13.48	6.42	49 (21%)	15	07
6.	44	18.18	85.35	14.65	5.83	44 (18%)	—	—

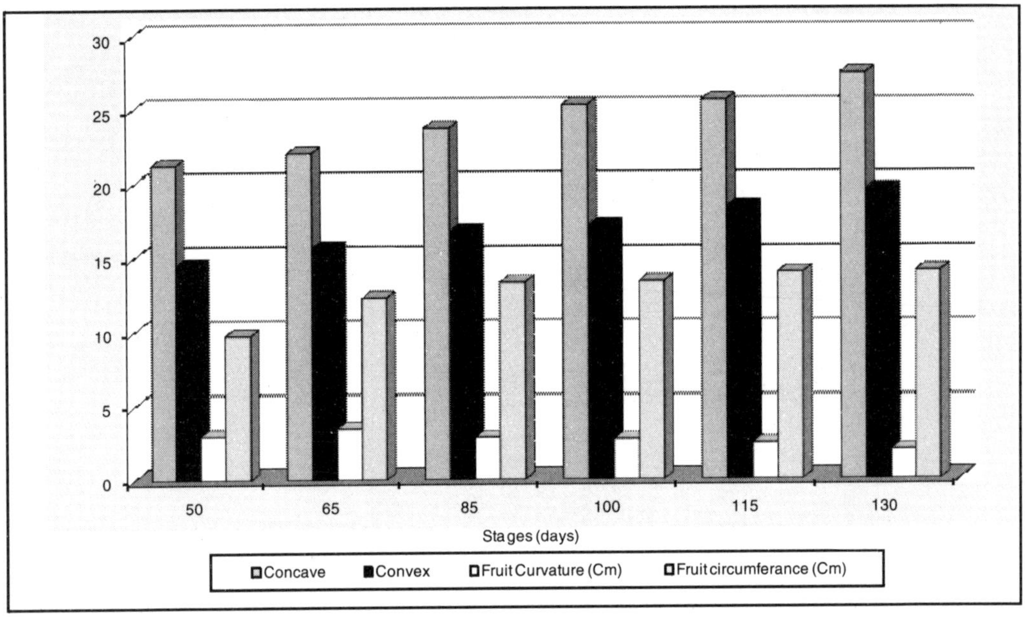

Fig.42.1. Fruit length, curvature, circumferance in banana Grand Naine

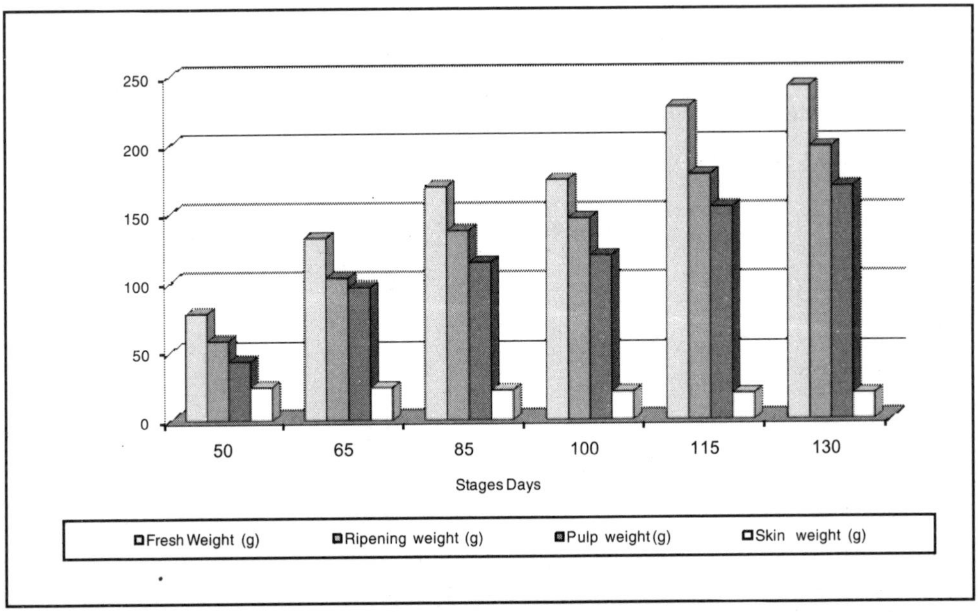

Fig.42.2. Fresh, ripen, pulp and skin weight of banana Grand Naine

References

Ahmad, S., Thompson, A.K. and Pervez, M.A. 2006. Effect of harvest maturity stage and position of hands on the ripening behaviour and quality of banana fruit. *Acta Horticulture* 741.

Laylieam, S., and Kosittrakun. 1998. Effect of harvest maturity on Banana Quality. *Journal of Food Quality,* **21**:539-44.

Mustaffa, Osmana, A., Yelsof, S., Mohanals, 1998. Physico chemical changes in canvendish banana at different posotion within a bunch during development and maturation. *J. Sci. Food. Agric,* **78**:201-207.

Panse, V.G. and Sukhatme, S.V. 1978. *Statistical Method for Agricultural Workers,* pp. 125-28, ICAR, New Delhi.

Ramma, I., Beni, Madhu S.P., Peerthum, Amas, S. 1999 Post-harvest quality improvement in banana. *Food and Agricultural Research Council Reduitmaritius,* **70**:112-15.

Trejo Marquez, M.A. Guillen, P.,Vendrell, M. 2000. Differences among bababa cultivarn Gran enanol harvested in canary islands,Spain and America *IV International Conference on Lost-harvest Science,* 553.

43

Post-Harvest Handling and Storage of Banana

Rema Menon, P. Jacob John, I.A. Chacko and A. Asha

Highly perishable nature of banana leads to substantial quantitative and qualitative losses brought about by handling at ambient temperature, poor transport, marketing and storage facilities. Therefore, improved pre-harvest and post-harvest practices are important. Under pre-harvest practices, timely cultivation, balanced use of fertilizers, adequate supply of moisture, plant-protection measures and bunch management are critical. One of the keys to minimizing post-harvest losses is to make use of the natural resistance of green banana fruits by preventing the initiation of ripening and storage at low temperature in an ethylene-free environment (Jones,2000). Nendran, the commercial plantain variety in Kerala, occupies 50% of the total area in the state and the bulk of the produce is marketed locally. A limited quantity is exported to Gulf countries, targeting the large Indian population and launching of Agri Export Zone in the state has further boosted the prospects. Presently, shipment carried out by air is constrained by the higher expenses involved. To arrive at a protocol for the sea shipment of Nendran over a time span of two weeks, a study on post-harvest handling and storage was carried out at the Banana Research Station, Kannara, in association with the Department of Processing Technology, College of Horticulture , Vellanikkara, and Vegetable and Fruit Promotion Council, Keralam (VFPCK), during 2006.

For procuring bunches of the required maturity (75-80%), a farmer's plot was selected based on survey of existing Nendran orchards. Bunches of 75-80% maturity were harvested from the selected plot. Harvesting done carefully to avoid bruises and injuries by covering the bunches at the time of harvesting with thin polythene sleeves and cutting peduncle with a sharp knife. The cut bunches were not allowed to touch the ground and individually transferred and loaded in a truck cushioned with sufficient dry leaves to protect from injuries during transportation. The bunches were transported to cold storage facility in Alathur unit of VFPCK in two hours, where the storage experiment was conducted. The bunches were hung vertically from a horizontal support to facilitate de handing. The hands were separated with a sharp knife to have a clean cut and an intact crown. The separated hands were spread on a clean surface to drain off the latex for 10 minutes.

Washing was accomplished in three stages: hands were washed twice for 10 minutes in clean water to remove any adhering dirt. They were washed in alum solution (500 ppm) for five minutes to remove any latex stains. And then, they were finally dipped in a fungicidal solution containing 500 ppm Bavistin for 10 minutes, to control rot in storage.

The hands were arranged on a clean surface and allowed to drain the adhering water completely for 30 minutes. For packing, 5 kg corrugated fibre board cartons were used. The following packaging treatments and storage conditions were tried (see *Box*).

Treatment	*Storage condition*	
	Temperature (C)	*Relative humidity (%)*
T_1, hands packed in cartons lined with polythene sheets of 100-guage and 2% ventilation.	13	90-95
T_2, hands packed in cartons lined with polythene sheets of 100-guage and 2% ventilation + $KMnO_4$ as ethylene absorbent	13	90-95
T_3, hands packed in polythene pouches of 100 guage and 2% ventilation and kept in cartons	13	90-95
T_4, hands packed in polythene pouches of 100 guage and 2% ventilation and kept in cartons + $KMnO_4$ as ethylene absorbent	13	90-95
T_5, clusters (4-5 fruits) packed in cartons lined with polythene sheets 100 guage and 2% ventilation	13	90-95
T_6, clusters (4-5 fruits) wrapped in cling film and kept in cartons	13	90-95
T_7, hands kept open in cartons (control 1)	13	90-95
T_8, hands kept open in cartons (control 2)	25	85
T_9, hands kept open in cartons + $KMnO_4$ as ethylene absorbent (control 3)	25	85
T_{10}, hands vacuum packed and kept in cartons	13	90-95
T_{11} - Hands vacuum packed and kept in cartons	25	85

A representative sample (1 carton) from each treatment was drawn 10 days after storage and observed for fruit quality .The fruits were also examined for quality after 14 days, the end of the storage period. Ripening was accomplished under two conditions with ethylene treatment (1000 ppm) for 24 hours. One carton from each treatment except T_8 and T_9 were kept for ripening in cold room. The temperature was increased to 18 °C during ripening. One carton representing all treatments except T_8 and T_9 were kept for ripening in an air-tight chamber. Observations on fruit quality were recorded 24 hours and 96 hours after ethylene treatment.

Irrespective of packing methods fruits stored at 13°C and 90-95% RH remained fresh, green and blemish-free even after 14 days of storage (Table 43.1).The fruits did not develop chilling injury and there was no physiological loss in weight. No spoilage due to post-harvest rots was observed. In contrast, fruits kept under ambient condition (25 °C and 85% RH) in the control treatments (T_7 and T_8) were completely ripe when observed after 10 days of storage. Vacuum packed fruits remained green for 14 days even under ambient conditions (T_{11}). Cooling the fruits as soon as possible after harvesting is critical to maintain their green life and to limit the development of crown rot (Stover ,1972). Temperature control to prevent fruit ripening is effectively used in international trade of Cavendish bananas, where fruits are stored and shipped at 13-14 °C. Stover and Simmonds (1987) noted that high humidity levels prevent the desiccation of crown, which favours tissue death and fungal colonization. In present study, the interval between harvesting and storage was 8 hours.

The temperature of 13 °C and relative humidity of 90-95% applied during storage period were effective in preventing ripening and development of post-harvest diseases.

Fruits could be uniformly ripen using ethylene treatment (1000ppm) for 24 hours at 18 °C and 90-95% RH and under ambient conditions. Colour development was normal during ripening at both the conditions, the peel attaining uniform yellow colour in 96 hours after ethylene treatment. Colour, texture and flavour of fruit pulp were also normal (Table 43.2). Vacuum packed fruits failed to ripen properly and hence pulp quality was not recorded. Fruits reached marketable stage 48 hours after ethylene treatment. Holding ripe fruits at 14 °C and 85 % RH was found optimum. Among packing methods, packing hands or clusters (4-5 fruits) in cartons lined with polythene sheet of 100-gauge thickness and 2% ventilation + $KMnO_4$ as ethylene absorbent is economical. Other treatments, viz. T_3, T_4, T_6 and T_{10} involve additional labour and inputs. Based on the study, a protocol for carrying out sea shipment of Nendran is developed (see Chart).

Harvest bunches at 75-80% maturity

↓

Transporting bunches wrapped in thin foam sheet to packing houses

↓

Dehanding by giving a clean cut with a sharp knife

↓

Delatexing by allowing the cut surface to drain latex for 10 minutes

↓

Washing the hands in clean water twice for 10 minutes for complete de latexing

↓

Washing in alum solution (500 ppm) for 5 minutes to prevent latex staining on fruits

↓

Dipping in fungicidal solution (Bavistin 500 ppm) for 10 minutes

↓

Drying

↓

Packing hands/clusters in corrugated fibre board cartons of 5 – 10 kg capacity lined with polythene sheets of 100-gauge thickness and 2% ventilation + $KMnO_4$ as ethylene absorbent

↓

Palletization

↓

Loading in precooled and sanitised reefer containers and maintaining temperature at 13 °C and 90-95% RH within 12 hours of harvesting

↓

Transportation at 13 °C and 90-95% RH

↓

Increasing the temperature to 18°C at destination and maintaining RH at 90-95%

↓

Ripening under controlled conditions ñ ethylene treatment at 1,000 ppm at 18 °C and 90- 95% RH for 24 hours

↓

Ventilation after 24 hours

↓

Holding ripe fruits at 13 – 14 °C and 85% RH

↓

Marketing in 2 – 3 days

Flow chart indicating harvest and post-harvest handling operations for sea shipment of banana

Table 43.1. Effect of packing and storage methods on fruit quality 14 days after storage

Treatment	*Storage conditions*		*Appcarance of ruits*	*Colour of peel*	*PLW**
	T (C)	*RH (%)*			
T_1	13	90-95	Fresh	Green	Nil
T_2	13	90-95	Fresh	Green	Nil
T_3	13	90-95	Fresh	Green	Nil
T_4	13	90-95	Fresh	Green	Nil
T_5	13	90-95	Fresh	Green	Nil
T_6	13	90-95	Fresh	Yellowish green in two clusters	Nil
T_7	13	90-95	Fresh	Yellowish green in two clusters	Nil
T_{10}	13	90-95	Fresh	Green	Nil
T_{11}	25	85	Fresh	Green	Nil

*Physiological loss in weight

T_8 and T_9 ripen 10 days after storage and hence not included

Table 43.2. Fruit and pulp characteristics after ripening (ethylene treatment at 1000 ppm) at 18 °C and 90-95% RH

Treatment	*After 24 hours appearance and peel colour*	*After 96 hours*				
		Peel colour	*Presence of spots*	*Pulp colour*	*Pulp texture*	*Taste*
T_1	Fresh and green	Uniform yellow	Nil	Orangeyellow	Firm	Normal
T_2	Fresh and green	Uniform yellow	Nil	Orangeyellow	Firm	Normal
T_3	Fresh and green	Uniform yellow	Nil	Orangeyellow	Firm	Normal
T_4	Fresh and green	Uniform yellow	Nil	Orangeyellow	Firm	Normal
T_5	Fresh and green	Uniform yellow	Nil	Orangeyellow	Firm	Normal
T_6	Fresh and yellowish green	Uniform yellow	Nil	Orangeyellow	Firm	Normal
T_7	Fresh and green	Uniform yellow	Nil	Orangeyellow	Firm	Normal
T_{10}	Fresh and green	Abnormal colour development in one hand	Present	–	–	–
T_{11}	Fresh and green	Green	Nil	–	–	–

References

Jones, D.R. 2000. *Diseases of Banana, Abacaí and Enset.* CAB International, Wallingford, UK, pp. 190-97.

Stover, R.H. 1972. *Banana, Plantain and Abaca Diseases.* Commonwealth Mycological Institute, Kew, Surrey, UK, 316 pp.

Stover, R.H and Simmonds, N.W.1987. *Bananas,* 3rdedn. Longman Scientific and Technical, Harlow, Essex, UK, 468pp.

44

Effect of Packaging on Shelf-Life and Quality of Banana

K.B. Sheela, M. Reni, Rema Menon and A. Suma

Despite high economic importance of banana, it has some special problems as a commercial fruit. Its shelf-life is comparatively short, hardly a week under normal tropical conditions. Short shelf-life creates serious difficulties for efficient marketing and results in heavy post-harvest loss. The use of modified atmosphere storage has considerable potential as an inexpensive means for prolonging shelf-life of banana. Hence, effect of modified atmosphere packaging on quality and shelf-life of banana Nendran was studied for developing a suitable and convenient retail packaging method for Nendran banana.

The popular clone of Nendran, Chengalikodan was used for study. Banana bunches of uniform maturity were harvested and hands of uniform size were separated using a sharp knife. The cut ends of banana hands were dipped in Bavistin (1000ppm) for one minute and then air dried and packed in retail packages. The treatments were: P_1 (cling film), P_2 (polythene cover) 100-gauge with ventilation, and P_3 (polythene cover 100-gauge without ventilation, P_4 sealed polythene cover with ethylene absorbent).

Ethylene absorbent was prepared by impregnating vermiculite with potassium permanganate. Saturated solution of potassium permanganate was poured into dried vermiculite and left for 30 minutes. The absorbent mixture was kept in muslin cloth bags and sealed in unglazed soft tissue paper sachets. The absorbent was used at the rate of 20 g/kg of fruits. The banana hands were packed in transparent low-density (100-gauge) polythene bags of 35 cm × 25 cm size.

Corrugated fibre board boxes (5 ply), of 30 cm × 25 cm × 20 cm size were used (P_5)

The other treatments were P_5 (corrugated fibre board boxes, P_6 CFB box with polythene lining, P_7 CFB box with polythene lining + ethylene absorbent, P_8 vacuum packaging in 150-gauge polythene bag and P_9 (control).

All the packages were kept 24-30 °C and relative humidity of 70-80%.

The data were recorded on physiological loss in weight during storage, shelf-life and days to 50 and 100% yellowing. The chemical constituents of ripe fruits, viz. TSS, acidity, total, reducing and non-reducing sugars were estimated. Organoleptic evaluation for assessing the appearance, taste, sweetness and overall acceptance of ripe fruits during storage was carried out by a trained panel of 10 members using a 10-point scale (0-2, poor; 3-5, satisfactory; 6-8, good and 9-10, excellent quality).

A gradual increase in PLW was noted throughout the storage period. Significant difference in PLW was observed among treatments, highest being in control and lowest in vacuum-packaged samples (Table 44.1). On 18th day of storage control fruits were discarded and maximum PLW was registered by P_5 (CFB boxes) up to 21 days of storage. Fruits packed in polyethylene covers (PE) without ventilation and PE cover sealed with ethylene absorbent also recorded lower PLW. The minimum PLW in vacuum packed fruits may be due to slowing down of enzymatic and respiratory activity and thereby reduced water loss. Nair *et al.* (1992) reported that bananas when stored in a hermetically sealed, tight polyethylene bag, the atmosphere around fruits rapidly becomes modified and drops to 10% oxygen within an hour after vacuum packaging of fruits.

The days taken for 50 and 100% yellowing was the longest in vacuum packaged fruits (34.5 and 37 days respectively) and shortest in the control (6.5 and 9.5 days respectively).

Analysis of variance showed significant difference for shelf-life among different treatments. Maximum shelf-life was recorded in vacuum-packed fruits (39.5 days) followed by fruits packed in polyethylene with ethylene absorbent (37.0 days). Minimum shelf-life was in the control fruits (15.0 days). Modified atmosphere packaging is known to extend the shelf-life of fresh produce by retarding the physiological metabolism leading to senescence, by increased CO_2 and decreased O_2 concentration in storage atmosphere, slowing down the rate of ethylene biosynthesis and its action subsequently helping in decreasing microbial contamination and by creating high humidity, resulting in less moisture loss and better quality retention.

Extension of storage life through vacuum packaging was reported by Nair *et al.* (1992); Rajeev (1996) and Gouthami (2001). Vacuum packaging slowed down the metabolic activity by limiting the oxygen supply and applying an elevated level of CO_2 (Gorris and Peppelenbos, 1992). Decrease in oxygen concentration also could have suppressed ethylene synthesis as conversion of ACC to ethylene is an oxidative reaction.

Use of ethylene absorbent inside the modified atmosphere packs resulted in an additive effect along with modified atmosphere in prolonging the green life by keeping the ethylene levels low inside the sealed packs. The KMO_4 has been reported to oxidize ethylene to ethylene glycol, thus partly delaying the ripening process (Mahajan and Chopra, 1994).

The fruits packed in CFB boxes recorded shortest green life and longest yellow life, indicating that maximum storage life after ripening was in CFB boxes.Yellow life was shorter in vacuum packed fruits as well as those packed in polyethylene covers.The use of CFB boxes for storage is being popular since they are lighter in weight, more convenient to handle,cheaper to transport, internationally acceptable and can be cycled as pulp or paper (Maini *et al.* 1993).

Significant difference was noticed in TSS, total, reducing and non-reducing sugars among different treatments on ripening. Control fruits registered highest TSS (31.5°Brix) and cling film wrapped fruits the lowest (23.5°Brix). The values for acidity ranged from 0.42% in P_5 to 0.51% in P_4 (unventilated polythene covers with ethylene absorbent). Maximum values for total and reducing sugars were registered in the control fruits (32% and 26.05% respectively) and minimum total sugar in cling film wrapped fruits (21.54%) and reducing sugar in vacuum

Table 44.1. Effect of packaging on PLW during storage in fruits of Chengalikodan under ambient conditions

Treatment	*3 DAS (%)*	*6 DAS (%)*	*9 DAS (%)*	*12 DAS (%)*	*15 DAS (%)*	*18 DAS (%)*	*21 DAS (%)*	*24 DAS (%)*	*27 DAS (%)*	*30 DAS (%)*	*33 DAS (%)*	*35 DAS (%)*	*38 DAS (%)*
P_1	2.25	2.55	2.95	3.60	3.60	3.60	4.00	4.00	5.1	5.35	ND	ND	ND
P_2	7.40	9.20	9.20	14.85	15.25	17.05	18.65	19.65	ND	ND	ND	ND	ND
P_3	1.80	2.05	2.05	2.40	2.90	3.90	4.25	4.25	4.45	4.45	ND	ND	ND
P_4	1.55	1.55	1.55	2.30	2.70	3.20	3.60	3.75	4.00	4.00	2.90	4.15	
P_5	7.50	11.00	11.00	23.35	25.70	27.40	30.30	ND	ND	ND	ND	ND	ND
P_6	3.65	5.35	5.35	15.30	16.50	17.95	21.55	22.95	ND	ND	ND	ND	ND
P_7	4.60	5.75	5.75	14.05	15.95	18.65	20.60	21.25	ND	ND	ND	ND	ND
P_8	0.65	0.65	0.65	0.65	1.10	1.10	1.35	1.35	2.05	2.05	4.15	4.35	4.6
P_9	12.55	12.55	18.10	27.75	30.80	ND	ND	ND	ND	ND	ND	ND	ND
CD (0.05)	2.776*	2.767*	2.619*	3.143*	3.024*	2.867*	1.001*	1.641*	NA	NA	NA	NA	NA

- Significant at 5% level; NA, not analysed statistically; ND, not determined due to spoilage

Table 44.2. Effect of packaging on chemical composition of ripe fruits of Chengalikodan

Treatment	*TSS (Brix)*	*Acidity (%)*	*Total sugars (%)*	*Reducing sugar (%)*	*Non-reducing sugar (%)*
P_1	23.5	0.465	21.540	15.215	6.375
P_2	28.0	0.457	24.155	17.580	4.755
P_3	26.5	0.458	23.165	15.195	7.970
P_4	27.5	0.509	22.335	16.670	7.485
P_5	26.5	0.420	25.650	20.115	5.535
P_6	27.0	0.483	28.860	23.130	5.730
P_7	28.5	0.500	27.015	21.145	5.870
P_8	25.0	0.470	21.855	14.735	7.120
P_9	31.5	0.496	32.000	26.050	6.000
CD (0.05)	2.151*	NS	1.315*	2.775*	1.529*

*- Significant at 5% level; NS, not significant

Table 44.3. Effect of packaging on green life and shelf-life of banana Chengalikodan

Treatment	*Days to 50% yellowing*	*Days to 100% yellowing*	*Shelf-life*
P_1	22.5(4.793)^{d}	23.5(4.897)	30.5(5.568)^{def}
P_2	12.0(3.533)^{e}	16.0(4.062)	24.5(5.000)^{hi}
P_3	24.5(4.998)^{c}	29.0(5.431)	33.5(5.830)^{cd}
P_4	29.5(5.477)^{b}	33.5(5.830)	37.0(6.123)^{ab}
P_5	12.5(3.605)^{e}	15.5(4.000)	21.5(4.690)^{j}
P_6	12.5(3.605)^{e}	15.0(3.937)	24.0(4.950)^{ij}
P_7	12.5(3.605)^{e}	16.5(4.123)	24.0(4.950)^{ij}
P_8	34.5(5.915)^{a}	37.0(6.123)	39.5(6.323)^{a}
P_9	6.5(2.644)^{f}	9.5(3.161)	15.0(3.937)^{l}

Figures in parentheses are transformed values; values with the same superscripts do not differ significantly

Table 44.4. Influence of packaging on sensory attributes of ripe fruits of Chengalikodan

Treatment	*Appearance*	*Taste*	*Sweetness*	*Overall acceptance*
P_1	7.30 (64.45)	6.9 (65.65)	6.7 (61.15)	6.7 (63.95)
P_2	7.2 (59.45)	7.0 (64.85)	7.0 (67.00)	6.0 (53.50)
P_3	5.3 (34.45)	3.9 (18.20)	3.7 (24.70)	3.9 (24.55)
P_4	6.7 (63.17)	4.0 (19.56)	3.6 (24.06)	3.6 (20.61)
P_5	5.4 (37.77)	6.9 (59.45)	6.6 (56.00)	6.8 (60.41)
P_6	3.6 (19.45)	7.1 (66.55)	6.9 (64.55)	6.7 (64.16)
P_7	6.9 (55.55)	7.0 (64.25)	7.6 (74.30)	7.8 (79.30)
P_8	8.1 (76.40)	3.7 (17.55)	2.8 (13.85)	3.3 (15.85)
P_9	3.2 (10.35)	4.7 (29.45)	3.6 (20.70)	3.8 (23.20)
Kruskalwallis statistic	56.896*	63.125*	67.628*	68.043*

*- Significant at 5% level

packed fruits (14.74%). Non-reducing sugars was highest in P_3 (unventilated polythene covers) (7.97%) and lowest in P_2 (ventilated polythene covers) (4.76%). Higher scores for appearance was obtained for vacuum packed fruits and taste, sweetness and overall acceptability for fruits packed in CFB boxes (Tables 44.2, 44.3 and 44.4).

References

Nair, H., Tung, H.F., Mohamad, W.W.O., Mohamad, R., Ahmad, S.H. and Chong, K.K. 1992. Low oxygen effect and storage of Mas bananas (Musa, AA group). *Acta Hort.*, **292**:209-15.

Mahajan, B.V.C. and Chopra, S.K. 1994. Effect of preharvest application of ethylene antagonists on the storage behaviour of apple. *Haryana J. Hort. Sci.*, **24**(3-4):163-70.

Gouthami, N. 2001. Post-harvest handling and packaging systems for banana *Musa* (AAB) ëNendraní. M.Sc. (Hort.) thesis, Kerala Agricultural University, Thrissur.

Rajeev, M. 1996. Studies on vacuum packaging of fruits and vegetables. ME (Agric. Process Engineering), thesis, Tamil Nadu Agricultural University, Coimbatore.

Gorris, L.G.M. and Peppelenbos, H.W. 1992. Modified Atmosphere and vacuum packaging to extend the shelf-life of respiring food products. *Hort. Technol.*, **2**(3):303-09.

Maini, S.B., Lal, B.B. and Anand, J.C. 1993. Fruit packaging. *Advances in Horticulture Vol. 4. Fruit Crops.* Chadha, K.L. and Pareek, O.P. (Eds). Malhotra Publishing House, New Delhi, pp.1775-93.

45

Carbohydrate Hydrolyzing and Antioxidative Enzymes in Banana Fruits

P. NANDHINI, I. RAVI AND C.K. NARAYANA

SURVEYS have revealed that a substantial portion of harvesting of banana is wasted annually due to improper harvesting and post-harvest practices, disease and lack of facilities and technology to extend storage life. Post-harvest losses have been estimated in developed countries to range from 5 to 25%, while in developing ones it is 20-50%, depending upon the commodity (Kader, 1992). This continues to cause heavy losses in revenue for growers, wholesalers, retailers and exporters. Therefore, to elucidate biochemical changes, antioxidative enzymes and carbohydrate hydrolyzing enzymes activity in hot water dip treated banana fruits stored at low temperature (10 °C) was necessary.

The fruits of triploid Robusta (AAA) were harvested to study chilling injury and biochemical mechanism. The hot water dip treatment by immersing them in hot water bath was done at 39 °C (T_1), 42 °C (T_2), 45 °C (T_3) for 30 minutes. The respective temperature was maintained constantly and control (no heat treatment) (T4) fruits were also taken. One set of all treatment samples were immediately processed as fruit peel and pulp for biochemical analysis. The remaining fruits of all the treatments were kept at 10 °C. The design was CRD. Each treatment had three replications and each replication consists of six fruits. The sampling was done after one day and sixth day after chilling treatment. The fruit samples were immediately processed for all biochemical analysis. For carbohydrate analysis, samples were stored in 80% ethanol and for all other analysis the samples were stored in cold acetone.

Chilling sensitivity varied among cultivars. Chilling injury results from exposure of 'Petite'bananas to temperatures lower than or equal to 10 °C for 7 or more days of storage or below 12.5 °C for 21 days of storage. Yangambi bananas were subjected to chilling injury when stored at temperates less than or equal to 10 °C for 7 days. Red Macabu bananas were subject to chilling injury when stored for 5 days at temperatures below 10 °C. Chilled fruits were more sensitive to mechanical damage and post-harvest decay.

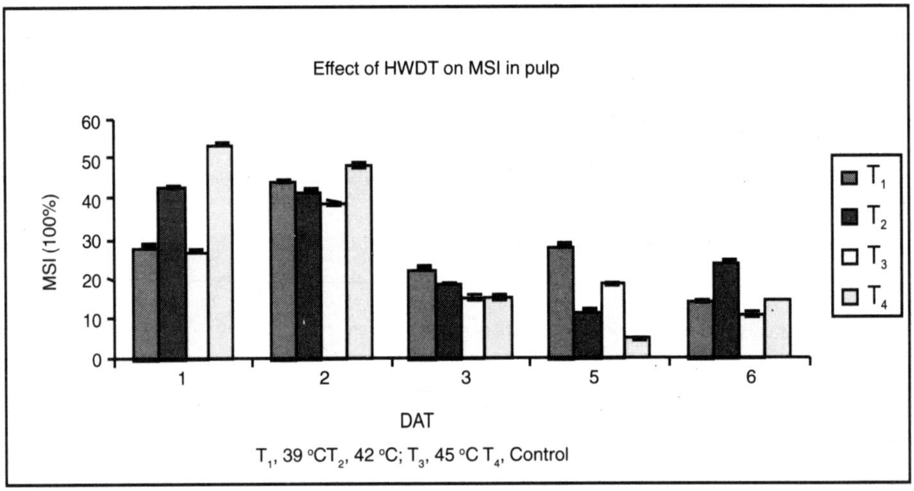

Fig. 45.1. Effect of hot water dip treated banana fruits stored at 10°C on membrane Stability Ind12ex (MSI) in pulp and peel of Robusta fruits.1

Table 45.1. Effect of chilling stress on RF Ascorbate peroxidase milli moles of acorbic acid consumed per mg of protein in pulp and peel of Robusta fruits

DAI	*Pulp*					*Peel*					*Grand X*
	T1	*T2*	*T3*	*T4*	*X*	*T1*	*L2*	*T3*	*T4*	*X*	
0	1.17	0.53	0.39	0.67	0.54	9.85	1.70	0.92	0.65	3.13	1.84
1	10.77	0.91	1.30	2.54	3.88	0.59	0.39	3.14	7.16	2.82	3.35
6	4.01	1.75	1.20	11.13	4.52	1.18	1.59	0.60	2.37	1.44	2.98
X	5.32	1.06	0.96	4.58	2.98	3.88	1.23	1.55	3.20	2.46	2.72

	SE	5% LSD
DAI	0.56	0.16
T	0.65	0.19
PL	0.46	0.13
DAI × T	0.11	0.32
DAI × P	0.79	0.23
P × T	0.92	0.26
DAF × P × T	0.16	0.45

Table 45.2 Effect of chilling stress on amylase units per mg of protein in pulp and peel of Robusta fruits

DAI	*Pulp*					*Peel*					*Grand*
	T1	*T2*	*T3*	*T4*	*X*	*T1*	*L2*	*T3*	*T4*	*X*	*X*
0	0.75	0.59	0.23	2.7	1.07	1.11	0.67	0.52	4.52	1.7	1.39
1	4.47	3.34	4.41	5.12	4.33	4.74	3.32	2.68	5.2	3.99	4.16
6	17.7	27.97	76.8	22.77	36.32	15.81	76.61	65.52	16.15	43.52	39.92
X	7.64	10.63	27.15	10.19	13.9	7.22	26.87	22.91	8.63	16.4	15.16

	SE	5% LSD
DAI	0.26	0.75
T	0.3	0.86
PL	0.21	0.61
DAI × T	0.52	0.15
DAI × P	0.37	0.11
P × T	0.42	0.12
DAF × P × T	0.74	0.21

Bananas can be stored for 2-3 weeks in good condition at low temperature 13.5 °C. However, chilling injury (CI) is a problem associated with lower than this critical temperature storage of bananas. Some symptoms of CI are sub-epidermal vascular browning, abnormal ripening and in severe cases, failure to ripen. Chilling sensitivity varies among cultivars. The present study revealed that when Robusta bananas were stored at 10 °C for six days after hot water dip treatment (HWDT) at 42 °C, the fruits maintain their function normally through better membrane stability Index (Rikin and Richmond 1979) and higher carbohydrate hydrolyzing enzyme, amylases and antioxidative enzymes activity and protein accumulation.

References

Kader, A. 2002. *Post-harvest Technology of Horticulture Crops.* UC Davuis Publication, 535 pp.

Rikin, A., Richmond, A.E. 1979. Factors affecting leakage from cucumber cotyledons during chilling stress. *Plant Sci. Lett.*, **14**:263-68.

46

Eco-Friendly Nematode Management in Banana

P. Sundararaju

Plant parasitic nematodes are a major constraint to sustainable banana production. Crop losses caused by nematodes to banana are very high, with an average annual yield losses of 20% worldwide (Sasser and Freckman, 1987). In India, nematodes have been identified as a major contributing factor in limiting production and goes as high as 45% crop losses due to nematodes in plantain and Cavendish group of banana which constitute more than 65% of the total banana industry in India. A total of 71 species belonging to 33 genera are associated with rhizosphere of banana. Among them, burrowing nematode (*Radopholus similis*), root-lesion nematode (*Pratylenchus coffeae*) and root-knot nematode (*Meloidogyne incognita*) are conomically important nematode pests on banana and are widely distributed in South India (Sundararaju, 1996; Jonathan and Rajendran, 2000). Crop losses caused due to *P. coffeae* in Nendran and *M. incognita* in Poovan banana were reported to be 44.4 and 30.9% respectively (Sundararaju and Cannayane, 2003; Jonathan and Rajendran, 2000). These nematodes can be controlled with chemicals, but they may have adverse environmental effects. Moreover, use of nematicides is too expensive and products are too dangerous for subsistence farmers. Indiscriminate use of nematicides has proved much disastrous not only to the environment but also to our health and much the worst, in eliminating natural enemies of plant pathogens and pests. The most economical and effective ways to control nematodes without affecting the produce is to use non-chemical methods. Different non-chemical methods for controlling nematodes on banana are discussed.

ECO-FRIENDLY NEMATODE MANAGEMENT

Cultural Methods

Effective prophylactic measures are of considerable importance to reduce the initial inoculum levels and to curb their progressive multiplication by appropriate cultural practices and intercropping. Nowadays, non-chemical methods received greater attention in view of the cost of chemical control and pollution of soil environment. The cultural practices are most reliable and easily accessible to adopt by banana farmers to control nematodes in banana. Fallowing for a period of three months after banana harvesting effectively suppressed the burrowing nematode population, while flood fallowing for five months destroyed not only burrowing nematode but also *Fusarium* sp. (Rajendran *et al.*, 1979). Crop rotation with rice, sugarcane, greengram or cotton suppressed the nematode population and increased the yield of banana. Intercropping of banana with *Crotalaria juncea* reduces *R.similis* population besides better growth and yield of banana in India (Venkitesan and Charles, 1983). Soil covers with black polyethylene, sugarcane and banana trashes reduce *R.similis* and *Pratylenchus* sp. in soil and roots of banana (Bhattacharya and Rao, 1984). Subramaniam and Selvaraj (1990) reported that application of Carbofuran at planting, intercropping with *Tagetes* sp. *Crotalaria* sp. or radish significantly reduced *R.similis* population with Carbofuran having greatest effect. Intercropping of various nematicidal plants, viz. sunnhemp, sesamum, mustard and marigold is effective method to control nematodes. Shanthi *et al.* (2001) reported that sunnhemp intercrop in banana field is effective in reducing *R.similis, P.coffeae* and *H. multicinctus* population to 38.4%, followed by marigold and cowpea which recorded 29.0 and 22.3% reduction respectively. However, effect of Carbofuran 3G and Monocrotophos (0.05%) is superior in increasing fruit yield (47.2 and 43.3%) consequent to highest reduction in nematode population by 56.4 and 46.4% respectively.

A field experiment conducted at farmer's field on banana Nendran (AAB) using intercrop *Tagetes erecta* in comparison with recommended cultural practices of paring of suckers and treated with Monocrotophos 36 EC dipping at 0.05 % at the time of planting showed significant reduction (85%) in root-lesion nematode population in banana field, where *Tagetes erecta* was grown as intercrop followed by Monocrotophos dipping treatment (75%). Maximum nematode population was recorded in untreated control plants. Further, yield of banana significantly increased when intercropped with *Tagetes erecta* compared to untreated control (Sundararaju, 2005).

Organic Cultivation

Amendment of soil with decomposable organic matter is recognized as most efficient method for changing the environment of soil and rhizosphere, thereby adversely affecting the life-cycle of nematodes and enabling the plants to resist the attack of these organisms to better vigour or alters the root physiology. It appears to be a good tool for reducing nematode density, infective capacity, host proneness and susceptibility. This method aims only to bring nematode population below economic injury level, to minimize their extent of damage rather than eradicating them. It is also true that organic amendments alone may not give satisfactory nematode management. It may give better results in integration with other methods of management. Jonathan *et al.* (2000) studied effect of organic amendments on management of root-knot nematode (*M. incognita*) and spiral nematode (*H. multicinctus*) on banana included significant reduction in nematode population and increased yield in plants treated with press mud (15 tonnes/ha) or neem cake (1.5 tonnes/ha).

The organic amendments are comparable with Carbofuran treatment. The effect of organics and inorganics for management of root-lesion nematode (*P.coffeae*) was studied in six commercial cultivars of banana, *viz.* Nendran, Karpuravalli, Rasthali, Robusta, Monthan and Poovan. The results revealed that significant reduction in *P.coffeae* population and yield increase was recorded in plants received 50% N applied through neem cake (Sundararaju and Kumar, 2003). An attempt was also made to study the analysis of best treatment and variety based on nematode population on banana using artificial neural networks (Sundararaju *et al.,* 2002). The population pattern of plant parasitic nematodes from field trial was carried out on locally available organic manures in banana Karpuravalli (ABB). Analysis of soil and root samples revealed the presence of four important nematode species, *viz. P.coffeae, M.incognita, H.multicinctus* and *H.oryzicola.* All the four nematodes were found to be significantly lower in plants which received distillery sludge @ 2.5 kg + vermicompost @ 1 kg + neem cake @ 1 kg + poultry manure @ 2.5 kg at 3, 5, 7 MAP compared to the control plants and it was on a par with distillery sludge @ 2.5 kg + neem cake 1 kg applied at the same interval. The root-lesion and root-gall indices were higher in control plants registering 4.0 and 3.7 respectively. Thus, there is a significant role of organic biological control strategies in banana (Sundararaju *et al.,* 2002).

Investigation carried out for the management of major nematode pathogen, namely *P.coffeae* and *M.incognita* infesting banana using different neem formulations (Econeem, Nimbicidine and Neemgold), plant growth promoter (Biovita) and compared with the standard treatment (Carbofuran 3 G) revealed that all the treatments are found effective in reducing nematode population with enhanced plant growth compared to the control plants. Among three neem formulations evaluated, Econeem and Nimbicidine showed maximum efficacy in reducing nematode population with increased plant growth. However, plants treated with Carbofuran 3G recorded maximum growth with siginificant reduction in nematode population (Sundararaju and Cannayane, 2003).

A field experiment carried out for the management of *P.coffeae* using press mud (a byproduct from sugar factory) recorded significant reduction in nematode population and increased plant growth as compared to too control. However, pressmud application was on a par with Carbofuran treatment. The use of press mud is economical and environmentally safe as compared to chemical nematicide (Sundararaju *et al.,* 2001).

The effect of neem-based formulations such as Neewin, Nimbecidine, Juerken and Neemgold were used as sucker dip treatment in different concentrations against root-knot nematode (*M. incognita*) on banana Robusta. The results of present investigation revealed that all the treatments are significantly superior to untreated control in terms of maximum plant growth with significant reduction in nematode populations. Among treatments, Nimbecidine or Juerken @ 15ml/litre of water for 30 minutes dipping or hot water treatment @ 50°C for 20 minutes dipping are most effective treatments, resulting in enhanced plant growth with significant reduction in nematode populations or absolute control of nematodes over untreated control plants and were on a par with each other (Sundararaju and Thenmozhi, 2006). Addition of organic amendments such as farmyard manure, oil cakes, green manure, press mud, etc. encouraged the multiplication of antagonistic microbes which in turn check plant parasitic nematodes. Organic acids such as formic, acetic, propionic and butyric acids are released in soil during microbial decomposition of organic amendments. Ammonia and hydrogen sulphate gases are released in soil during decomposition. These organic acids and gases are toxic to nematodes. Since nematodes are very well-managed by using organic amendments in different horticultural and plantation crops, a study was therefore initiated on banana. Nendran by using different organic amendments against the control of root-lesion nematode (*Pratylenchus coffeae*) infesting Nendran banana. The results revealed that application of neem cake @ 200g/plant at planting, 3, 5 and 7 months after planting recorded maximum plant growth parameters with significant reduction in nematode population which could be recommended for the management of nematodes on banana (Sundararaju and Sudha, 2006).

Botanical Biopesticides

Vijayalakshmi *et al.* (1979) screened 50 indigenous plant products for their nematicidal properties against the second stage juveniles of *M.incognita* under laboratory condition. Datura *(Datura metel)* leaves at 10% concentration in 24h exposure at 5% concentration in 48 hour exposure gave 100% mortality and appear to be the most promising of all the materials tested. Among 10 plant extracts, *Azadirachta indica* showed maximum mortality of *P. coffeae* when exposed to 20 hour in 80% plant extract, followed by *Vitex negundo* and *Crotolaria juncea* (Sundararaju and Cannayane, 2002). Fresh leaves of eight plants, viz, *Acalypha indica, Cassia fistula, Cassia auriculata, Crotolaria juncea, Calotropis gigantea, Abutilon indicum, Azadirachta indica* and *Solanum tarvu*m were tested for its nematicidal activity against root-knot nematode, *M. incognita* infesting banana. At 50 and 100% concentrations, all plant extracts showed cent per cent mortality of adults and larvae of *M.incognita* when exposed to 72 hours. However, plant extracts, *Abutilon indicum, Azadirachta indica* and *Solanum tarvum* showed high degree of nematicidal action recording cent per cent mortality even at 24 hour (Sundararaju and Saritha, 2005).

Fresh leaves of 15 botanicals which are commonly found in nearby banana gardens were collected, dried and extracts were prepared. They were tested for their nematicidal and nematostatic properties against *M. incognita.* The results indicated that all the plant species showed nematicidal activity ranging between 46.6 and 98.7% mortality at 24 hour exposure period. The best five plant extracts, viz. *Tagetes erecta, Azadirachta indica, Tithonia diversifolia, Acalypha indica* and *Solanum tarvum* were further diluted at 20, 40, 60 and 80% concentration and tested. The maximum mortality rate of *M. incognita* recorded was 98.7, 96.5, 89.3, 88.0 and 85.7% when exposed to 24 hour at 80% concentration with respect to *T.erecta, A. indica, T. diversifolia, A. indica* and *S. tarvum* respectively. However, cent per cent mortality was recorded at 80% concentration when exposed to 72 hours and thus these plants could be effectively utilized as sucker dip treatment to control *M.incognita.* No mortality was recorded in distilled water (Sundararaju and Madhula, 2007)

Green Manuring

Green manuring plants are grown and ploughed *in-situ* which provide nutrition to the main crop after proper decomposition. This process not only changes the soil environment during the decomposition but also affect the soil micro-fauna. Linford and Yap (1938) and Lindford *et al.* (1938) demonstrated for the first time that incidence of root-knot nematode could be significantly reduced by incorporating chopped pineapple leaves @ 50-200 tonnes/ha soil. Patel and Desai (1964) found *Melilotus alba* variety *annua* and *Sorghum vulgare* effective in reducing root-knot nematode populations. Singh (1965) applied chopped leaves of Karanj to soil infected with *M. javanica* and obtained 50% reduction in tomato root-knot gall intensity. Singh and Sitaramaiah (1967) reported that application of 5-10% of leaves of *Azhadirachta indica, Mellia azederach, Cassia fistula, Crotolaria juncea* and *Sesbania aculeate* to infested soil reduced the incidence of root-knot on tomato and okra. Yuhara (1971) studied that plant materials of *Crotolaria* or marigold amended to soil gradually reduce *M.hapla* population. The plants of *Azadirachtas indica, Vitex negundo* and *Crotolaria juncea* using their dry and fresh leaves were tested further against root-lesion nematode, *P. coffeae,* in banana Nendran and Rasthali under field conditions. All the botanicals are effective in reducing nematode population and significantly increased the plant growth and yield compared to untreated control. Among different botanicals tried, *A. indica, Calotropis procera, Datura stramonium, C. juncea* and *V. negundo* were found to be superior and effective in reducing the nematode population, increasing the yield significantly (Sundararaju *et al.,* 2003).

Biocontrol Agents

Management of banana nematodes with synthetic nematicides pose several problems like groundwater pollution, effect on beneficial organisms, resurgence of pest etc. Biomolecules having antagonistic effect on target nematodes is one such approach to solve the aforesaid ill effects of chemical nematicides. Fungi that colonize the healthy roots exhibiting antagonistic effect to invaded nematodes are termed as endophytic fungi, regarded as the best alternative to manage the banana nematodes. Therefore studies were undertaken to isolate endophytic fungi from 12 accessions belonging to different genomic groups of banana and to evaluate its bioefficacy on *P. coffeae* and *M. incognita* under *in vitro* conditions. The endophytic fungi isolated from banana were identified as *Fusarium* spp. and their biocidal effects on *P. coffeae* and *M. incognita* were studied. The culture filtrates of endophytic fungi isolated from diploids exerted higher nematicidal effect than those isolated from triploids *in vitro.* The nematicidal effects of endophytic fungi on *P. coffeae* and *M. incognita* juveniles increased with increasing exposure period to the culture filtrates (Sundararaju *et al.*, 2002).

An attempt was made to mass-produce the nematode egg parasitic fungal bioagent, *P. lilacinus* on banana wastes (leaf, petiole and pseudostem), castor and pungam leaves and their combinations and a standard check with pre-soaked sorghum grains. Highest spore load of 8 x 10^6 *P. lilacinus* per g of substrate was harvested in sorghum grains, banana petiole and castor leaf mixture. Banana pseudostem and pungam leaves were comparable in their substrate nature to the fungus (Sundararaju and Cannayane, 2002). Field and greenhouse experiments were conducted by using different doses of bioagent, *T. viride* on susceptible cultivars, viz. Virupakshi and Rasthali. Significant yield increase and reduction in nematode population was noticed in the treated plants. The effect of *T. viride* on growth of Rasthali banana infected with root-lesion nematode, *P. coffeae* and fungus, *F. oxysporum* f. sp. *cubense* in pots under greenhouse conditions showed a significant reduction in nematode population. Two application of biocontrol agent, *T. viride* at 20g/plant one at the time of planting and second 3 months after planting was found effective in controlling nematodes (*P. coffeae* and *M. incognita*) as well as reduced the incidence of panama wilt disease in Rasthali and Virupakshi (Sundararaju *et al.*, 2001).

An experiment on effect of *Verticillium chlamydosporium* culture filtrates on second stage juveniles and eggs of *M. incognita* was carried out under *in-vitro* conditions. The results revealed that the marked deleterious effect of culture filtrates (25, 50, 75 and 100 %) was noticed on *M. incognita* second stage juveniles. Among different dilutions of *V. chlamydosporium* culture filtrates, 100% concentration of fungus culture filtrate showed highest (78.67%) mortality of *M. incognita* juveniles. The nematode egg parasitic fungus, *V. chlamydosporium* was also multiplied on locally available organic substrate like farmyard manure (FYM), neem cake, wheat bran, cumbu grains, banana wastes (petiole, leaf and pseudostem) and sorghum grains.

It was observed that neem cake has shown profuse growth of the fungus when compared to rest of the substrates (Sundararaju *et al.*, 2003).

The root-knot nematode, *M. incognita* was recorded for the first time in an ornamental banana, *Ensete superbum*. The eggs of root-knot nematode were parasitized by the fungus naturally and the fungus was identified as *P. lilacinus*. Though *P. lilacinus* has been reported to be a potential egg pathogen of root-knot nematodes but the present isolate has shown its antagonistic effect on root-knot nematode (Sundararaju *et al.*, 2003).

Jonathan and Rajendran (2000) reported that application of *P.lilacinus* at 15 or 20 g per plant significantly reduced the root gall index, number of egg masses, eggs per mass, females and soil population of *M.incognita* in banana Poovan under greenhouse conditions. The effect of this treatment was comparable with 40 g Carbofuran per plant. *Paecilomyces lilacinus* applied @ 30 g/kg of soil at planting (30 days before nematode inoculation) was comparatively effective in controlling the population of *R.similis* in banana (Devarajan and Rajendran, 2001). Devarajan and Rajendran (2002) reported that highest reduction in population of *M.incognita* infesting banana Robusta was obtained with the highest level of application of *P.lilacinus* (30 g/kg of soil) during planting compared to 10, 20 g / kg of soil during planting or 30 or 60 days after planting respectively.

Integrated Nematode Management

Chennabasappa (1994) reported that integration of neem cake (400 g/plant), Carbofuran (20 g/plant), mycorrhiza (500 chlamydospores/plant) and *Pasteuria penetrans* (100 g soil containing an average of 20 *P.penetrans)* infected J_2 of *M.incognita* race I (each of which had an average of 15 spores attached to their cuticle) was effective in improving growth parameters of banana plant besides highest reduction in nematode population in root and soil as well as lesion index. Parvatha Reddy *et al.* (1996) found that by integrating eco-friendly components such as oilseed cakes of castor, karanj and neem at 400 g/plant with endomycorrhiza, *Glomus mosseae* (containing 25 chlamydospores per g of inoculum) reduced the population of *R.similis* both in soil and roots increasing plant growth parameters and fruit yield of banana.

Vidya and Reddy (1998) reported that integration of neem cake (neem extracts), Carbofuran, *P.penetrans* and *G.fasciculatum* was the most effective in enhancing the plant growth and yield of banana, raising the cost: benefit ratio to 1:2.65, reducing the *R.similis* population both in soil (95.16%) and roots (89.27%).

Investigations were carried out on the management of burrowing nematode, *R.similis*, infesting banana Dwarf Cavendish, by integrating eco-friendly components such as oil cakes (neem and pongamia, each at 500 g/plant), a biological control agent (*Trichoderma viride* at 10 g/plant) and a nematicide (Carbofuran 3G at 40 g/plant). Among different treatments, combination of neem cake, Carbofuran and *T.viride* was most effective in reducing the nematode population, improving plant growth and increasing fruit yield (76.3 tonnes/ha) with wider cost: benefit ratio (1:2.9) (Harish and Gowda, 2001). Kumar (2001) reported that *R.similis* population in soil and roots significantly reduced when banana plants are treated with Carbofuran 3G at 40 g/sucker followed by *P.fluorescens* @ 6.25 kg/ha with FYM 500 kg/ha and neem cake 500 kg/ha. The fruit yield increased by 165.40 kg/40m^2 in Carbofuran 3G at 40g / sucker followed by *T.viride* 2.5 kg/ha with FYM 500 kg/ha (147.25 kg/m^2).

The combined application of *P.lilacinus* (25 kg/ha) and Carbofuran (33 kg/ha) recorded significantly lowest root-knot index (1.8) with higher plant height and better growth, and also gave highest fruit yield (42017.8 kg/ha) (Vyas *et al.*, 2001). Sundararaju and Kiruthika (2007) investigated an egg parasitizing biocontrol agent, *Paecilomyces lilacinus* alone, and in combination with neem cake and botanicals such as *Tagetes* sp. (leaf extracts), *Tagetes* sp. (flower extracts), and *Solanum torvum* against root-knot nematode, *M. incognita* on banana Robusta. The results indicated that an individual treatment of *P.lilacinus* or *Tagetes* spp. (leaf and flower extracts) alone or in combination with neem cake or *Tagetes* spp. (flower extracts) or *S.torvum* are most effective treatments in increasing plant growth with significant reduction in nematode populations. Thus, integration of *P. lilacinus* with neem cake or any one of botanicals namely, *Tagetes* spp., (leaf or flower extracts) or *S.torvum* can be effectively used in management of root-knot nematode in banana, since use of single bioagent or botanical cannot be very effective in management of nematode induced disease complex (*see* Box). Using VAM (*Glomus fasciculatum* and *G.mosseae)* @20g/plant one at the time of planting and second three months after planting to control root-knot and root-lesion nematodes in all commercial cultivars of banana has been developed (Sundararaju,2006).

Host reaction of different genotypes: Reaction of *Musa* clones for nematode resistance

Musa clone	*Host response*	*Nematode*	*Reference*
Kadali, Padalimoongil, Ayiranka Poovan, Peykunnan, Vennettukunnan, Tongat, Anaikomban	R	*R.similis*	Rajendran *et al.*, 1979
Yelakkibale	R	*M.incognita*	Ravichandran and Krishnappa (1986)
Anaikomban	R	*R.similis*	Sathiamoorthy (1987)
Bantala, Dwarf Cavendish	MR	*R.similis*	Ray and Parija (1987)
Padali Moongil, Kunnan, Ayiramka Poovan	R	*R.similis*	Saeed *et al.* (1988)
Athia	T	*H.dihystera*	Choudhury and Phukan (1990)
Kunnan, Poom Kadali, Nalli Poovan	R	*R.similis*	Anitha *et al.* (1996)
Palayankodan	MR	*R.similis*	Sudha and Sundararaju (1996)
Anaikomban, Adakkakunnan, Kunnan	T	*R.similis*	Johnson and Sathiamoorthy, 1999
Matti, Kullan, Chinali, Nallabontha, Burharia	MR	*R.similis, P.coffeae* and *H.multicinctus*	Shanthi *et al.*, 2001
Anaikomban, Namarai, H 59, H 65 Tongat, H 84, H 109, H 110	MRR	*R.similis* and *P.coffeae*	Devarajan and Rajendran, 2001
Agnieswar, Mottapoovan, CO.1, Neyvazhai, Vannan, Ladan Padathi, Melakali, Cheenabale, Nendrakunnan, Lambi, Batheesa Ashy, Alshi,	T	*R.similis, P.coffeae* and *H.multicinctus*	Shanthi *et al.*, 2002
Neyvannan, Sambarani, Boothibale, Rajavazhai, Kothia and Muthia Beula, Kanthali, Kothia and Muthia	T	*R.similis, P.coffeae* and *H.multicinctus*	Shanthi *et al.*, 2002Beula, Kanthali,
Pisang Mas, Pisang Berlin, Kanaibansi, Namarai, cultivar Rose	T	*P. coffea*	Sundararaju *et al.*, 2002
Namarai, Hatidat	R	*H.multicinctus*	
Cultivars Burrow Cemsa and hybrid E.A.0322. FHIA-01 in Pome group	R	*R.similis, P.coffeae, M. incognita H.multicinctu*	Sundararaju and Uma, 1999
Singhlal, Sakkarachayna, Malai Kali, Manik Champa, Thiruvananthapuram, Kartobiumtham and FHIA601 and FHIA-23	R	*P.coffeae*	Sundararaju *et al.*, 2007

SUMMARY

Although chemical nematicides resulted in higher reduction of nematodes, due to economical and environmental issues, it is essential to switch over to other avenues. The use of biocontrol agents in management of nematodes is gaining importance worldwide. Though it is gaining momentum, there are several impediments which need further research perfection for improving their bioefficacy under field conditions. The lack of proper delivery systems which has to be standardized depending on crop duration, annual/perennial nature of crop and frequency of application of bioagents, is a major hurdle. Some of the botanicals and organic amendments are proved to possess nematicidal properties. The principle compound Azhadirachtin present in neem showed better reduction of nematode hatching and reproduction. The organic amendments increased population of microorganisms in soil, which release ammonia by organic amendments decomposition also positively reduces the nematode population in soil. Hence, exploitation of these properties of botanicals, organic amendments and biocontrol agents will be helpful for eco-friendly nematode management strategies in banana.

References

Anitha, N., Rajamony, L. and C.Radhakrishnan, T. 1996. Reaction of banana clones against the major biotic stress. *Planter,* **72:**315-21.

Bhattacharya, R.K. and Rao, V.N.M. 1984. Effect of soil covers and soil moisture regimes on nematode population in soil and in root of banana. *J. Res.* Assam Agric. Univ., **5**:206-09.

Chennabasappa, B.S. 1994. M.Sc. (Ag). thesis. University of Agricultural Sciences, Bangalore.

Choudhury, B.N. and P.N. Phukan, 1990. Distribution and occurrence of certain plant parasitic nematodes in different cultivars of banana. *Curr. Nematol.*, **1**:153-56.

Devarajan, K. and Rajendran, G. 2001. Effect of fungus *Paecilomyces lilacinus* on the burrowing nematode, *Radopholus similis* in banana. *Pest Management in Horticultural Ecosystems,* **7**:171-173.

Devarajan, K. and Rajendran, G. 2002. Effect of fungal egg parasite, *Paecilomyces lilacinus* (Thom.) Samson on *Meloidogyne incognita* in banana. *Indian J. Nematol.*, **32**:88-90.

Devarajan, K. and Rajendran, G. 2001. Reaction of some banana clones to the burrowing and lesion nematodes. *Indian J. Nematol.,* **31**:159-78.

Harish, M. and D.N. Gowda. 2001. Management of the burrowing nematode, *Radopholus similis* (Cobb,1893). Thorne, 1949 infesting banana. *Indian J.Nematol,* **31**:23-25.

Johnson, R.N. and Sathiamoorthy, S. 1999. Reaction of diploid banana cvs. to the burrowing nematode, *Radopholus similis. South Indian Hort.*, **47**:170-171.

Jonathan, E.I. and Rajendran, G. 2000. Assessment of avoidable yield loss in banana due to root-knot nematode, *Meloidogyne incognita. Indian J. Nematol.,* **30**:162-164.

Jonathan, E.I. and Rajendran, G. 2000. Biocontrol potential of the parasitic fungus *Paecilomyces lilacinus* against the root-knot nematode *Meloidogyne incognita* in banana. *J. Biol. Control.*, **14**:67-69.

Rajendran G., Naganathan, T.N. and Vadivelu, S.1979. Studies on banana nematodes.*Indian J. Nematol.,* **9**:54.

Ravichandran, N.G. and Krishnappa, K. 1986. Reaction of banana cultivars popularly grown in Karnataka to the burrowing nematode, *Radopholus similis. Indian J. Nematol,* **15**:240.

Saeed, M., A.Khan and Qamar, F. 1988. Reaction of two commercial varieties of bananato root-knot nematode, *Pak. J. Sci. Indul. Res.,* **31**:632-633.

Sasser, J.N and Freckman, D.W. 1987. A world perspective on nematology: the role of the society. pp. 7-14. *In*: *Vistas on Nematology.* Veech, J.A. and Dickson, D.W. (Eds). Society of Nematologiest Inc., Hyattsville, USA.

Sathiamoorthy, S. 1987. Studies on male breeding potential and certain aspects of breeding bananas. Ph.D. (Hort.) thesis, Tamil Nadu Agricultural University, Coimbatore.

Shanthi, A., Rajendran, G., Kumar, N. and Soorianathasundaram, K. 2001. Screening of Banana cultivars for nematode resistance. *In: Proceeding of National Seminar on EmergingTrends in Pests and Diseases and their Management, held* at TNAU, during 11-13 October 2001.

Shanthi, A., Rajendran, G. Kumar, N. and Soorianathasundaram, K. 2002. Evaluation of promising cultivars of banana against major nematodes. *In: Proceeding National Symposium on Biodiversity and Management of nematodes in Cropping systems for Sustainable Agriculture,* Durgapura, Jaipur (Rajasthan), held during 11-13. November 2002.

Shanthi, A., Poornima, K. Rajendran, G. Kumar, N. and K. Soorianatha Sundaram. 2001. Management of nematodes in banana. *In: Proceeding National Seminar on Emerging Trends in Pests and Diseases and their Management,* held at CPPS, TNAU, during 11-13. October pp. 326-30.

Subramaniam, S. and P.Selvaraj. 1990. Effect of antagonistic intercrops on burrowing nematode in Robusta banana. *South Indian Horti.*, **38:**216-17.

Sudha, S and Sundararaju, P. 1996. Reaction of some banana cultivars to the burrowing nematode, *Radopholus similis. In: Proceeding Banana Improvement, Production and Utilization.* Singh, H.P. and Chadha, K.L. (Eds). National Research Centre for Banana, Trichy., pp. 341-43.

Sundararaju, P and Uma, S. 1999. Evaluation of promising hybrids and cultivars of banana against major nematode pathogens. *In: Proceeding Third International Symposium of Afro-Asian Society of Nematologist,* (TISAASN). Usha, K.Mehata, (Ed.). Sugarcane Breeding Institute Coimbatore, pp. 122-25.

Sundararaju, P. 1996. Nematode pests of banana and their management. ‵*Challenges for Banana Production and Utilization in 21st Centrury*′ held at NRC for Banana, Trichy, 24-25 September 1996, pp. 17-19.

Sundararaju, P and Cannayane, I. 2002. Production of nematode egg parasitic fungus, *Paecilomyces lilacinus*, on banana wastes and certain plant leaves. *Indian J. Nematol.,* **32**:188-89.

Sundararaju, P and Cannayane, I.2002. Antinemic activity of plant extracts against *Pratylenchus coffeae* infecting banana. *Indian J Nematol,* **32**:121-24.

Sundararaju, P., Cannayane, I. and Sathiamoorthy, S. 2002. Evaluation of promising diploids banana against major banana nematodes. Global Conference on Banana and Plantain, 28-31 October 2003, Bangalore, pp.176.

Sundararaju, P. and Kumar, V. 2003. Management of root-lesion nematode, *Pratylenchus coffeae* in six commercial cultivars of banana through organic and inorganic amendments. *Infomusa,* **12:**21-25.

Sundararaju, P., Padmanaban, B and Sathiamoorthy, S. 2003. Efficacy of certain botanicals against root-lesion nematode, *Pratylenchus coffeae* in banana. *Nematol. medit.,* **31**:201-5.

Sundararaju, P. Cannayane, I. and Sathiamoorthy, S. 2003. Occurrence of *Meloidogyne incognita* on *Ensete superbum. Infomusa.,* **12:**43-44.

Sundararaju, P., Mustaffa, M.M., Kumar, V. Cannayane, I. and Tanuja Priya, B. 2002. Effect of organic farming on plant parasitic nematodes infesting banana cv. Karpuravalli, *Current Nematology,* **13**:39-43.

Sundararaju, P., Selvarajan, R. and Sathiamoorthy, S. 2001. Management of nematode- Fusarium wilt complex in banana through bioagent, *Trichoderma viride* (Bio-T). *In: Fifteenth Nematological Congress,* at Skukuza Kruger National Park, South Africa, during 20-24 May, 2001.

Sundararaju, P.2005. Effect of marigold, *Tagetes errecta* intercropped with banana against root- lesion nematode, *Pratylenchus coffeae. Indian J. Nematol.,* **35**:123-26.

Sundararaju, P and Saritha, V. 2006. Effect of leaf extracts of *Acalypha indica, Cassia fistula* and *Solanum torvum* on *Pratylenchus coffeae. Indian J. Nematol.,* **36**(1):144-45.

Sundararaju, P. and Thenmozhi, R. 2007. Efficacy of neem based pesticides used as sucker treatment against root-knot nematode infesting banana cv.Robusta. III National Symposium on Plant Protection in Horticulture: *Emerging Trend and Challenges,* organized by Association for Advancement of Pest Management in Horticultural Ecosystems (AAPMHE) at IIHR, Bangalore, 7-9, March, 2007, p.51.

Sundararaju, P. and Sudha, R. 2007. Efficacy of organic amendments on root-lesion nematode, *Pratylenchus coffeae* infecting banana in cv. Nendran. Ibid p.06.

Sundararaju, P., Thangavelu, R. and Cannayane, I. 2002. Management of *Pratylenchus coffeae* and *Meloidogyne incognita* on banana by using endophytic fungi. *Current Nematology,* **13**:77-81.

Sundararaju, P., Leela Devi, R. and Manimegalai, M. 2002. Analysis of best treatment and variety based on nematode population on banana using artificial neural networks. *Indian J. Nematol.,* **32:**84-85.

Sundararaju, P., Thangavelu, R. and Cannayane, I. 2002. Management of *Pratylenchus coffeae* and *Meloidogyne incognita* on banana by using endophytic fungi. *Current Nematology,* **13**:77-81.

Sundararaju, P and I.Cannayane. 2003. Evaluation of different biopesticides against major nematode pathogens infesting banana cv. Nendran. *National Symposium on Organic Farming in Horticulture for Sustainable Production.* 29-30, August, 2003, CISH, Lucknow. 66-67p.

Sundararaju, P., Sasikala, T. and Cannayane, I. 2003. Biocontrol Potential of *Verticillium chlamydosporium* against *Meloidogyne incognita* infesting banana. *Current Nematology,* **15**(2):72-75.

Sundararaju, P. and Kiruthika, P. 2007. Effect of compatible biocontrol agents along with neemcake and botanicals on *Meloidogyne incognita* infecting banana. *In: National Conference on Banana* at Trichirapalli 25-28, during October, 2007.

Sundararaju, P. and Madhula, V. 2007. Effect of compatible biocontrol agents along with neemcake and botanicals on *Meloidogyne incognita* infecting banana. Ibid.

Venkitesan, T.S. and J.S.K.Charles. 1983. Sucker dip treatment for banana nematode control. *In: Third Nematology Symposium*, Solan, India. 24-26 May,1983.

Vidya, K. and Reddy, B.M.R. 1998. Integrated management of *Radopholus similis* infecting banana using plant product, nematicide and biocontrol agents. *Mysore J. Agric. Sci.*, **32**:186-90.

47

Management of Root Mealybug – A New Pest on Banana

M.S. Smitha, P. Mathew Maicykutty

Root mealybug posed a threat to the cultivation of Nendran banana in several areas of Palakkad, Thrissur and Ernakulam districts (Verghese *et al.*, 2002) in Kerala. Since infestation found on roots, it goes unnoticed and results in complete devastation of the crop. Considering its widespread occurrence and damages caused to commercial cultivation of Nendran banana, a study was undertaken on identification its biology and management.

Adult mealybugs were collected and preserved in 70% alcohol in small vials of 5ml size. These were sent to Agricultural Research Organization, Bet Dagan, Israel, for identification.

Mealybug population was assessed at monthly intervals from July 2005 to June 2006 in two panchayats, *viz.* Thachampara and Karakurissi in Palakkad. From each *panchayat*, four gardens each with a minimum of 100 banana plants of uniform age were selected randomly. Compact soil samples of 15cm^3 size were dug out and mealybug intensity was rated by counting the number of mealybug colonies in unit soil sample (15cm^3 soil). The data on mean number of root mealybugs in each *panchayat* was subjected to completely randomized block design (2 factor) ANOVA and the means were separated by the Duncans' Multiple Range Test to study the peak period of occurrence of root mealybug.

The root mealybugs were cultured on potato sprouts and biology was studied at the College of Horticulture, Vellanikkara. The study was conducted at susceptibility of popular banana cultivars. Most popular cultivars of banana in Kerala, *viz.* Rasthali (Poovan), Robusta, Njalipoovan, Kodappanillakunnn and Palayankodan were evaluated under field conditions for susceptibility to root mealybug. The experiment was laid out at farmers' fields in Thannerpanthal area of Palakkad district in a randomized block design with four replications. Four suckers were maintained for each variety. The data on intensity of population was collected by counting the number of mealybug colonies. The observations were taken at bimonthly intervals, viz. two, four and six months after planting of suckers. The mean data on number of root mealybug colonies on different cultivars were subjected to ANOVA 2 analysis and the means were separated by Duncan's Multiple Range Test.

MANAGEMENT OF ROOT MEALYBUG

Management study was conducted in two steps. In first step, treatment materials under different category were screened under field condition. In second step, 15 different combinations were designed with four best treatments from each screening category and the experiment was laid out in a randomized block design with two replications and nine plants were maintained for each replication. Sodium silicate was applied one month after planting and all the other treatments were applied as drench at bimonthly intervals for seven months commencing from one month after planting of suckers. Five litres of drenching solution was applied for each plant. One control plot was also maintained, which was drenched with five litres of water at bimonthly intervals. Data on mealybug intensity and bunch weight were collected.

The identification report from coccid taxonomist indicated the presence of two species belonging to genus *Geococcus*. The species identified were: *Geococcus citrinus* Kuwana, 1923 and *G. coffeae* Green, 1902. The *G. citrinus* was reported from Kerala for the first time and on banana for the first time in India.

The root mealybug infestation on banana was recorded from 10 districts in Kerala, *viz.* Pathanamthitta, Kottayam, Idukki, Ernakulam, Thrissur, Palakkad, Malappuram, Kozhikkode, Wynad and Kannur. The occurrence of root mealybug as a new pest on banana was reported by Verghese *et al.* (2002), where they have recorded the pest from three districts, *viz.* Ernakulam, Thrissur and Palakkad. The present study shows that the pest has spread to seven more districts, *viz.* Pathanamthitta, Kottayam, Idukki, Ernakulam, Thrissur, Palakkad, Malappuram, Kozhikkode, Kannur and Wynad.

The highest population of root mealybug was recorded in July (Table 47.1) and was significantly superior compared to remaining months. This was followed by June and August. There might be positive correlation between moisture present in soil and root mealybug population, which may be the reason for the occurrence of mealybugs in high numbers during rainy season.

Table 47.1. Population dynamics root mealy bug in Thachampara and Karakurissi *panchayats* during July '05 to June '06.

Month	**No. of colonies/15cm³soil*		
	Thachampara	*Karakurissi*	*Mean population*
July 2005	7.35 ^{a}(2.80)	8.73 ^{a} (3.04)	8.04 ^{a}(2.92)
August 2005	4.48 ^{b}(2.23)	4.28 ^{b}(2.19)	4.38 ^{b}(2.21)
September 2005	3.06 ^{c}(1.89)	2.48 ^{c}(1.73)	2.77 ^{c}(1.81)
October 2005	2.95 ^{c}(1.86)	2.80 ^{c}(1.82)	2.88 ^{c}(1.84)
November 2005	3.38 ^{c}(1.97)	3.00 ^{c}(1.87)	3.19 ^{c}(1.92)
December 2005	1.78 ^{d}(1.57)	2.05 ^{cd}(1.60)	1.91 ^{d}(1.55)
January 2006	1.45 ^{d}(1.40)	1.38 ^{e}(1.37)	1.41 ^{d}(1.38)
February 2006	1.65 ^{d}(1.47)	1.73 ^{d}(1.49)	1.69 ^{d}(1.48)
March 2006	1.93 ^{d}(1.56)	1.85 ^{d}(1.53)	1.89 ^{d}(1.55)
April 2006	1.65 ^{d}(1.47)	1.43 ^{e}(1.39)	1.54 ^{d}(1.43)
May 2006	1.53 ^{d}(1.42)	1.55 ^{e}(1.43)	1.60 ^{d}(1.45)
June 2006	4.50 ^{b}(2.24)	3.73 ^{b}(2.06)	4.11 ^{b}(2.15)

* Mean of 40 observations

Means followed by the same alphabets did not differ significantly (P=0.01)

Figures in parentheses are square root transformed values

Eggs were elongate-oval in shape and pearly white in colour. Eggs were observed in masses. On an average, 79.4 eggs were observed in an egg mass. The incubation period was 8 days during March-April and 11.6 days during November-December. Three nymphal instars were identified.

Adult female of *G. citrinus* was creamy-white colour, elongate-oval in shape and measured 2.5-3 mm in length and 1.5 mm in width. The anal hooks were light brown in colour. Adult female of *G. coffeae* was snow-white, elongate-oval and 2-2.5 mm in length and 0.8 mm in width. The anal lobes were reddish-brown and strongly curved. Though there were reports on occurrence of *Geococcus* sp. as early as in 1902, so far nobody have studied its biology.

Of the six varieties tested, Palayankodan and Kodappanillakunnan were completely free from infestation. The results revealed that out of four cultivars, Nendran (4.38 colonies/15 cm^3 soil) (Table 47.2) recorded significantly highest mealybug population followed by Njalipoovan (2.55 colonies/sample). Poovan recorded the least number of colonies (0.42/sample).

Table 47.2. Intensity of root mealy bug infestation on different cultivars of banana

Month	** Mealy bug colonies/15cm³ soil*						
	Njalipoovan	*Poovan*	*Robusta*	*Nendran*	*Palayan -kodan*	*Kodappanilla - kunnan*	*Mean for months*
February 2006	2.38 ^{d}(1.70)	0.25 ^{g}(0.87)	0.38 ^{fg}(0.93)	4.50 ^{b}(2.24	0.00 ^{h}(0.71)	0.00 ^{h}(0.71)	1.88 ^{B}(1.54)
April 2006	2.13 ^{d}(1.62)	0.38 ^{fg}(0.93)	1.25 ^{e}(1.32)	3.38 ^{c}(1.97)	0.00 ^{h}(0.71)	0.00 ^{h}(0.71)	1.79 ^{B}(1.51)
June 2006	3.13 ^{c}(1.90)	0.63 ^{fg}(0.16)	1.38 ^{e}(1.37)	5.25 ^{a}(2.40)	0.00 ^{h}(0.71	0.00 ^{h}(0.71	2.60 ^{A}(1.76)
Mean for varieties	2.55 ^{B}(1.74)	0.42 ^{D}(0.95)	1.00 ^{C}(1.21)	4.38 ^{A}(2.20)	0.00 ^{E}(0.71)	0.00 ^{E}(0.71)	

* Mean of three replications
Figures followed by the same alphabets did not differ significantly (P=0.01)
Figures in parentheses are square root transformed values

The use of Chlorpyrifos alone and its combinations showed higher reduction of the pest. This effect might be due to the fumigant action of Chlorpyrifos (Table 47.3). Many workers reported the use of Chlorpyrifos in controlling root mealybugs. Hara *et al.* (2001) reported that in general, drenching of dursban (Chlorpyrifos) twice at two weeks interval controlled root mealybugs. The treatment with Chlorpyrifos recorded highest bunch weight (11.43kg) and was on a par with CV (11.27kg) and were significantly superior.

SUMMARY

Two species of root mealybug pests, *Geococcus coffeae* and *G. citrinus,* were identified infesting on Nendran banana in Kerala. The survey revealed the occurrence of this pest in 10 districts of Kerala. The study on population dynamics for a year showed that it reached a peak in July and thereafter declined. The adult female laid eggs in masses with 79.4 eggs in an egg mass. Three nymphal stages could be identified in its life-cycle. One coccinellid predator, *Scymnus* (*Pullus*) sp., and an entomopathogenic fungi, *Hirsutella* sp., could be observed naturally among mealybug colonies in field. Fourteen species of weed hosts belonging to eight plant families were recorded as alternative hosts for root mealybug. Palayankodan and Kodappanillakunnan were completely free from the infestation, but Njalipoovaan was equally

Table 47.3. Reduction of population (%) in different treatments evaluated on root mealybug, *Geococcus* spp. infesting banana.

Treatment	Reduction over control						
	2 MAP	3 MAP	4 MAP	5 MAP	6 MAP	7 MAP	8 MAP
SNCV	34.27 ^{cde}(0.621)	32.46 ^{cde}(0.606)	37.85 ^{bc}(0.661)	43.75 ^{cde}(0.722)	51.07 ^{abc}(0.796)	59.02 ^{ab}(0.877)	61.57 ^{ab}(0.902)
SNC	48.09 ^{abc}(0.766)	39.35 ^{bcd}(0.674)	46.53 ^{ab}(0.750)	46.88 ^{bcd}(0.752)	58.45 ^{ab}(0.871)	62.55 ^{a}(0.912)	65.13 ^{ab}(0.940)
SNV	26.32 ^{def}(0.533)	22.18 ^{defg}(0.486)	28.82 ^{cd}(0.561)	34.38 ^{def}(0.623)	31.55 ^{d}(0.594)	43.33 ^{bcde}(0.718)	58.12 ^{abc}(0.868)
SCV	58.53 ^{a}(0.874)	57.52 ^{a}(0.861)	61.46 ^{a}(0.902)	65.63 ^{a}(0.945)	60.95 a(0.896)	63.82 a(0.926)	71.47 ^{a}(1.009)
SN	38.09 ^{bcd}(0.665)	22.43 ^{defg}(0.493)	32.29 ^{bcd}(0.604)	43.75 ^{cde}(0.722)	33.93 ^{cd}(0.619)	53.14 ^{abc}(0.817)	58.12 ^{abc}(0.868)
SC	58.97 ^{a}(0.876)	54.89 ^{ab}(0.834)	64.58 ^{a}(0.934)	62.50 ^{ab}(0.913)	58.45 ^{ab}(0.871)	62.35 a(0.910)	64.79 ^{ab}(0.937)
SV	26.32 ^{def}(0.533)	24.81 ^{defg}(0.520)	31.94 ^{bcd}(0.598)0.598	37.5 ^{def}(0.658)	29.05 ^{d}(0.564)	51.47 ^{abc}(0.800)	58.01 ^{abc}(0.867)
S	28.82 ^{def}(0.559)	27.44 ^{cdef}(0.551)	46.53 ^{ab}(0.750)	46.88 ^{bcd}(0.754)	31.43 ^{d}(0.589)	52.94 ^{abc}(0.815)	54.89 ^{abc}(0.835)
NCV	48.09 ^{abc}(0.766)	34.59 ^{cde}(0.626)	49.65 ^{ab}(0.782)	50.00 ^{abcd}(0.785)	41.19 ^{bcd}(0.694)	52.94 ^{abc}(0.815)	53.28 ^{bc}(0.818)
NC	41.03 ^{abcd}(0.695)	37.72 ^{bcd}(0.661)	38.19 ^{bc}(0.666)	40.63 ^{def}(0.691)	38.81 ^{cd}(0.671)	52.00 ^{abcd}(0.785)	51.56 ^{bc}(0.801)
NV	24.27 ^{def}(0.515)	20.05 ^{efg}(0.464)	22.92 ^{cd}(0.488)	28.13 ^{ef}(0.558)	29.17 ^{d}(0.570)	35.59 ^{cde}(0.638)	41.55 ^{cd}(0.700)
N	20.88 ^{ef}(0.465)	15.04 ^{fg}(0.398)	34.72 ^{bcd}(0.627)	34.38 ^{def}(0.623)	24.29 ^{d}(0.514)	32.65 ^{de}(0.608)	39.93 ^{cd}(0.684)
CV	56.03 ^{ab}(0.847)	37.22 ^{bcd}(0.655)	61.46 ^{a}(0.902)	59.38 ^{abc}(0.881)	60.83 ^{a}(0.895)	60.88 ^{ab}(0.895)	63.29 ^{ab}(0.920)
C	58.97 ^{a}(0.876)	42.73 ^{abc}(0.712)	61.46 ^{a}(0.902)	62.50 ^{ab}(0.913)	58.45 ^{ab}(0.871)	60.68 ^{ab}(0.893)	66.52 ^{ab}(0.955)
V	15.88 ^{f}(0.407)	12.66 ^{g}(0.361)	20.14 ^{d}(0.458)	25.00 ^{f}(0.524)	26.79 ^{d}(0.544)	26.37 ^{e}(0.539)	31.54 ^{d}(0.596)

MAP- Months after planting
*Mean of two replications
Figures in parentheses are angular transformed values

susceptible as Nendran. Management studies revealed that treatment combinations containing Chlorpyrifos reduced the mealybug population to a higher extent.

References

Hara, A.H., Nino-DuPonte, R.Y. and Jacobsen, C.M. 2001. Root mealy bugs of quarantine significance in Hawaii. Cooperative Extension Service, CTAHR, University of Hawaii, Manoa (US). *Insect Pests,* **6**.

Verghese, A., Ajith, C.B. and Priyamol, S. 2000. Mealy bugs - threat to banana cultivation. *Krishiyankanams,* **6**:5-6.

48

Depredatory Birds of Banana and Their Eco-Friendly Management

MANI CHELLAPPAN, V. VASUDEVA RAO AND JIM THOMAS

BANANA is a major fruit crop in Kerala. Despite being one of the major man-made fruit plantations, its avifaunal pattern is poorly understood. The bird community in banana plantations belonged to different residency status. Periodical observations in banana plantation indicated that about 20 species of birds are frequent to the field. The birds of banana plantations could be assigned to different guilds based on their food and feeding habits. Since birds cause damage even to un-ripen fruits, present study was undertaken to understand banana plantation avifauna composition, extent of damage caused and ecological management of these vertebrate depredators.

ECO-FRIENDLY MANAGEMENT OF BIRDS

Five banana plantations were chosen for the study. The data were collected from 2003 onwards. Birds were counted regularly using line transect method within the field. About 2-3 censuses were carried out per month. Details of food eaten by birds in banana plantations were recorded by direct observation, using binoculars from a close range. Bird diversity in each study area was calculated using modified Simpson index (Rottenberry, 1978). The bunch damage was calculated based on per cent feeding. Birds frequent banana plantations for food or used the standing crop as a shelter or cover or as a site for nesting or to roost at night. Survey at different stages of banana crop, viz. vegetative, flowering, fruiting and fruit maturity has shown that birds like the bee eater (on insects), myna and babblers (on insects and fruits) frequent banana plantations in all these stages for feeding and resting. Many birds visiting plantation for feeding on insects and for mere cover but only few birds like the small green barbets, bulbuls, koel, parakeet, babbler and raven cause economic damage to the crop at the fruiting stage (Table 48.1). Among 22 species, 45% birds actually cause damage to banana bunches. The initial damage is being done by small green barbet, which attracts other birds to feed on fruits (Mani Chellappan *et al.*, 2002). Based on gravity of damage, a preference study was conducted for the small green barbet.

Table 48.1. Diversity of banana plantation avifauna

Stage of crop	*Avifauna diversity*		*Relative abundance (%)*
	Common name	*Scientific name*	
Vegetative	Babbler	*Turdoides affinis*	21.0
	Bee eater	*Merops orientalis*	9.0
	Black drongo	*Dicrurus* sp.	8.0
	Coucal	*Centropus* sp.	15.0
	Fly catcher	*Muscicapa* sp.	12.0
	Hoopoe	*Upupa epops*	9.0
	Myna	*Acridotheres tristis*	15.0
	Jungle myna	*A. fuscus*	7.0
	King fisher	*Halcyon* sp.	5.0
	Spotted dove	*Streptopelia chinensis*	2.0
	Tree pie	*Dendrocita vagabunda*	4.0
	Warbler	*Bradypterus* sp.	2.0
Flowering stage	Babbler	*Turdoides affinis*	19.0
	Bee eater	*Merops orientalis*	14.0
	Black drongo	*Dicrurus* sp.	7.0
	Coucal	*Centropus* sp.	6.0
	Fly catcher	*Muscicapa* sp.	9.0
	Hoopoe	*Upupa epops*	10.0
	Myna	*Acridotheres tristis*	15.0
	Jungle myna	*A. fuscus*	5.0
	King fisher	*Halcyon* sp.	7.0
	Malabar trogon	*Harpactes fasciatus*	2.0
	Tree pie	*Dendrocita vagabunda*	4.0
	Warbler	*Bradypterus* sp.	1.0
Fruiting	Babbler	*Turdoides affinis*	18.0
	Bee eater	*Merops orientalis*	5.0
	Bulbul-black	*Hypsipetes leucocephalus*	20.0
	Bulbulñred vented	*Pycnonotus cafer*	18.0
	Coucal	*Centropus* sp.	5.0
	Koel	*Eudynamys scolopacea*	6.0
	Raven	*Corvus corax*	8.0
	Jungle myna	*Acridotheres fuscus*	11.0
	King fisher	*Halcyon* sp.	9.0
Fruit maturity	Barbet (small)	*Megalaima viridis*	18.0
	*Barbet	*M. lineate*	8.0
	*Babbler	*Turdoides affinis*	14.0
	*Bee eater	*Merops orientalis*	2.0
	Bulbul whitecheeked*	*Hypsipetes leucocephalus*	14.0
	Bulbulñred vented	*Pycnonotus cafer*	6.0
	Myna	*Acridotheres tristis*	13.0
	Koel	*Eudynamys scolopacea*	10.0
	*Jungle myna	*Acridotheres fuscus*	6.0
	*Raven	*Corvus corax*	2.0
	Rose ringed parakeet	*Psittcula krameri*	8.0

* Causes economic damage; pooled observations from six locations

THE RESULT

The result showed that bird had a particular preference to banana Nendran (Table 48.2).

Table 48.2. Feeding preferences of small green barbet

Food material	*Quantity fed (g)*	*Feeding (%)*
Banana *Nendran*	40g out of 60g kept	66.66
Banana *Poovan*	5 g out of 155g kept	3.20
Banana *Palayankodan*	50g out of 150g kept	45.45
Luvi - luvi	107.51g out of 130g kept	82.70
Mango	120g out of 300g kept	40.00
Cherry	52.87g out of 80g kept	28.20
Guava	19.92g out of 120g kept	16.66
Jamun	8.51g out of 250g kept	2.81
Papaya	231.95g out of 250g kept	92.78

Population of barbet fluctuated throughout the year and two peaks were observed (Fig. 48.1) during March-April and July-September (ManiChellappan and JimThomas, 2004). The birds excavate rotting trees, viz. rain tree, mango and jack, and construct the nest deep inside with zig-zag tunnel. Nesting holes could be seen at any height from 2m above the ground level. Multiple holes were recorded in tree branches and sometimes the holes were never occupied. On an average, two chicks were present in the nest.

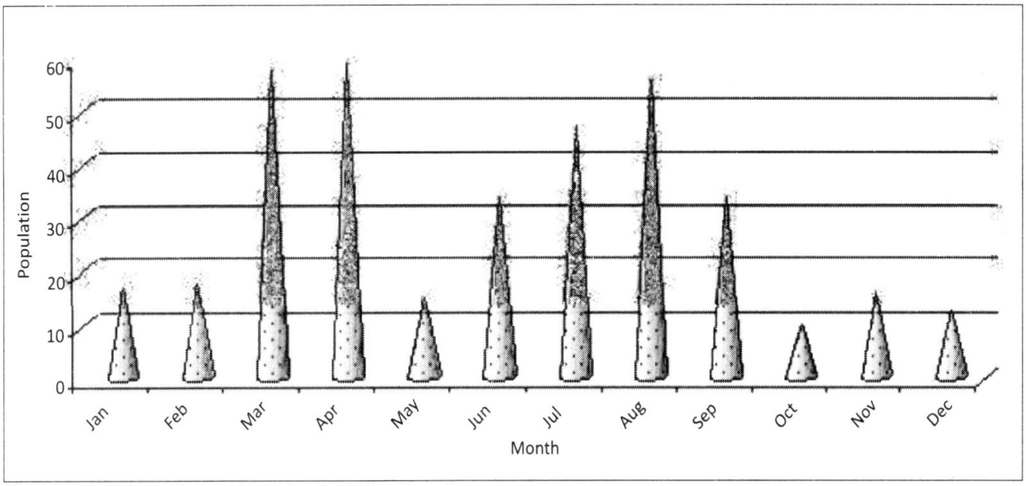

Fig. 48.1. Population dynamics of small green barbet *(Megalaima viridis)*

To minimize the damage caused by small green barbet and other depredatory birds, the bunches were protected with polythene covers of varying colours (transparent, white, blue, black and yellow), jute bag, dried banana leaf trash covers and compared with unprotected ones. Covering the bunches with polyethylene covers completely protected the bunches from the bird damage (Table 48.3). The exposed fingers out of cover were damaged by the bird. There was no correlation between colours of covers and the damage. Covering with non-porous covers hastened the maturity of bunches during hot summer months, compared to dried banana trashes. However, those covers gave better protection against the bird damage.

Table 48.3. Effect of bunch protection against depredatory birds in banana

Type of bunch protection	*Bunch damage (%)*
Transparent polyethylene cover	damage (nil)
Yellow polyethylene cover	damage (nil)
Blue polyethylene cover	damage (nil)
Black polyethylene cover	damage (nil)
White polyethylene cover	damage (nil)
Manure bags (white)	damage (nil)
Jute bag	damage (nil)
Banana trash	5.0
Unprotected bunch	20.5

Pooled observations from six locations

SUMMARY

The covering of bunches fully with polyethylene covers completely protected them from the bird damage. The birds damaged only the exposed fingers out of the covers. There was no correlation between colour of covers and bird damage. Covering with non-porous covers hastened the maturity of bunches during hot summer months, compared to dried banana trashes. The gunny bag cover attracted the rodents to nest inside and cause subsequent damage. In plastic covers, bunches had fungal infection due to higher humidity. Bio-bird repellent spray on bunches was effective for a brief period and required to be repeated more than once owing to length of duration of crop. Among different methods tried, poly covers gave better protection against the bird damage and could be reused if properly secured at end of the season. The cost: benefit ratio was also more for the polythene covering than other methods of protection.

References

Mani Chellappan, Jim Thomas and Anitha, S. 2002. Bio-ecological and depredatory role of small green barbet, *Megalaima viridis*, in banana plantations. *In: third National Symposium on Avian Biodiversity – Issues and Conservation Strategies*, ANGRAU, Hyderabad, India, pp 63.

Mani Chellappan and Jim Thomas. 2004. Bioecology of small green barbet, *Megalaima viridis. In: International Conference on Bird and Environment,* Gurukula Kangri University, Haridwar, India, pp. 112.

49

Biochemical Interactions in Banana Hybrids to Root-Lesion Nematodes

P.S. Kavitha, T.N. Balamohan and K. Poornima

Banana is an important food crop in tropical countries. Among pests and diseases that significantly reduce yield, nematodes are considered as one of the main soil-borne factors inducing root damages. Plants when attacked by nematodes show selective changes in their metabolism due to host-pathogen interaction, inducing an immune response of host to the parasite. Several researchers have emphasized the role of phenols as an expression of defense mechanism by host plants. The objective of the present work was to study the levels of biochemical alterations in banana hybrids inoculated with lesion nematodes and to correlate it with their level of resistance.

BIOCHEMICAL INTERACTIONS IN BANANA HYBRIDS TO NEMATODES

The root sample of 100 mg was chopped into small bits and plunged in 80% hot ethanol kept in water bath for 10 minutes and cooled in running tap water. The tissues were crushed thoroughly in a pestle and mortar, macerate was squeezed through two layers of cheese cloth and the extract was collected in a beaker. The residue was again extracted with 80% hot ethanol and squeezed through the cheese cloth. Both the extracts were pooled together and centrifuged at 5000 rpm for 10 minutes. The supernatant was collected and allowed to evaporate completely. The residue left was dissolved in 10 ml distilled water and used for estimation. Total and ortho-dihydroxy phenols (OD) were estimated by Folin Ciocalteau method (Malik and Singh, 1980) and reported as µg/g of roots. The lignin content of banana roots was gravimetrically estimated following the method of Chesson (1978). The results were expressed in terms of lignin content (%) on dry-weight basis of tissues.

The biochemical estimation was done at three stages of hybrids under field conditions (7 months after planting, shooting and harvesting). The total and ortho-dihydroxy phenols (OD) were estimated by Folin Ciocalteau method (Malik and Singh, 1980) and the lignin content of banana roots was gravimetrically estimated following the method of Chesson (1978). With a view to elicit information on presence of lignified cells and phenols accumulated cells in roots of susceptible and resistant banana hybrids, histological studies were carried out (Johansen, 1940).

THE RESULTS

Besides phytoalexins, most of the plants synthesis toxic compounds like phenols, proline, and lignin as part of the normal development and are called phytoanticipins (Van Etten *et al.*, 1995). Their roles in resistance mechanism have been reported earlier by many workers (Sariah *et al.*, 1999; Reuveni *et al.*, 1992; Reuveni and Karchi, 1987; Fogain and Gowen, 1996).

The hybrid, H-03-19, registered highest total phenol content of 235.60, 268.70 and 250.60 μg/g at 7MAP, shooting and harvesting respectively; the lowest being 36.40, 46.20 and 42.50 μg/g in H-03-12 at 7MAP, shooting and harvesting respectively. The hybrid H-03-19 recorded maximum OD phenol content of 1.61, 1.70 and 1.74 μg/g at 7MAP, shooting and harvesting respectively, while H-03-12 recorded minimum content of 0.85, 0.92 and 0.95 μg/g at 7 MAP, shooting and harvesting respectively. The maximum lignin content of 1.22, 1.36 and 1.40% was observed in H-03-05 at 7MAP, shooting and harvesting respectively and the minimum being 0.35, 0.38 and 0.40% in H-03-07 at 7MAP, shooting and harvesting respectively.

Vidhyasekaran (1988) described the occurrence of many kinds of phenolics in plants. Among them, total phenols play a unique role in response to pathogen and nematode invasion. There was a significant increase in phenol content in resistant (H-03-05 and H-03-19) and tolerant hybrids (H-03-13 and NPH-02-01) *vis-à-vis* susceptible ones. The accumulation of phenol may be due to the excess production of hydrogen peroxide by increased respiration (Farkas and Kiraly, 1962) or due to activation of hexose monophosphate (HMP) shunt pathway, acetate pathway and release of bound phenols by hydrolytic enzymes (Goodman *et al.*, 1967). Similar results were also observed by Gunavathi (2000) for *Fusarium* wilt resistance and Fogain and Gowen (1996) and Krishnamoorthy (2002) for nematode resistance in banana.

Lignin content perceptibly increased in resistant hybrids, H-03-19, H-03-05 and H-03-13, whereas in susceptible hybrids it was very low. This is in line with the findings of Binks (1996). Lignin is deposited in cell walls in response to damage but is also present at high levels in plant roots prior to infection, suggesting a pre-infectional defense mechanism. In present study, increasing lignin content was also observed in resistant cultivars, indicating the possibility of such a mechanism in operation to induce resistance. The increased synthesis of phenols followed by oxidation of them by higher rate of polyphenol oxidase would have helped in increased conversion of them into strong biochemical compound like lignin. Presence of more number of lignified cells was noticed in resistant cultivars than susceptible ones (Fogain, 1996; Giebel, 1982).

Higher levels of total phenols were recorded in resistant hybrids which ranged 235.60, 268.70 and 250.60 μg/g at 7MAP, shooting and harvesting respectively in hybrid H-03-19 and lowest of 36.40, 46.20 and 42.50 μg/g in susceptible hybrid H-03-12 at 7MAP, shooting and harvesting respectively (Table 49.1). Similarly, maximum OD phenol content was found in H-03-19 which recorded 1.61, 1.70 and 1.74 μg/g at 7MAP, shooting and harvesting respectively.

Similar trend was found in lignin content also. The maximum lignin content of 1.22, 1.36 and 1.40% was observed in H-03-05 at 7MAP, shooting and harvesting respectively and minimum of 0.35, 0.38 and 0.40% was observed in susceptible hybrid H-03-07 at 7MAP, shooting and harvesting respectively (Table 49.1). According to Muller and Beckman (1974) and Matielle (1994), these phenolic cells may have been pre-formed and may have a role in resistance. Increase in concentration of phenols following infection by nematodes was reported by Ganguly and Dasgupta (1984).

Table 49.1. Biochemical contents of phase II hybrids under field condition

Hybrids	*Reaction level*	*Total phenol (µg/g)*			*OD phenol (µg/g)*			*Lignin (%)*		
		7MAP	*Shooting*	*Harvest*	*7MAP*	*Shooting*	*Harvest*	*7MAP*	*Shooting*	*Harvest*
H-03-04	T	188.50	220.75	200.60	1.20	1.46	1.40	0.66	0.80	0.88
H-03-05	R	169.70	198.50	180.20	1.09	1.10	1.09	1.22	1.36	1.40
H-03-06	T	209.50	248.00	236.50	1.23	1.49	1.42	1.12	1.25	1.30
H-03-07	S	39.50	68.90	55.20	1.00	1.08	1.06	0.35	0.38	0.40
H-03-08	S	88.20	110.50	100.60	1.13	1.26	1.28	0.41	0.56	0.65
H-03-09	HS	46.80	54.75	50.70	0.90	1.00	1.06	0.54	0.60	0.64
H-03-10	R	110.75	143.50	132.60	1.26	1.38	1.40	0.69	0.78	0.81
H-03-11	R	163.20	185.75	180.20	1.35	1.47	1.52	0.72	0.80	0.84
H-03-12	HS	36.40	46.20	42.50	0.85	0.92	0.95	0.45	0.52	0.54
H-03-13	R	175.70	198.50	186.00	1.23	1.34	1.46	1.25	1.36	1.42
H-03-14	R	198.50	222.00	216.50	1.33	1.44	1.51	0.97	1.14	1.18
H-03-15	R	140.60	172.75	165.00	1.08	1.17	1.20	0.95	1.10	1.20
H-03-16	S	63.40	70.60	67.00	0.98	1.18	1.22	0.85	0.92	0.94
H-03-17	R	202.60	248.30	240.00	1.52	1.68	1.70	1.32	1.44	1.53
H-03-18	S	38.60	44.50	42.50	0.82	0.90	0.92	0.52	0.56	0.58
H-03-19	R	235.60	268.70	250.60	1.61	1.70	1.74	1.70	1.88	2.00
H-03-20	S	80.50	88.75	86.40	1.01	1.18	1.16	0.72	0.78	0.81
H-03-21	HS	40.60	48.60	44.20	0.89	1.00	0.98	0.46	0.54	0.59
NPH-02-01	T	212.50	249.00	240.20	1.52	1.60	1.64	1.80	1.97	2.05
Cultivar										
Pisang Lilin	R	253.25	288.00	267.75	1.70	1.81	1.84	1.78	1.97	2.06
Rasthali	HS	68.60	75.40	70.00	0.95	1.07	1.10	0.64	0.73	0.79
SEd		5.5597	15.2962	5.7886	0.0443	0.1142	0.0500	0.0443	0.1092	0.0456
CD (0.05%)		11.1378	30.6427	11.5963	0.0887	0.2287	0.1001	0.0887	0.2188	0.0913
CD (0.01%)		14.8278	40.7947	15.4382	0.1180	0.3044	0.1332	0.1180	0.2913	0.1215

MAP, months after planting ; R, resistant; T, tolerant; S,susceptible; HS, highly susceptible

More number of phenolic cells and lignified cells were found in resistant/tolerant hybrids, viz. H-03-17, H-03-19 and NPH-02-01 than susceptible hybrids, H-03-12 and H-03-20. The reference resistant cultivar, Pisang Lilin had more number of phenolic cells and lignification than susceptible cultivar Rasthali. More cavity space due to entry of nematodes was found in susceptible cultivars, whereas cells were compact in case of resistant cultivars. No lignified cells were found in cortical parenchyma of either resistant/susceptible hybrids. Most of the lignified cells were found in stele. The lignification along the endodermis was seen in all hybrids. An increase in number of lignified cells in tolerant cultivars compared to susceptible cultivars of banana has already been reported (Fogain, 1996) and their role in tolerance mechanisms also reported by Sariah *et al.* (1999) and Reuveni and Karchi (1987).

SUMMARY

Increased lignin content was observed in resistant banana cultivars, indicating possibility of defense mechanism in operation to induce resistance. Lignin is deposited in cell walls in response to damage but is also present at high levels in plant roots prior to infection, suggesting a pre-infectional defense mechanism. The increased synthesis of phenols followed by oxidation of them by higher rate of polyphenol oxidase would have helped in increased conversion of them into strong biochemical compound like lignin.

References

Binks, R.H. 1996. Aspects of biochemical resistance in *Musa* spp. *Nematropica,* **26**(3):1-5.

Chesson, A. 1978. The maceration of line flax under anaerobic condition. *J. Appli. Bacteriol.,* **45:**219-30.

Farkas, G.L. and Kirlay, Z. 1962. Role of phenolic compounds in the physiology of plant disease resistance. *Phytopath. Z.,* **44:**105-50.

Fogain, R. 1996. Evaluation of *Musa* spp. for susceptibility to nematodes and study of resistance mechanisms. *Acta Hort.,* **540:**215-24.

Fogain, R. and S. R. Gowen. 1996. Investigations on possible mechanisms of resistance to nematodes in *Musa. Euphytica,* **92:**375-81.

Ganguly, A.K. and Dasgupta, D.R. 1980. Purification and some properties of peroxidase from resistant and susceptible varieties of *Lycopersicum esculentum* infested with root-knot nematode, *M. incognita. Indian J. Nematol.,* **21:**13-22.

Giebel, J. 1982. Mechanism of resistance to plant nematodes. *Ann. Rev. Phytopathol.,* **20**:257-79.

Goodman, R.N., Kiraly, E. and Ziatlin, M. 1967. *The Biochemistry and Physiology of Infectious Plant Diseases.* D. Van. Nosti and Co., Princeton, New Jersey, pp.354.

Gunavathi, J. 2000. Studies on resistance to Panama wilt *(Fusarium oxysporum* f. sp. cubense) in certain banana hybrids. M.Sc. thesis, Tamil Nadu Agricultural University, Coimbatore.

Johansen, D.A. 1940. *Plant Microtechnique.* McGraw-Hill, New York.

Krishnamoorthy, V. 2002. Breeding for resistance to Sigatoka leaf spot and nematodes in banana (*Musa* spp.). Ph.D. thesis, Tamil Nadu Agricultural University, Coimbatore.

Malik, C. P. and M. B. Singh. 1980. *In: Plant Enzymology, Histo Enzymology.* Kalyani Publication, Delhi, pp-53.

Mateille, T. 1994. Biochemical reactions provoked by three phytoparasitic nematodes in roots of *Musa acuminata* (AA Cavendish group) cv. Poyo and Gros Michel. *Fundamental and Applied Nematol.,* **17**(3):283-90.

Mueller, W. C. and C. H. Beckman. 1974. Ultrastructure of phenol strong cells in cotton roots. *Canadian J. Bot.,* **54:**2074-82.

Reuveni, R. and J. F. Karchi. 1987. Peroxidase activity : A possible marker for resistance of melon against downy mildew (Abstract). *Phytopathology,* **77**:1724.

Reuveni, R., M. Shimmoni and I. R. Crute. 1992. An association between high peroxidase activity in lettuce (*Lactuca sativa*) and field resistance to Downy mildew (*Bremia lactuca*). *J. Phytopathol.,* **132:**312-18.

Sariah, M., Lim, C.L. and Tariq, S.A. 1999. Use of biochemical marker as a measure of resistance towards *Fusarium* wilt of banana. *Proceedings of the International Workshop on the Banana Fusarium Wilt Disease,* 18-20 October 1999, Malaysia, pp. 243-49.

Valette, C., Andary, C. Geiger, J.P., Sarah, J.L. and Nicole, M. 1998. Histochemical and cytochemical investigations of phenols in roots of banana infected by the burrowing nematode *Radopholus similis. Phytopathol.,* **88:**1141-48.

Van Etten, H.D., Sandrock, R.W., Wasmam, C.C., Soby, S.D., Mellusky, K and Wang, P. 1995. Detoxification of phytoanticipins and phytoalexins by phytopathogenic fungi. *Canadian. J. Bot.,* **73** (suppl.1):518-25.

50

Role of Enzymes and Phenols in Banana Hybrids against Nematode

P.S. Kavitha, T.N. Balamohan, B. Senthamizh Selvi and M. Kavitha

Plant parasitic nematodes are widespread and important pests damaging bananas. Among nematodes affecting banana and plantain, burrowing nematode (*Radopholus similis*) is more destructive. Development of nematode resistance with quality and food safety, therefore assumes priority in banana breeding programmes in many parts of the world (Simmonds, 1966). When a pathogen infects the host tissue, a small number of specific genes are induced to produce mRNA's that permit synthesis of similar number of specific proteins (Vera and Conejero, 1989). Many of these proteins are enzymes such as phenylalanine ammonia lyase, polyphenol oxidase, peroxidase and β-1-3 glucanase (Vidhyasekaran, 1993). These are involved in synthesis of low-molecular weight substances such as phytoalexins, phenols and lignin, which are inhibitory to invading pathogens (Vidhyasekaran, 1993). Vidhyasekaran (1988) described the occurrence of many kinds of phenolics in plants. Among them, total phenols play a unique role in response to pathogen and nematode invasion. Hence, estimation of these biochemical markers, which provide mechanism for resistance to pathogens, is highly essential. The banana hybrids were assessed for biochemical content (total phenol) and enzyme activities (peroxidase and polyphenol oxidase).

In the labelled pots, with three replications and 10 suckers/replication, 45 days after planting, suckers were inoculated with infective juveniles of *Radopholus similis* at the rate of 1/g of soil, which were artificially cultured by carrot disc culture method. After inoculation, soil was lightly watered. The data on nematode population in soil and root were collected 45 and 90 days after inoculation. The lesion index in roots and corm observed 90 days after inoculation. Samples for estimation were taken 90 days after inoculation under pot culture. Total phenols were estimated by Folin Ciocalteau method (Malik and Singh, 1980) and reported as μg/g of roots. Peroxidase activity was assayed at 30 °C as described by Sindhu *et al.* (1984) using guaiacol as substrate. The reaction was initiated and increase in absorption was measured at 470 nm. The polyphenol oxidase activity was done Jayaraman *et al.* (1987) and changes in absorbance were recorded at 495 nm at 30 second interval for 3 minutes.

Table 50.1. Enzyme activities and phenol content of banana hybrids inoculated with *Radopholus similis* 90 days after iococulation under pot culture

Hybrid	*Peroxidase (abs/min/g)*			*Polyphenol oxidase (abs/min/g)*			*Total phenol (µg/g)*		
	C	*I*	*% diff*	*C*	*I*	*% diff*	*C*	*I*	*% diff*
H-03-04	1.080	1.128	4.44	0.025	0.032	28.00	300.00	304.00	1.33
H-03-05	1.065	1.200	12.68	0.035	0.040	14.29	420.30	592.50	4.97
H-03-06	1.300	1.398	7.54	0.028	0.041	46.43	346.75	375.00	8.15
H-03-07	0.485	0.522	7.63	0.018	0.020	11.11	172.60	175.00	1.39
H-03-08	0.530	0.693	30.75	0.036	0.045	25.00	118.00	120.50	2.12
H-03-09	0.372	0.410	10.22	0.018	0.020	11.11	161.50	172.50	6.81
H-03-10	0.852	0.936	9.86	0.055	0.075	36.36	296.75	322.00	8.51
H-03-11	1.378	1.420	3.05	0.080	0.098	22.50	257.50	275.00	6.80
H-03-12	0.415	0.518	24.82	0.021	0.041	95.24	148.00	154.00	4.05
H-03-13	1.120	1.195	6.70	0.088	0.100	13.64	514.00	560.00	8.95
H-03-14	1.118	1.150	2.86	0.010	0.014	40.00	349.50	378.00	8.15
H-03-15	1.090	1.272	16.70	0.080	0.094	17.50	433.00	456.60	5.45
H-03-16	0.516	0.676	31.01	0.026	0.038	46.15	130.00	138.00	6.15
H-03-17	1.250	1.389	11.12	0.040	0.060	50.00	445.00	524.00	17.75
H-03-18	0.440	0.500	13.64	0.049	0.074	51.02	166.75	175.50	5.25
H-03-19	1.892	1.950	3.07	0.091	0.109	19.78	541.00	572.00	5.73
H-03-20	0.575	0.654	13.74	0.018	0.030	66.67	178.00	185.00	3.93
H-03-21	0.408	0.426	4.41	0.027	0.052	92.59	158.75	164.50	3.62
NPH-02-01	1.240	1.310	5.65	0.089	0.125	40.45	464.60	482.00	3.75
Cultivars									
YKm#5	1.419	1.529	7.75	0.125	0.135	8.00	488.75	527.60	7.95
Rasthali	0.330	0.367	11.21	0.025	0.026	4.00	127.50	135.00	5.88
Sources	G	T	GT	G	T	GT	G	T	GT
SEd	0.046	0.012	0.066	0.004	0.001	0.005	11.111	2.918	15.714
CD (0.05%)	0.091	0.024	0.129	0.008	0.002	0.010	22.008	5.779	31.125
CD (0.01%)	0.121	0.032	0.171	0.010	0.003	0.013	29.101	7.642	41.155

DAI, Day after inoculation; C, control; I, inoculated; % difference over control

Expression of enzyme activities and total phenol content in roots of banana hybrids on 90 days after inoculation under control and inoculated plants varied significantly among hybrids, treatments and for their interaction. The peroxidase activity is highest in H-03-11 (1.378 abs/min/g under-control and 1.420 abs/min/g under inoculated) and the lowest in H-03-09 (0.372 abs/min/g under control and 0.410 abs/min/g under inoculated) among hybrids. The difference was maximum (30.75 %) in H-03-08 and minimum (2.86%) being in H-03-14.

The polyphenol oxidase activity is maximum in H-03-19 (0.091 abs/min/g under control and 0.109 abs/min/g under inoculated) and minimum being H-03-14 (0.010 abs/min/g under control and 0.014 abs/min/g under inoculated). The difference is highest in H-03-12 (95.24%) and lowest (11.11%) in both hybrids, H-03-07 and H-03-09.

The highest total phenol content was found in H-03-19 (541.00 µg/g under control and 512.00 µg/g under inoculated) and lowest in H-03-08 (118.00 µg/g under control and 120.50 µg/g under inoculated). Thus, there is significant increase in phenol content in resistant (H-03-05 and H-03-19) and tolerant hybrids (H-03-13 and NPH-02-01) *vis-à-vis* susceptible ones.

Vidhyasekaran (1988) described occurrence of many kinds of phenolics in plants. The total phenols play an unique role in response to pathogen and nematode invasion. The accumulation of phenol may be due to excess production of hydrogen peroxide by increased respiration (Farkas and Kiraly, 1962) or due to the activation of hexose monophosphate (HMP) shunt pathway, acetate pathway and release of bound phenols by hydrolytic enzymes (Goodman *et al.*, 1967). Similar results were also observed by Gunavathi (2000) for *Fusarium* wilt resistance and Fogain and Gowen, (1996) and Krishnamoorthy (2002) for nematode resistance in banana.

SUMMARY

Total phenols and activity of enzymes like peroxidase and polyphenol oxidase were higher in nematode (*Radopholus similis)* resistant and tolerant banana hybrids which indicate them as defense mechanism in response to nematode invasion. The tolerant hybrids, H-03-06, H-03-13 and NPH-02-01, and resistant ones, H-03-19 and NPH-02-01, produced higher peroxidase and poly phenol oxidase activity than susceptible ones.

References

Farkas, G.L. and Kirlay, Z. 1962. Role of phenolic compounds in the physiology of plant disease resistance. *Phytopath. Z.,* **44:**105-50.

Fogain, R. and Gowen, S.R. 1996. Investigations on possible mechanisms of resistance to nematodes in *Musa. Euphytica,* **92:**375-81.

Goodman, R.N., Kiraly, E. and Ziatlin, M. 1967. The Biochemistry and Physiology of Infectious Plant Diseases. D. Van. Nosti and Co., Princeton, New Jersey, pp.354.

Gunavathi, J. 2000. Studies on resistance to Panama wilt *(Fusarium oxysporum* f. sp. *cubense*) in certain banana hybrids. M.Sc. (Hort.) thesis, Tamil Nadu Agricultural University, Coimbatore.

Jayaraman, K.S., Ramanuja, M.N. Vijayaraghavan, P.K. and Vaidyanathan, C.S. 1987. Oxidative enzyme in pearl millet. *Food Chem.,* **24:**203.

Krishnamoorthy, V. 2002. Breeding for resistance to Sigatoka leaf spot and nematodes in banana (*Musa* spp.). Ph.D. (Hort.) thesis. Tamil Nadu Agricultural University, Coimbatore.

Malik, C.P. and Singh, M.B. 1980. *In: Plant Enzymology, Histo Enzymology.* Kalyani Publication, Delhi, pp-53.

Simmonds, N. W. 1966. *Bananas.* Longman, London.

Sindhu, J.S., Ravi, S. and. Minocha, J.L 1984. Peroxidases isozyme patterns in primary trisomics of pearl millet. *Theor. Appl. Genet.,* **68:**179-82.

Vera, P. and Conejero, V. 1989. The induction and accumulation of pathogenesis -related P69 proteinase in tomato during citrus exocortis viroid infection and after chemical treatments. *Physiol. Mol. Plant Pathol.,* **34:**323-34.

Vidhyasekaran, P. 1988. *Physiology of Disease Resistance in Plants.* vol.I, CRC Press, Boca Raton, Florida, p. 149.

Vidhyasekaran, P. 1993. Defense genes for crop disease management. *Genetic Engineering, Tissue Culture, and Molecular Biology for Crop Pest and Disease Management.* Vidhyasekaran, P. (Ed.). Daya Publishing House, New Delhi, pp.17-30.

51

Management of Banana Bunchy Top Disease by Tissue Culture Technique

D.S. AMBIKA AND A.S. BYADGI

BANANA is economically profitable plantation crop in India. Fungi, bacteria, virus and nematode attack banana. Bunchy top viral disease was first reported from Fiji in 1889 (Stover 1972) and in Kerala in India in 1940 (Mehta 1964). The disease is prevailing in almost all banana growing areas of the country, causing losses up to 100%. Hence, present study was undertaken.

Tissue culture techniques are now a days widely used for production of disease-free planting material. A rapid multiplication of banana and plantain by *in-vitro* shoot tip culture has been developed by Cronauer and Krikorian (1984). Wu and Su (1991) reported that BBTV infected tissue is cultured for three months at 35 °C. Some of the buds produce healthy looking plantlet. Five out of eleven cultures produced healthy looking plantlets in six months. Crude extracts of these plantlets did not react with monoclonal antibody prepared against BBTV, but uneven distribution and low concentration of virus after high temperature treatments lead to BBTV-free primordial cells which in turn developed healthy plantlets. Diekman and Putter (1996) standardized protocol for obtaining BSV free material through a combination of thermotherapy and other methods of tissue culture techniques. Helloit *et al.* (2001) proposed the use of meristems excised from *in-vitro* plants, *in-vivo* plants and from highly proliferating meristems for the elimination of virus disease from *Musa*. Carpia (2002) revealed that mass production of healthy planting material is the control approach to virus disease. Kenganal (2003) found that meristem culture alone or incombination with the thermotherapy could not eliminate the BSV from the infected meristems.

MANAGEMENT OF BANANA BUNCHY TOP DISEASES

The experiment was conducted at the Department of Plant Pathology and Department of Horticulture, University of Agricultural Sciences, Dharwad, during 2003-04. Novel meristem culture technique was adopted for production of healthy plants from diseased meristems. The diseases suckers of Grand Naine banana collected from the field were used for study along with healthy suckers as the control.

Preparation of Ex-plants

The suckers were cleaned and washed with water. The corms were trimmed by removing outer leaves and corm tissue were trimmed carefully using a knife at the size of 4-5 cm long and 1 cm diameter. The meristem buds were washed using tween-20 to remove the adhering soil particles followed by three times washing with ordinary water. The buds were sterilized with 0.1% Bavistin solution for 3 minutes, followed by three times washing with sterile double distilled water. The buds were next surface sterilized with 70% alcohol for 30 seconds followed by three washings with sterile double distilled water. The later two surface sterilizations were done under Laminar Air Flow Chamber (LFC).

Initiation of Culture

Outer leaf sheaths from meristem buds are removed until meristem retains one or two layers of leaf sheath along with small base in a petriplate containing small amount of 0.1% ascorbic acid. The meristems were then inoculated on B1 Medium (MS basal supplemented with 2 mg/l Benzyl Adenine) and kept for shoot proliferation at 23$\pm$2 °C with 16 hour photoperiod. The cultures are subcultured whenever browning of medium is observed due to exudation of phenolics.

Multiplication of Plantlets

After proliferation of shoots they are cut longitudinally and inoculated for multiplication on B 2 medium (MS basal supplemented with 5mg/litre Benzyl Adenine and 30 mg/litre adenine sulphate along with 0.8% agar as a solidifying agent) for shoot multiplication. Sub-culturing is done at weekly intervals.

***In-vitro* Rooting of Shoots**

Multiple shoots formed are separated individually and inoculated in B-3 rooting medium (MS basal supplemented with 3% sucrose, 0.8% agar as solidifying agent and supplemented with NAA (1 mg/litre). The cultures are incubated for 20 days. The rooted plantlets are transferred to hardening and those non-rooted are again sub-cultured to the same medium for another 15 days.

Hardening of Plantlets

The rooted plantlets are taken out from bottles after rooting and washed in water to remove any adhered medium on basal portion. The plantlets are kept immersed in water for 12 hours and later treated with 0.1% Bavistin solution for one hour and finally rinsed in tap water. The plantlets are then transferred to net pots containing sterilized sand and placed in greenhouse for a period of one week, after which they are transferred to polybags containing potting medium (soil, sand, compost, coir dust in a 1:1:1:1 ratio) without damaging the root system and kept under the greenhouse for another one month.

Detection of BBTV by DAS-ELISA

The detection of virus is carried out using DAS-ELISA. Four samples are subjected for the test. Samples from tissue culture seedling raised from meristem of diseased mother plant. Samples from tissue culture seedling raised from meristem of healthy mother plant. Samples from field with distinct symptoms. Samples from healthy plants.

Reagents

- **Coating/sample buffer:** (sodium carbonate buffer) pH 9.6.59 g Na_2CO_3, 2. 93g $NaHCO_3$, 0.2g NaN_3, 0.02m DIECA dissolved in distilled water and volume is made to 1 litre.
- **Phosphate buffer saline:** (PBS) pH 7.4 (stock buffer). 8g $NaCl_2$, 0.2 g KH_2PO_4, 1.4 g $Na_2HPO_22H_2O$, 0.2 g KCL, 0.2 g NaN_3dissolved in 1 litre distilled water.
- **Wash buffer:** (PBS-T) Phosphate buffer containing 0.5 ml Tween 20/litre. Antibody buffer/sample buffer PBS-T containing 2% PVP (polyvinyl pyriolodine)
- **Conjugate buffer:** PBS-T containing 2% PVP and 0.2% ovalbumin substrate buffer 97 ml diethanolamine dissolved in 800 ml distilled water and pH adjusted to 9.8 with 6 N HCl and volume made to 1 litre
- **Blocking buffer:** PBS-T containing 2% PVP and 4% skimmed milk powder.

Procedure

The 200 µl of coating buffer containing anti BBTV γ globulin diluted to 1:100 was added in each well. It was incubated at 37 °C for 2 hours.

- Plate was washed with wash buffer thrice at three minutes interval.
- 200– µl of sample including buffer, healthy and positive control diluted to 1:10 in sample buffer was added in each well.
- Incubated at 37 °C for 2.5 hours.
- Plate was washed with wash buffer (PBST) thrice at three minutes interval.
- 200-µl conjugate buffer containing conjugated antibodies diluted to 1:100 were added in each well.
- Incubated at 37 °C for 2.5 hours.
- Plate was washed with wash buffer (PBST) thrice at three minutes interval.
- 200-µl of freshly prepared substrate (pP- nitro phenyl phosphate) was added in each well.
- Incubated at room temperature, observed for the development of colour.
- Reaction was stopped by adding 50 µl of 3 M sodium hydroxide solution in each well.
- Plate was read in ELISA reader (ELX 800 MS) at a wavelength of 405 nm.
- Reactions were tabulated.

The inoculated meristem took 35 days for shoot proliferation. The multiplied shoots are subjected for second cycle of multiplication, where results are found non-consistence. Multiplied shoots inoculated on rooting medium, the root induction is observed after 25 days of inoculation. The rooted plantlets are subjected to hardening and plantlets are maintained for one year in nursery.

The plantlets produced from infected mother plants do not show any visual symptoms and they were looking healthy. Further, samples from tissue culture seedling raised from meristem of diseased mother plant, samples from tissue culture seedling raised from meristem of healthy mother plant, samples from field with distinct symptoms and samples from healthy plants are subjected to ELISA. The BBTV infected samples collected from field and meristem culture plantlets obtained from suckers of infected mother plant show positive reaction to ELISA. Though plantlets obtained from meristem culture are apparently healthy, they were found to be carrier of BBTV.

Fig. 51.1. Initiation of meristem culture

Fig. 51.2. Proliferation of shoots

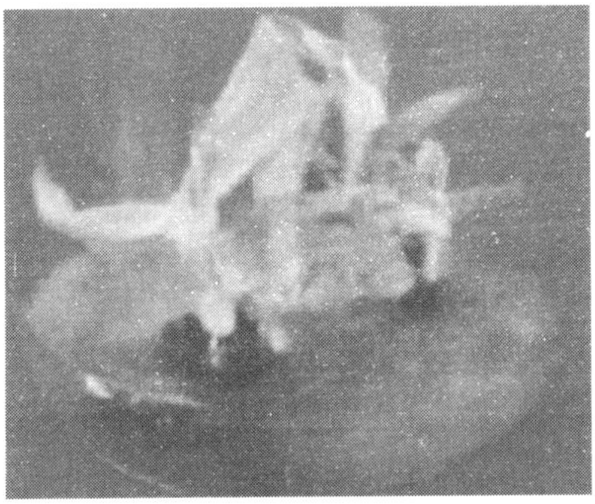

Fig. 51.3. Multiplication of shoots

Fig. 51.4. Stages of hardening

The meristem culture technology has been employed for elimination of virus on consideration factor that low concentration of virus in highly multiplying cells, presence of less number of plasmodesmata also meristematic cell constituents inhibitory for multiplication of virus. Hence attempts are made to know the possibility of elimination of virus by meristem culture techniques, which could be employed as a management strategy for bunchy top disease. Laxmikanth and Nataraja (1989) also successfully regenerated banana plantlets using meristem as explants to produce healthy planting material. Similar efforts were made by Wu and Su (1991) and they used BBTV infected tissue for elimination of the virus in their study. The regenerated plantlets do not exhibit any visual symptoms even up to one year but carried banana bunchy top virus. Such plants which tested positive in ELISA, indicated failure of meristem tissue culture technique in elimination of BBTV. Drew *et al.* (1992) observed very severe symptoms of bunchy top disease in all regenerated plantlets through meristem tissue culture technique 12 months after hardening. This confirms our results. It is evident that BBTV cannot be eliminated by meristem culture technique. It may be because of the difficulty in getting meristem, which could be free from virus. As banana meristem is very deep seated and small, hence care should be taken while selecting the explants for mass multiplication. Indexing the mother plants also would help in eliminating the latent infection, which is not expressed in the suckers.

SUMMARY

The regenerated plantlets do not show any visual symptoms even up to one year. Plantlets are subjected to DAS-ELISA technique. Plantlets produced from infected suckers gave positive results for BBTV, whereas plantlets of healthy suckers showed negative reaction to disease. This indicated that meristem culture technique does not eliminate BBTV and it appear that the viruses are present even in meristematic cells.

References

Carpio. T. 2002. Impact of virus free banana plantlets in rehabilitation of Banana Bunchy to affected areas in the Philippines, Paper presented in Global Conference on Banana and Plantain 28-31, October. Bangalore, India, p 195.

Cronauer, S.S and Krikorian, A.D., 1984. Rapid multiplication of bananas and plantains by In vitro shoot tip culture. *Horticulture Science,* **19:**234-35.

Diekmann, D.K. and Putter, C.A.J 1996. *Musa* 2nd edn. FAO/IPGRI Technical guidelines for the safe movement of germplasm. No 5, Food and Agriculture Organization of the United Nations, Rome/ International Plante Genetic Resources Institute, Rome: pp. 28 DNA in range of musa cultivars. *Molecular Plant Pathology,* **2:**207-13.

Drew, R., Smith, M.K and Aderson, D.W., 1992. Field evaluation of micro propagated bananas derived from plats containing banana bunchy top virus. *Plant Cell, Tissue and Organ Culture,* **28:**203-5.

Kenganal, M.Y. 2003. Further studies on banana streak virus. M.Sc (Agric) thesis, Marathwada Agricultural University, Parbhani.

Lasmikanth, D, M and Nataraja, K., 1989. Mircopropogation of banana through shoot tip culture. *Current Science,* **58:**140-41.

Mehta, P.K.1964. Bunchy top a serious disease of banana in India. *Science and Culture.,* **30:**259-63.

Murashige T. and Skoog, F. 1962. A revised medium for rapid growth and bioassays with tobacco tissue cultures. *Physiologia Plantarum,* **15:**473-79.

Stover, R.H. 1972. Banana Plantain and Abaca diseases. Common Wealth Mycological Institute, Kew, England.

Wu, R.Y and Su, H.J. 1991. Regeneration of healthy banana plantlets from banana bunchy top virus infected tissue cultured at high at high temperature. *Plant pathology,* **40:**4-7.

52

Virus Indexing in Production of Quality Planting Material

R. Selvarajan

Banana is one of the most important staple food and fruit crop in the world. It plays a major role for millions of resource poor people for their livelihood and food security. The sucker-borne obligate pathogens are major cause in destroying its plantations. The planting material should be healthy and vigorous in order to obtain a good yield. Quality of a planting material plays a major role in banana production and productivity. In order to plant the entire banana-growing area with tissue cultured planting material we need to produce approximately 2,000 million plants. Even if we want to cover 50% area under quality tissue culture planting material, we need approximately one billion plants. Our country produces approximately 30 million tissue culture plants only. So potential of tissue culture industry growth is enormous. Of late, viral pathogens are known to spread through tissue culture plants because on non-indexing of mother plants used in propagation. Recently DBT has accredited Molecular Plant Virology Lab and Plant Tissue Culture and Molecular Biology Labs of NRC for Banana, for virus testing and genetic fidelity testing of tissue culture raised plants of banana respectively to ensure quality plants as per the standards set by the DBT and ICAR. In this chapter, virus indexing part in banana tissue culture propagation has been dealt in detail.

VIRUS INDEXING OF BANANA TISSUE CULTURE PLANTS

Diagnosis is important step in managing banana viral diseases. In order to make proper management decisions, early and correct diagnosis of viruses is important. This helps in checking further spread of pathogens especially to a new area. So, early detection by means of sensitive diagnostic methods is the main way to control them. The viral pathogens are a major impediment in tissue culture industry of banana. These viral pathogens cannot be eliminated through shoot tip culture which is the technique adopted by industries for mass propagation. Latency is an another point which has to kept in mind since most of banana viruses reside in host in latent form, i.e. without exhibiting any visual symptoms in the host for some period which can be even 2 years. If non-indexed plants are used as mother plants, virus spread and its effect on yield is very high. The tissue culture technology in banana is one of the widely adopted technologies in horticulture. Hence in order to sustain and reap the benefits of tissue culture, we have to assure the industry quality mother plants or virus indexing technology for producing healthy and high-yielding planting material. Different molecular diagnostic techniques have been used to detect banana viruses in tissue culture plants as well as mother plants used by the tissue culture industries. NRC for Banana is offering contract services for virus testing to tissue culture industries and for banana growers.

Diagnostic Tests

At Molecular Virology Laboratory of NRCB, ELISA/DIBA based serodiagnosis kits are available for testing against BBTV and CMV. The PCR is routinely being done for detection of BBTV and BSV. Reverse transcriptase –PCR is performed for RNA genome containing viruses such as CMV and BBrMV. Keeping the cost factor in view, a technique has been developed for simultaneous detection of more than two viral pathogens by PCR. Duplex PCR was developed to detect BSV and BBTV simultaneously in PCR. A multiplex PCR has also been developed for the detection of DNA and RNA viruses, viz. BBTV, BSV and BBMV. This technique is sensitive and cost-effective for indexing of tissue culture mother plants. For handling more number of samples at a time, a non-radioactive probe based nucleic acid spot hybridization (NASH) technique has also been developed for detecting all four banana viruses (see Box).

Indexing techniques available for banana virus diagnosis at NRC for Banana

Viruses	Test methods developed and used			
	PCR/RT-PCR/ IC-PCR/ DB-PCR/ IC-RT-PCR	ELISA/DIBA	Nucleic acid spot hybridization (NASH) using non-radioactive DNA / RNA probes	Southern and northern blots using non-radioactive probes
Banana Bunchy Top Virus (BBTV)	"	"	"	"
Banana Bract Mosaic Virus(BBrMV)	"	-	"	"
Banana Streak Virus	"	-	"	"
Cucumber Mosaic Virus	"	"	"	"

Role of DBT

In order to get quality banana tissue culture plants from tissue culture industries, the DBT along with ICAR has developed specific standards for banana tissue culture for certification (Fig.52.1). The DBT has identified accredited labs for virus testing and also genetic fidelity of banana tissue culture plants. Molecular Plant Virology Lab and Plant Tissue Culture and Molecular Biology Labs of NRC for Banana, Trichy, have been accredited to take up virus indexing and also testing genetic fidelity of banana tissue culture plants in August 2007. The DBT has set eligibility requirements for the tissue culture industries involved in banana as per the NCS-TCP. Following are the requirements for banana tissue culture production as per the standards set by DBT.

- Laboratory and greenhouse facilities used for production of plantlets shall be maintained

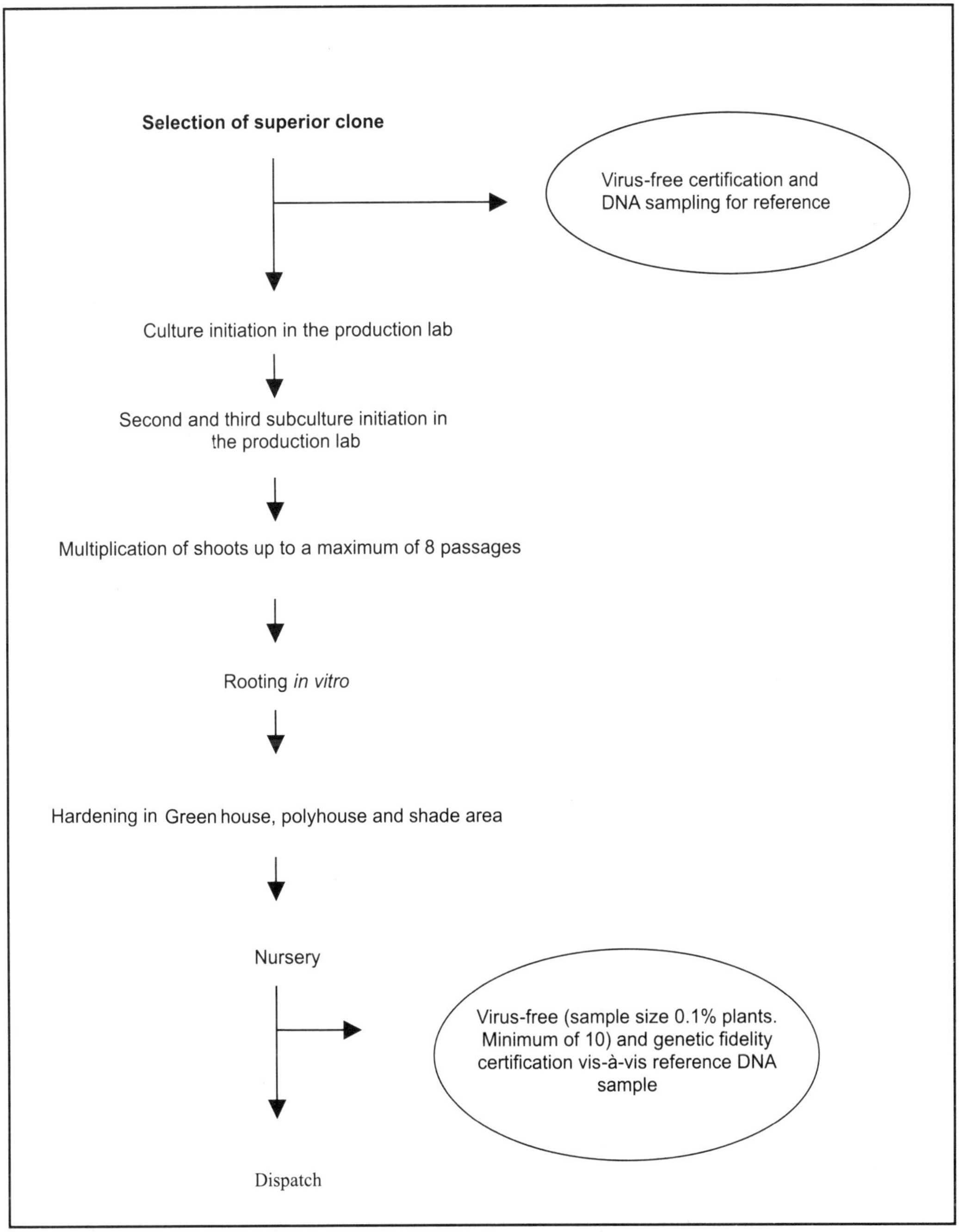

Fig. 52.1. Procedures and standard parameters for production of banana by tissue culure

free of pests of vectors of banana pathogens. Failure to keep such pests under control may cause rejection of all lots maintained in the facility. All potting or growth media shall be sterile. Water sources used in the laboratory or greenhouses operation shall be treated or otherwise rendered free of all pathogens by the applicant.

- Hygienic conditions shall be strictly observed during micropropagation, potting, planting, irrigating, movement and use of equipment and other laboratory and greenhouse practices to guard against the spread of diseases or pests in the facilities used for banana plant multiplication.
- The greenhouse (protected environment) must be insect-proof and be equipped with a double-door entrance, provision for disinfection prior to entering the protected environment and insect-proof ventilation screening on intakes and exhaust openings.
- The material being initiated must be of a notified variety and confirmed identity. It must be duly documented with respect to origin.
- All samples of banana varieties being initiated should be tested in an accredited laboratory and be free of viruses (Banana Bunchy Top Virus, Cucumber Mosaic Virus, Banana Bract Mosaic Virus, Banana Streak Virus).
- The basic material for sub-multiplication need to be obtained afresh from the nodal organization as soon as the maximum permitted number of passages (as confirmed by DNA fingerprinting) of shoot multiplication with old cultures has been completed.
- On application for inspection, the mother cultures as developed above are eligible for certification. The micropropagation facility to be inspected must have been approved by the competent Authority. All stocks must have a valid variety identification and disease testing report at any time during multiplication process.
- *In vitro* multiplication of an imported variety or a non-notified variety can be taken up by the industry exclusively for export purposes. Such varieties, however, should be introduced following the approved guidelines of Government of India.

Source of Seed

- The facility should use recognized aseptic initiation and propagation procedures.
- The initiating facility must maintain following information on each variety for review and audit by the competent authority at least once in a year: variety identification, date of initiation, origin and testing results from accredited laboratory.
- Tests must be carried out on a minimum of 0.1% (at least ten) plantlets for each variety by an accredited laboratory. Such tests will be valid so long as cultures of that particular batch are under production (subject to a maximum of 8 passages). No plant should contain (Banana Bunchy Top Virus, Cucumber Mosaic Virus, Banana Bract Mosaic Virus and Banana Streak Virus).
- Valid pathogen testing results are required at the two-thirds subculture stage prior to the bulking up of the cultures.

Minimum Quality Standards

The minimum quality standards set by DBT for growing of plants inside greenhouse/ polyhouse are:

- Effective sanitation practices including insect and disease monitoring and prevention must be adhered to.
- No field-produced banana plants can be grown in the protected environment (greenhouse/ polyhouse) along with tissue cultured plants.
- Varieties must be separated by physical barriers (such as proper tagging), which will prevent varietal mixtures.
- Before dispatch to the farmers, the tissue cultured plants growing in the nursery should be tested for the absence of the viruses (Banana Bunchy Top Virus, Cucumber Mosaic Virus, Banana Bract Mosaic Virus, Banana Streak Virus) and clonal uniformity. For establishing clonal fidelity, the sample size should be 0.1% of the batch size with a minimum of 10 plants.
- If testing performed by an accredited laboratory reveals the presence of banned viruses, fungus or bacteria, the tissue-cultured plants should not be dispatched from the premises of the production lab and entire material should be destroyed.
- The concerned laboratory/ agency producing the tissue culture raised material should issue a certificate to the effect that BTC have been produced as per guidelines.

CONCLUSION

To ensure quality banana tissue culture plants, all the stake holders of banana tissue culture industry, viz. banana farmers, banana tissue culture plant suppliers, scientific organizations involved in banana research, KVK's and state horticulture departments have to work in co-ordination to sustain the banana production and productivity. As tissue culture banana is one of the successful technologies which played a major role in increasing the productivity of Cavendish bananas, there will be a good future in export markets for banana through this technology. Hence regular and strict virus indexing of mother and tissue culture plants of banana and avoidance of somaclonal variants by following correct protocols of tissue culture are very much necessary and the standard set by DBT have to be followed to obtain quality banana tissue culture for the betterment of banana growers and industries.

53

Molecular Characterization of BSV Infected Banana Using ISSR Markers

R. Selvarajan, K.Santhanalakshmi, V. Balasubramanian, S.Mohana, S.Indumathi, P.Suganya Devi and M.M. Mustaffa

Banana streak virus (BSV) disease has become a major threat to Mysore (AAB) group of banana. It has been reported in nearly all countries including Africa, Europe, Asia, North, Central and South America, Australia and Pacific Island (Diekmann and Putter, 1996; Lockhart and Jones, 2000). This includes all known plant pararetroviruses. Non-enveloped bacilliform particles of size 30x 130-150 nm, containing a circular double stranded DNA genome of 7.4-8.0 kbp are characteristics of the virus. The BSV infection causes characteristic chlorotic and necrotic leaf streak symptoms. Depending on infecting BSV species, highly susceptible banana cultivars can develop more severe symptoms, such as pseudostem splitting and necrosis, eventually leading to death of infected plants (Lockhart and Jones, 2000; Daniels *et al.*, 2001). It is currently the main viral constraint to *Musa* germplasm movement, genetic improvement and mass propagation (Diekmann and Putter, 1996). Transmitted in a semi-persistent manner by mealy bugs, *Planococcus citri* and *Ferrisia virgata* (Jones and Lockhart, 1993). The BSV genomic sequences are integrated into genomic DNA of *Musa*, which could give rise to episomal BSV infection (Harper *et al.,* 1999; Ndowara *et al.,* 1999).

Integration of BSV genome in the host banana genome and episomal infection arising from integrated sequences has been reported. Most of the cultivated bananas are man-made selections derived from interspecific natural hybridization containing *acuminata* (A) genome and *balbisiana* (B) genome.

The PCR based technique has been used to recognize the BSV sequences integrated into *Musa* genome. This was confirmed by Fluorescent *in-situ* hybridization (FISH) technique (Harper and Hull, 1998). The BSV integrants were detected in the Indian *Musa* germplasm by PCR (Selvarajan *et al.*, 2004). Geering *et al.* (2005) have reported on a large group of badnavirus integrations in various *Musa* species, termed Banana Endogenous Viruses (BEVs), some of which may be unique to a particular *Musa* species and some are more dispersed throughout the genus. In recent years, a variety of molecular markers based on microsatellite or Inter Simple

Sequence Repeats (ISSR) have become a marker of choice that is frequently used in fingerprinting. Therefore, an attempt was made to use ISSR primers to differentiate BSV integration among populations of Poovan (AAB) banana.

MOLECULAR CHARACTERIZATION OF BSV-INFECTED BANANA

Banana Poovan having BSV symptoms and also non-symptomatic plants were used for study. The total DNA was isolated from 100 Poovan leaves samples as per the protocol described by Gawel and Jarret (1991) with little modification. Finally, DNA was dissolved in sterile double distilled water, quantified and used for PCR amplification. Six ISSR primers-UBC_{822}, UBC_{823}, UBC_{824}, UBC_{826}, UBC_{810} and UBC_{828} developed by Biotechnology Laboratory, University of British Columbia, Canada, were used in this study (Godwin *et al.*, 1997). These primers were synthesized at Bangalore Genei Pvt.Ltd, Bangalore. The PCR reaction mixture in a volume of 25 μl contained 2.5 μl of 10X PCR buffer with 15mM $MgCl_2$, 2.0μl of 10mM dNTP's, 0.4 μM of primers 0.25 μl (5 units/ μl) of Taq DNA polymerase and 5.0 μl of total DNA. The PCR was performed in an Eppendorf Master Cycler gradient with the following parameters. Initial denaturation at 94 °C for 4 minute was followed by 30 cycles of denaturation at 94 °C for 1 minute, annealing at 50 °C for 1 minute and extension at 72 °C for 2 minute with a final extension at 72 °C for 10 minute. After reaction, 1.0 μl of 6x gel loading dye was mixed with PCR product and loaded in 1.5% agarose gel and electrophoresed. The gel was documented using Alpha Imager.

THE RESULT

To distinguish among population of Poovan bananas having integration of banana streak viral genome, six ISSR primers were tested. These primers could differentiate Poovan bananas based on the banding pattern obtained by PCR amplification. The ISSR primer UBC_{823} produced polymorphic bands in samples. The amplified fragment of 300-500bp were not found in some samples. The same samples do not show any viral disease symptoms and UBC_{828} ISSR primer also produced polymorphic bands in healthy samples and not in infected samples which might be due to BSV genome integration.

The UBC_{822} produced polymorphic bands in the infected samples; the amplicons size was between 300bp and 500bp which were absent, indicating BSV integration. The results reveal that ISSR primers can be employed for differentiating infected and healthy banana cultivars of Poovan. Random amplified polymorphic DNA markers (RAPD) techniques has been successfully used to distinguish diverse *Musa* germplasm, molecular linkage map development and to differentiate between closely related to *Musa* genome (Bhat and Jarret, 1995). The ISSR markers can be used to identify high level of polymorphisms and development of ISSR DNA fingerprinting key. The ISSR assay also used to identify the cultivar specific markers that have a potential of being used as diagnostic tools for cultivar identification or could be developed for cultivar specific sequences analysis (Arnau *et al.*, 2003). Variable Number of Tandem Repeats (VNTR) of microsatellite has been reported to detect abundant polymorphism in many systems including *Musa* (Crouch *et al.*, 1991).

Venkatachalam *et al.* (2007) reported that ISSR markers could be used for genetic fidelity test in banana Nanjangud Rasable. They reported that very long period of clonal propagation through tissue culture technique did not have any genetic change as proved by ISSR and RAPD markers. The BSV genome sequences have been known to be integrated in banana genome (Harper *et al.,* 1999; Ndowara *et al.,*1999). These integrants have terminal repeats such that they can recombine illegitimately and form episomal virus and cause infection. Non-expression of symptoms, mild symptoms might be due to integration of dead sequences incapable of giving rise to episomal virus from eprv's. Immuno Capture Polymerase Chain Reaction (IC-PCR) may confirm the non-activation of eprv's into episomal form of virus in such non-symptomatic plants. Thus, banana Poovan might contain endogenous pararetroviral sequences that might also vary among populations. The mild and severe symptoms expression might be due to activation of integrants dispersed in host genome.

Therefore, it is recommended to further work such as southern blot analysis to get clear idea about BSV and its integration pattern. Among molecular markers, ISSR has been widely used to phylogenetic analysis and clonal identity. The ISSR markers are capable of identifying banana having BSV integrants.

SUMMARY

The ISSR primers can be employed for differentiating plants of banana Poovan having different integration pattern. However, southern blot analysis has to be done to confirm the viral genome integration in Poovan banana.

References

Arnau, G., Lallemand, J., Bourgoin, M. 2003. Fast and reliable starwbery cultivar identification using inter simple sequence repeat (ISSR) amplification. *Euphytica,* **129:**69-79.

Bhat, K.V.and Jarret, R.L. 1995. Random Amplified Polymorphic DNA and genetic diversity in Indian Musa germplasm. *Genetic Resources and Crop Evolution,* **42:**107-118.

Crouch, J.H., Crouch, H.K., A.Tenkouano and Ortiz, R. 1991. VNTR based diversity analysis of 2x and 4x full-sib *Musa* hybrids. *Electronic Journal of Biotechnology,* **2:**99-108.

Daniells, J.W, Geering, A.D.W, Bryde, N.J and Thomas, J.E. 2001. The effect of banana streak virus on the growth and yield of dessert banana in tropical Australia. *Ann. Appl. Biol.,* **139:**51-60.

Diekmann, M., and Putter, C.A.J. 1996. FAO. IPGRI Technical guidelines for the safe movement of germplasm. Musa, 2nd edn. Food and Agriculture Organization of the United Nations, Rome/International Plant Genetic Resources Institute, Rome.

Gawel, N.J. and Jarret, R.L. 1991. A modified CTAB DNA extraction procedure for *Musa* and *Ipomea. Plant. Mol. Bio. Rep.,* **9:**262-66.

Geering, A.D.W., Olszewski, N.E., Harper, G., Lockhart, B.E.L., Hull, R. and Thomas, J.E. 2005. Banana contains a diverse array of endogenous badna viruses. *J. Gen. Virol.,* **86:**511-20.

Godwin, I.D., Aitken, E.A.B. and Smith, L.W. 1997. Application of inter-simple sequence repeat (ISSR) markers to plant genetics. Electrophoresis, **18:**1524-28.

Harper, G. and Hull, R.1998. Cloning and sequence analysis of banana streak virus DNA. *Virus Genes,* **17:**271-78.

Harper, G., Osuji, J.O., Heslop-Harrison, J.S., and Hull, R. 1999. Integration of banana streak badnavirus into the *Musa* genome: molecular and cytogenetics evidence. *Virology,* **255:**207-13.

Jones, D.R., and Lockhart, B.E.L.1993. *Banana Streak Disease.* International Network for the Improvement of Banana and Plantain, Montpellier, France.

Lockhart, B.E.L, Jones, D.R. 2000. Banana mosaic and banana streak, diseases caused by viruses. *In: Diseases of Banana, Abaca and Enset.* Jones, D.R. (Ed.), CAB Int, Wallingford, UK, pp. 241-93.

Ndowora, T., Dahal. G., La Fleur, D., Harper. G., Hull, R., Olzewski, N.E., Lockhart, B.E. 1999. Evidence that badnavirus infection in *Musa* can originate from integrated Para retroviral sequences. *Virology,* **255:**214-20.

Selvarajan, R., Balasubramanian, V., Dayakar, S., Uma, S., Ahlawat, Y.S. and Sathiamoorthy, S. 2004. Integration of banana streak virus genome in Musa germplasm with B genome. pp. 102. *In*: *Proceedings of First International Congress on Musa: Harnessing Research to Improve Livelihoods, held during* 6-9 July 2004 at Penang, Malaysia.

Venkatachalam, L., Sreedhar, R.V. and Bhagyalakshmi, N. 2007. Molecular analysis of genetic stability in long term micropropagated shoots of banana using RAPD and ISSR markers. *Molecular Biology and Genetics*, **10** (1).

54

Genetic Diversity among *Fusarium oxysporum* f.sp. *cubense* Race1 Isolates

R.Selvarajan, P.Suganya Devi, R.Thangavelu, R.Kumudha and M.M.Mustaffa

Fusarium wilt disease caused by *Fusarium oxysporum* f.sp. *cubense* (*Foc*) is one of most destructive disease and considered to be a major constraint in banana production worldwide. Being the center of origin of banana, India is also the center of diversity of banana (Valmayor, 2000) and hence genetic diversity of *Foc* is also likely to be more as the *Foc* co-evolved with banana (Moore *et al.,* 1995). Hence, genetic diversity of *Foc* is also likely to be high as *Foc* found to be co-evolved with banana. Variability in pathogen may cause breakdown of resistance in host cultivars (Moore, 1994). The study on pathogenic and genetic variation within *Foc* isolates is therefore required before the development of any resistant varieties to combat the disease. Several workers have proposed the use of molecular markers as a fast and safe alternative to differentiate and characterize fungal isolates.

Molecular markers can be used to evaluate different levels of genetic diversity and to identify particular races and pathotypes. The ISSR markers have great potential for studies of natural populations (Wolfe *et al.,* 1998b). The ISSR markers are obtained through PCR amplification using simple sequence repeat (SSR) primers. These markers have the reproducibility of SSR markers because of longer length of their primers (Bornet *et al.*, 2001). They are very informative because of co-dominance and high polymorphism found in them (Luana *et al.,* 2001). India being one of the center of origin for bananas, variability among *Foc* isolates expected to high owing to genetic variability of host and other soil factors. Therefore, 10 *Foc* isolates collected from different bananagrowing states were analyzed for variability using 19 ISSR primers.

GENETIC DIVERSITY IN FUSARIUM

The *Foc* cultures preserved on dried filter papers were cut into small pieces using a sterilized forceps and scissors. These cultures were initiated by plating on weakhalf strength PDA medium (Anisworth, 1971) amended with antibacterial agent (eg. Streptomycin sulfate @ 1.2 ml/250ml PDA). All these were carried out under aseptic conditions. The plates were incubated at room temperature (28 ± 2 °C) for 7 days and observed for fungal growth. These then subcultured in half strength PD broth for 7 days at 25 °C from which DNA can be extracted.

The DNA was extracted as per the protocol of Amy *et al.* (1997). The amplification was performed in a Gradient Thermal Mastercycler PCR machine (Eppendorf) with some modifications in PCR conditions of Prasad, (2004). Each reaction contained 20ng of genomic DNA, 0.4 μM final concentration of primer, 0.2 mM each of four dNTPs (Eppendorf), 2 units of Taq DNA polymerase (Bangalore Genei., India), and PCR buffer (50 mM KCl, 10mM Tris [pH 8.3], and 2 mM (MgCl2) in a final volume of 25 μl. The PCR reactions were performed for 35 cycles consisting of initial denaturation of template DNA at 94 °C for 2 minutes, 20 seconds denaturation at 94 °C, 50 seconds annealing at 50 °C, 1 minutes 30 second elongation at 72 °C culminating with a final extension step of 7 minutes at 72 °C followed by incubation at 4 °C. Then the PCR products were analysed by 1.5% agarose gel.

The PCR amplified products of individual isolates were scored as 1 or 0 depending on presence or absence of a particular band respectively. Data analysis was performed using the NTSYSpc. The UPGMA clustering was done to draw the dendrogram.

THE RESULTS

To optimize the concentration of primers, Taq DNA polymerase and dNTP's a PCR was performed using 4 ISSR markers for 4 different isolates of *Foc* (Fig 54.1). In PCR amplification, amplicons ranging size from 100bp to 5Kbp have been obtained. After optimization of PCR conditions, 19 ISSR primers were tested with the same 4 isolates used in initial PCR optimization experiment. The 19 primers, only 7 gave consistent scorable bands. The sequences of 7 primers are listed in Table 54.1. The ISSR primers are of di and tri nucleotide repeats of CA, GA, AC, and CAC. Initially, 24 isolates of *Foc* (Table 54.2) were amplified using 75PC primer and later only 10 of 24 *Foc* isolates were selected based on the origin (location) of isolates the variety it infects for testing polymorphism using 7 ISSR primers. For these 10 representative isolates, the DNA extraction and ISSR-PCR analysis using 75PC primer (Fig 54.2) was repeated twice and the reproducibility was confirmed. With these 10 representative isolates all other 7 primers were analyzed and each of which produced scorable amplification products.

The PCR amplification products were scored as either '1' for presence or '0' for absence and a combined cluster analysis was done using NTSYS software application for fingerprints obtained with 7 ISSR primers for 10 *Foc* isolates. This resulted in two major clusters which grouped most of Rasthali, Hill banana and Karpooravalli isolates in one cluster and Ney Poovan and Rasthali collected from Tamil Nadu in other cluster with 50% similarity among them. The Ney poovan isolate collected from two different districts were 100% similar. The similarity between two main clusters is only 15%. It is concluded that a lot of variability exists among Race-1 isolates of *Foc* infecting different cultivars collected from different locations.

Similar studies were carried out by Prashant *et al.* (2006). High genetic diversity is found within the population of *Fusarium pseudograminearum* isolated from wheat grains from Canadian provinces of Alberta and Saskatchewan with 8 ISSR primers. Analysis of pathogen diversity is therefore, a prerequisite to successful control and both traditional and molecular methods are now available. Among molecular markers, ISSR have proven to be a rapid, simple and inexpensive way to assess the genetic diversity (Kantety *et al.*, 1995).

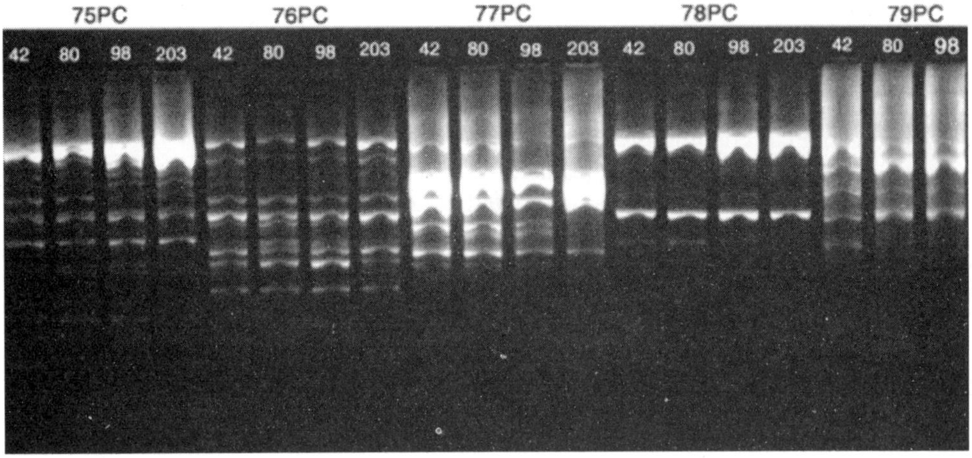

Lane 1-4: Products amplified using Primer 75 PC, Lane 5-8: Products amplified using Primer 76 PC, lane 9-12: products amplified using primer 77 PC, lane 13-16: products amplified using primer 78 PC, lane 17-20: products amplified using primer 79 PC.

Fig. 54.1. ISSR profile of 4 isolates of Fusarium oxysporum f.sp cubense using 19 primers

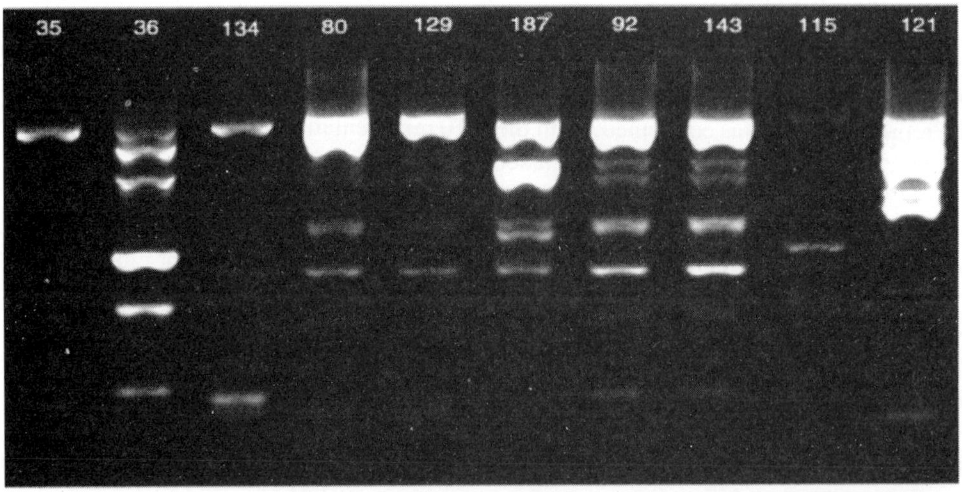

Lane 1-2: Rasthali Foc collected from Tamil Nadu, lane 3: Rasthali Foc collected from Andhra Pradesh, lane 4: rasthali Foc collected from Kerala, Lane 5: Rasthali Foc collected from Karnataka, Lane 6: Rasthali Foc collected from Assam, Lane 7&8: Neypoovan Foc collected from Tamil Nadu, Lane 9: Hill banana Foc collected from Tamil Nadu and lane 10: Karpooravalli isolate from Karnataka

Fig 54.2. ISSR profile of 10 isolates of Foc using the Primer 75PC

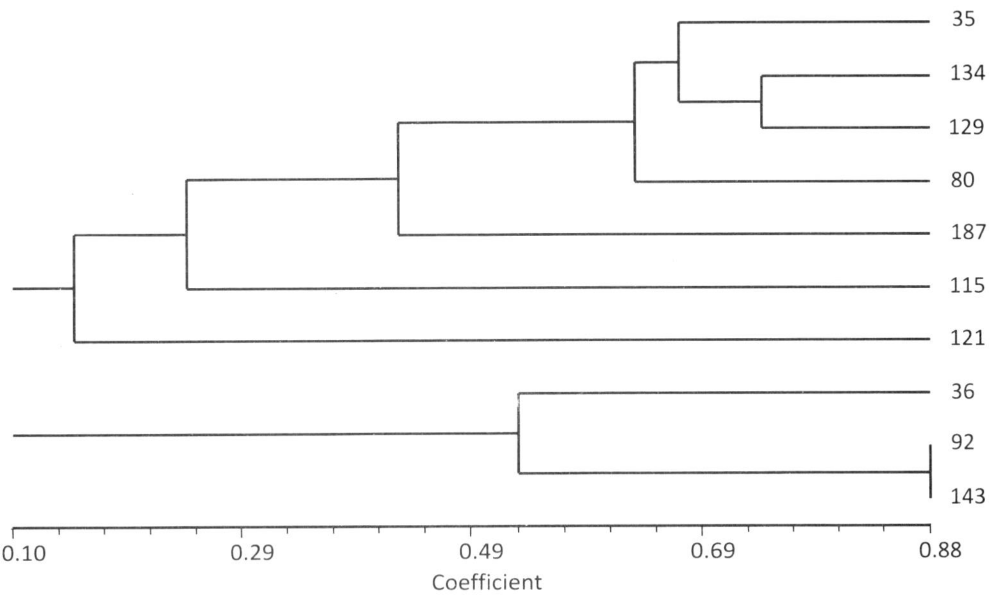

Foc Isolate No 35 and 36: Rasthali Foc collected from Tamil Nadu, 134: Rasthali *Foc* collected from Andhra pradesh, 80: Rasthali *Foc* collected from Kerala, 129: Rasthali Foc collected from Karnataka, 187: Rasthali *Foc* collected from Assam, 92 and 143: Neypoovan *Foc* collected from Tami! Nadu, 115: Hill banana *Foc* collected from Tamil Nadu and 121: Karpooravalli isolate from Karnataka

Fig. 54.3. A combined dendrogram for 10 *Fusarium oxysporum* f. sp. *cubense* isolates using similarity coefficients of markers generated by 7 primers.

Table 54.1. List of primers and their sequences used in study

Code	*Primer sequence 5í-3í*	*Number of bases*
75PC27A3	CCCGCATCC$(CA)_9$	27
76PC27A4	CCCGCATCC$(GA)_9$	27
77PC17G5	ACACACACACACACACG	17
78PC19A6	CGAGAGAGAGAGAGAGAGA	19
79PC15C7	CACCACCACCACCAC	15
84PC17T5	ACACACACACACACACT	17
85PC17C6	ACACACACACACACACC	17

Table 54.2. list of isolates and their place of collections used in study

Isolate No.	*Variety from which pathogen isolated*	*Place of collection*
7	Monthan	Periyakulam.
14	Rasthali	Sebathiapuram.
35	Rasthali	Valliyur
36	Rasthali	Valliyur, Vadakadu.
42	Neypoovan	Thorakalputhur, Suseendram, N-Koil
80	Rasthali	Thrissur
84	Neypoovan	Edalore, Thiruvananthapuram
92	Neypoovan	Thottiyam.
98	Karpooravalli	Kasipalayam, Erode
107	Rasthali	Sathyamangalam
115	Hill Banana	Thandikudi
117	Hill Banana	Joseph farm, Perumalmalai
121	Karpooravalli	Dharmapuri town.
124	Neypoovan	Mr. Nagaraj, Mysore
129	Rasthali	Nanjankodu
134	Mortaman	East Godavari
142	Karpooravalli	Jagdevpetta, Nellore
143	Neypoovan	Cuddalore
152	Karpooravalli	Karadialla
182	Monthan	Vaadipatti T.K
187	Malbhog	Posinbaragaon, Guwahati, Assam.
190	Malbhog	-DO-
194	Manohar	Aharimale, Chaitia, Sunipur, Assam
203	Manohar	Jarnapani, NEH-ICAR, Research complex, Nagaland

SUMMARY

The results indicated two major clusters which grouped most of Rasthali, Hill Banana and Karpooravalli isolates in one cluster and NeyPoovan and Rasthali collected from Tamil Nadu in other cluster with 50% similarity among them. The similarity between two main clusters is only 15%. It is concluded that a lot of variability existed among Race-1 isolates of *Foc* infecting different cultivars collected from different locations.

References

Amy, J., Nelson, Karol,. S. Elias, Arvbalo, G., Darlington, Lee C., and Bailey, Bryan, A. 1997. Genetic characterization by RAPD analysis of isolates of *Fusarium oxysporum f. sp. Erythroxylic* associated with an emerging epidemic in Peru. *Phytopathology,* **87:**1220-25.

Ainsworth, G.C. 1971. '*Ainsworth and Bisbyís Dictionary of Fungi'*, 6th edn. Commonwealth Mycological Institute, Kew, Surrey, England, pp. 663.

Bornet, B. and Branchard, M. 2001. 'Nonanchored Inter Simple Sequence Repeat (ISSR) Markers: Reproducible and Specific Tools for Genome Fingerprinting', *Plant Mol Biol Rep.,* **19:**209-15.

Kantety, R., Zeng, X., Bennetzen, J. and Zehr, B. 1995. 'Assessment of genetic diversity in Dent and popcorn (*Zea mays* L.) inbred lines using Inter-simple sequence repeat (ISSR) amplification'. *Mol. Breed.,* **1:**365-73.

Luana, M., Alessandro, P., Marcos, A.G., Cesar, B., Celina, F.V. and Celso, L.M. 2001. 'Identification of ISSR (Inter Simple Sequence Repeat) codominant markers in *Eucalyptus grandis*', *Proceedings of the Ninth Annual Conference of the Plant and Animal Genome,* Town & Country Hotel, San Diego, CA, January, pp. 13-17.

Moore, N.Y., Bentley, S., Pegg, K.G. and Jones, D.R. 1995. '*Fusarium* wilt of Banana', *Musa disease fact sheet*, INBAP, Montpellier, France, **5**:4.

Moreno, S., Martin, J.P and Ortiz, J.M. 1998. inter simple sequence repeats PCR for characterization of closely related grapevine germplasm. *Euphytica,* **101**(1):117-25.

Prasad, R.D. 2004. Wilt of crops with special reference to cultural, morphological, molecular characterization and pathogenic variability of isolates in India, *Annual Report*, pp 48-55. ICAR Network Project.

Prashant, K. Mishra, Jalpa P. Tewari, Randall M. Clear, T. Kelly Turkington 2006. 'Genetic diversity and recombination within populations of *Fusarium pseudograminearum* from western Canada', International Microbiology.

Wolfe, A.D., Xiang, Q.Y. and Kephart, S.R. 1998b. 'Assessing hybridization in natural populations of Penstemon (*Scrophulariaceae*) using hypervariable Inter-Simple Sequence Repeat markers', *Molecular Ecology,* **7**:1107-25.

Valmayor, R.V. 2000. 'Cooking bananas-classification', Production and utilization in South East, *Asia Infomusa,* **9**:28-30.